Hyperbranched Polymers

Macromolecules in between Deterministic Linear Chains and Dendrimer Structures

RSC Polymer Chemistry Series

Editor-in-Chief:
Professor Ben Zhong Tang, *The Hong Kong University of Science and Technology, Hong Kong, China*

Series Editors:
Professor Alaa S. Abd-El-Aziz, *University of Prince Edward Island, Canada*
Professor Stephen Craig, *Duke University, USA*
Professor Jianhua Dong, *National Natural Science Foundation of China, China*
Professor Toshio Masuda, *Shanghai University, China*
Professor Christoph Weder, *University of Fribourg, Switzerland*

Titles in the Series:
1: Renewable Resources for Functional Polymers and Biomaterials
2: Molecular Design and Applications of Photofunctional Polymers and Materials
3: Functional Polymers for Nanomedicine
4: Fundamentals of Controlled/Living Radical Polymerization
5: Healable Polymer Systems
6: Thiol-X Chemistries in Polymer and Materials Science
7: Natural Rubber Materials: Volume 1: Blends and IPNs
8: Natural Rubber Materials: Volume 2: Composites and Nanocomposites
9: Conjugated Polymers: A Practical Guide to Synthesis
10: Polymeric Materials with Antimicrobial Activity: From Synthesis to Applications
11: Phosphorus-Based Polymers: From Synthesis to Applications
12: Poly(lactic acid) Science and Technology: Processing, Properties, Additives and Applications
13: Cationic Polymers in Regenerative Medicine
14: Electrospinning: Principles, Practice and Possibilities
15: Glycopolymer Code: Synthesis of Glycopolymers and their Applications
16: Hyperbranched Polymers: Macromolecules in between Deterministic Linear Chains and Dendrimer Structures

How to obtain future titles on publication:
A standing order plan is available for this series. A standing order will bring delivery of each new volume immediately on publication.

For further information please contact:
Book Sales Department, Royal Society of Chemistry, Thomas Graham House, Science Park, Milton Road, Cambridge, CB4 0WF, UK
Telephone: +44 (0)1223 420066, Fax: +44 (0)1223 420247
Email: booksales@rsc.org
Visit our website at www.rsc.org/books

Hyperbranched Polymers
Macromolecules in between Deterministic Linear Chains and Dendrimer Structures

Albena Lederer
Leibniz Institute of Polymer Research, Dresden, Germany
Email: lederer@ipfdd.de

Walther Burchard
Albert-Ludwigs-University Freiburg, Freiburg, Germany
Email: walther.burchard@makro.uni-freiburg.de

ROYAL SOCIETY OF CHEMISTRY

THE QUEEN'S AWARDS FOR ENTERPRISE: INTERNATIONAL TRADE 2013

RSC Polymer Chemistry Series No. 16

Print ISBN: 978-1-84973-394-6
PDF eISBN: 978-1-78262-246-8
ISSN: 2044-0790

A catalogue record for this book is available from the British Library

Published by The Royal Society of Chemistry,
Thomas Graham House, Science Park, Milton Road,
Cambridge CB4 0WF, UK

Registered Charity Number 207890

For further information see our web site at www.rsc.org

Printed and bound by CPI Group (UK) Ltd, Croydon, CR0 4YY

Preface

Hyperbranched macromolecules are an integral part of nature. Since Flory's first successful effort to calculate such structures it took several decades before they could be synthesized. In the 1980s the first success in preparing polymers with AB_2 monomers was reported.

Some 40 years ago uncontrolled branching was classified as an undesired side-effect of polymerization, causing a delay in network formation. At present the variety of defined branched architectures and their application potentials seem to be endless. Countless numbers of functional groups fascinated chemists, permitting modifications which are used for different applications, such as processing or coatings. They also possess significant potential in sensorics, diagnostics, catalysis and therapeutics. In addition, the compact structure causes a decrease of the viscosity when applied as additives to melts.

Challenging requirements for suitable tailoring of material properties generates the need to understand the molecular characteristics of hyperbranched macromolecules. The discovery potential in this area is huge, taking into account the statistical character and manifold modification possibilities. Academic interest in hyperbranched molecules has continuously increased during the past 20 years since the discovery by Kim and Webster. More than 700 research articles per year regarding hyperbranched polymers are reported, with a continuously growing tendency.

Since Flory's fundamental work major efforts have been undertaken to synthesize dendritic structures. With the appearance of hyperbranched molecules many new routes for applications have opened. Yet, only a few research groups attempt to understand their properties by suitable theories.

Preparative physical chemists commonly possess a widespread experimental experience but lack knowledge in modern theory. *Vice versa*,

RSC Polymer Chemistry Series No. 16
Hyperbranched Polymers: Macromolecules in between Deterministic Linear Chains and Dendrimer Structures
By Albena Lederer and Walther Burchard

Published by the Royal Society of Chemistry, www.rsc.org

theoreticians tend to have insufficient insight into the feasibility of polymer sythesis for special structures. When the present two authors came together they recognized that their experience overlapped. Collaboration over several years led to the idea to assemble this knowledge in a book.

In this book we are attempt to present the knowledge of physical chemistry characterization in a suitable mathematical scheme. The main problem of hyperbranched structures arises from statistical processes combined with the formation of large structures. As far as we know, this book is the first attempt to bridge the experimental and theoretical views in a consistent way.

We hope that our book will support a new trend in future studies of hyperbranched polymers.

Contents

Acknowledgements **xiii**

Abbreviations **xv**

Chapter 1 Introduction **1**

1.1 Overview 1
1.2 Branched Polymers 3
1.2.1 Dendrimers 6
1.2.2 Hyperbranched Polymers 8
1.2.3 Dendritic Hybrids 11
1.3 Why Solution Properties? 13
References 16

Chapter 2 Degree of Branching **20**

2.1 Definition of Branching 20
2.1.1 Contraction Factors 22
2.1.2 Degree of Branching 24
2.1.3 Correlation Between the Degree of Branching and Contraction 27
2.2 Determination of the Degree of Branching 28
2.2.1 Analysis by Spectroscopic Approaches 28
2.2.2 Influence of the Synthetic Approach on the Degree of Branching 30
2.2.3 Side Reactions and the Degree of Branching 33
References 35

RSC Polymer Chemistry Series No. 16
Hyperbranched Polymers: Macromolecules in between Deterministic Linear Chains and Dendrimer Structures
By Albena Lederer and Walther Burchard

Published by the Royal Society of Chemistry, www.rsc.org

Chapter 3 **Control over the Conformation of Dendritic Polymers** 39

3.1 Introduction 39
3.2 Tuning the Degree of Branching 42
3.2.1 The $AB_2 + AB$ Approach 42
3.2.2 The $AB_2 + ABB^*$ Approach 47
3.2.3 Pseudo-dendrimers by AB_2 Modification 50
3.3 Control over Molar Mass 56
3.4 Variation of End-group Functionality 61
References 66

Chapter 4 **Separation and Determination of Molar Mass** 69

4.1 Determination of Absolute Molar Mass 69
4.2 Suitability of the Separation Technique 76
4.3 Novel Separation Methods for Dendritic Polymers 78
4.3.1 Separation According to Degree of Branching 81
4.3.2 Elution Properties Depending on Functionality 83
References 86

Chapter 5 **Solution Viscosity** 88

5.1 General Remarks 88
5.2 Determination of the Kuhn–Mark–Houwink–Sakurada Dependency 93
5.3 The Viscosity Behaviour of Dendritic Polymers 97
5.3.1 Dendrimers 97
5.3.2 Hyperbranched Polymers 99
5.3.3 Pseudo-dendrimers 102
5.3.4 Dendronized Polymers 105
5.4 Influence of Chemical Structure on Viscosity Behaviour 107
5.4.1 Influence of the Backbone Character 107
5.4.2 Influence of Functionality 110
5.5 Influence of Segmental Density on Viscosity and Scaling Characteristics 115
5.5.1 Gradually Increasing Statistical Branching 115
5.5.2 Pseudo-dendrimers with Increasing Generation 124
References 132

Chapter 6 **Size Determination by Scattering Techniques** 136

6.1 Introduction 136

6.2 The Search for Non-invasive Structure Characterization 137
6.2.1 Microscopy 138
6.2.2 Rheology 139
6.2.3 Sedimentation by Ultracentrifugation 140
6.2.4 Application of Light 140
6.3 Static Light Scattering 141
6.3.1 Scattered Light from Point-like Scattering Units: Correlation to Density and Concentration Fluctuations 142
6.3.2 Effect of Polarization 144
6.3.3 Scattering from Particles of Dimensions $> \lambda/20$ 145
6.3.4 Molar Mass and Radius of Gyration 151
6.3.5 Summary: Static Light Scattering 154
6.4 Global Parameters 154
6.4.1 Mean Square Radius of Gyration 154
6.4.2 Fractal Behaviour 158
6.4.3 Concentration Dependence 159
6.4.4 Some Properties of the Second Virial Coefficient 161
6.4.5 Self-similarity 162
6.5 Dynamic Light Scattering 164
6.5.1 Introduction 164
6.5.2 A Short Outline of the History of the Development of Optics 165
6.5.3 Brownian Motion and its Relation to Stochastic Processes 167
6.5.4 Time Correlation Functions 168
6.5.5 Comment on Averages and the Effect of External Forces 169
6.5.6 The Time Correlation Function in Dynamic Light Scattering 171
6.5.7 Flexible Particles: Linear and Branched Macromolecules in Solution 176
6.5.8 Small Particles: Proteins 177
6.5.9 Polydispersity and the Effects of Cluster Formation 179
6.5.10 Electrophoretic/Dynamic Light Scattering 180
6.5.11 Concentration Dependence 183
6.5.12 Summary: Dynamic Light Scattering 185
6.6 Effects of Heterodyne Light Scattering 185

6.7 Appendix: Aspects of Modern Theory of Segmental Dynamics 186
6.8 Recommended Literature 189
References 189

Chapter 7 Model Calculation of Different Hyperbranched Structures 193

7.1 Introduction 193
7.2 Conformation Properties of Branched Polymers in the Light of a General Branching Theory 194
7.2.1 Some General Remarks 194
7.2.2 Basic Properties of the General Branching Theory 196
7.2.3 Generating Functions of Probable Distributions 196
7.2.4 Randomly Branched Polymers 199
7.3 Angular Dependence of Scattered Light 204
7.3.1 The General Scheme 204
7.3.2 Radius of Gyration 206
7.3.3 Hydrodynamic Radius 206
7.3.4 AB_2 Hyperbranched Samples 207
7.3.5 Degree of Polymerization and Global Structure Parameters 208
7.3.6 Summary: Unperturbed Conformations 209
7.4 Effects of Excluded Volume 210
7.4.1 Introduction: What Does Volume Exclusion Mean? 210
7.4.2 The Flory Expansion Factor 211
7.4.3 Size Distribution of Real Chains 212
7.4.4 The Particle Scattering Factor, $P(q)$ 213
7.4.5 Fractal Approach to Fits of Measured Particle Scattering Factors 215
7.4.6 The Radius of Gyration and Hydrodynamic Radius of Branched Macromolecules in Good Solvents 220
7.5 Summary 224
References 225

Chapter 8 Scaling Conception and Interpretation of Structural Parameters and Scattering Experiments 227

8.1 Molecular Properties in a Global Scale 227
8.1.1 General Comments 227
8.1.2 Molecular Parameters 228

8.1.3 Generalized Ratios of Molecular Parameters 234
8.1.4 Contraction Parameters Related to Branching 244
8.1.5 Molar Mass Distribution 249
8.1.6 Pending Problems to be Solved 251
8.1.7 The Fractal Approach 251
8.1.8 Percolation on a Lattice 255
8.1.9 Characteristics of Size Exclusion Chromatography Diagrams 257
8.1.10 Angular Dependence of Scattering Curves from Measurements and Theoretical Prediction 260
8.2 Addendum 264
8.3 Summary 265
References 266

Chapter 9 Correlation of Molecular and Bulk Properties 269

9.1 Bulk Properties in Flow and in Thin Films 269
9.2 Thermal Properties 271
9.2.1 Influence of the Degree of Branching 272
9.2.2 Influence of the Molar Mass 272
9.2.3 Influence of the End Groups 275
References 277

Subject Index 280

Acknowledgements

Walther Burchard expresses his warmest thanks to Professor Ziya Akcasu, University of Michigan, USA, who 30 years ago brought him into contact with the modern theory of dynamic light scattering. He thanks Professor Andrzej Dworak, Centre of Polymer and Carbon Materials, Polish Academy of Sciences at Zabrze, Silesian District, Poland, who kindly suggested using any opportunity to write something on the history of how dynamic light scattering was developed. This task became a real challenge but in the end it was a very enjoyable effort! He feels indebted to Professor Murugappan Muthukumar at University of Massachusetts, Amherst, USA, for his very helpful suggestions regarding how excluded volume interactions could be treated. Part of this is now included in this book. Thanks are expressed to Professor Brigitte Voit, Director of the Leibniz-Institute of Polymer Research, Dresden, Germany, for financial support throughout this work.

Albena Lederer is deeply grateful to Brigitte Voit for initiating the work on dendritic polymers some 14 years ago. Professor Peter Schoenmakers, University of Amsterdam, the Netherlands, and Dr Wolfgang Radke, Polymer Standard Services in Mainz, Germany, inspired the search for advanced separation of very complex dendritic macromolecules. Warmest acknowledgements go to several colleagues who contributed significantly to our progress in understanding the physical chemistry of dendritic polymers. Here we mention the major contributors: Dr Hartmut Komber, Dr Peter Lindner, Dr Ralf Schweins, Dr Dietmar Appelhans and all the members of the Polymer Separation Group at the Leibniz-Institute of Polymer Research in Dresden. Bojan Bojkov and Johannes Windisch provided great technical assistance during the compilation of this book. We thank Dr Peter Friedel for several MD simulations including the one shown on the cover picture of this book.

RSC Polymer Chemistry Series No. 16
Hyperbranched Polymers: Macromolecules in between Deterministic Linear Chains and Dendrimer Structures
By Albena Lederer and Walther Burchard

Published by the Royal Society of Chemistry, www.rsc.org

Abbreviations

$[\eta]$	intrinsic viscosity
2D-LC	two-dimensional liquid chromatography
A	functional group of type A
A_2, A_3	second, third virial coefficient
a_{A2}; a_η	scaling exponents
AF4	asymmetric flow field flow fractionation
AFM	atomic force miscroscopy
arbPS	arborescent polystyrene
b	bond length
B	functional group of type B
c	concentration
c^*	overlap concentration
D	diffusion coefficient
D	dendritic units
Đ	molar mass distribution, polydispersity index
d_{app}	apparent density
DB	degree of branching
d_{f}	fractal dimension
DLS	dynamic light scattering
DMF	dimethylformamide
DMSO-d_6	dimethylsulfoxide
dn/dc	refractive index increment
DP	degree of polymerization
DP_n	number average polymerization degree
DP_w	weight average polymerization degree
DSC	differential scanning calorimetry
ELS	electrophoretic light scattering

RSC Polymer Chemistry Series No. 16
Hyperbranched Polymers: Macromolecules in between Deterministic Linear Chains and Dendrimer Structures
By Albena Lederer and Walther Burchard

Published by the Royal Society of Chemistry, www.rsc.org

f	functionality
FB	branching fraction
FTIR	Fourier-transform infrared spectrometry
G	generation
g	contraction factor, branching parameter, radius contraction factor
g'	viscosity contraction factor
$g(r)$	radial distribution function
hb	hyperbranched
KMHS	Kuhn–Mark–Houwink–Sakurada
L	linear units
LAC	liquid adsorption chromatography
LCB	long-chain branched
LCCC	liquid chromatography under critical conditions
LCST	lower critical solution temperature
LDPE	low density polyethylene
lin	linear
LLDPE	linear low density polyethylene
l_p	persistence length
LS	light scattering
M	molar mass
M_0	molar mass of the monomer unit
MALDI-TOF-MS	matrix assisted laser desorption/ionization–time of flight–mass spectrometry
MALLS	multi-angle laser light scattering
MALS	multi-angle light scattering
M_{app}	apparent molar mass
M_c	molar mass of the core
MD	molecular dynamic
M_n	number average molar mass
M_t	molar mass of the terminal units
M_w	weight average molar mass
M_z	z-average molar mass
n_0	refractive index of the solvent
N_b	functionality of the branching site
N_c	functionality of the core
n_e	number of end-groups in a dendrimer
NMR	nuclear magnetic resonance
p_A	probabilty of reaction/conversion of functionality A
p	pressure
$P(q)$	particle scattering factor
p.g.f.	probability generating function
PAMAM	polyamidoamine
PDI	polydispersity index
PEA	polyetheramide
PS	polystyrene

q	scattering vector
RAFT	reversible addition–fragmentation chain transfer
R_g	radius of gyration
RG theory	renormalization group theory
R_h	hydrodynamic radius
RI	refractive index detection
r_{jk}	distance vector
r_m	monomer ratio AB/AB_2 in the feed
R_N	end-to-end distance of the chain
r_p	monomer ratio AB/AB_2 in the product
R_θ	Rayleigh ratio
SANS	small-angle neutron scattering
SAW	self-avoiding walk
SAXS	small-angle X-ray scattering
SCVP	self-condensing vinyl polymerization
SEC	size exclusion chromatography
SLS	static light scattering
SY	TBDMS capped end group
t	time
T	terminal units
TBDMS	*tert*-butyldimethylsilyl group/ether
TCF	time correlation fucntion
TEM	transmission electron microscopy
T_g	glass transition temperature
THF	tetrahydrofuran
V	volume
X_{ABB*}	molar fraction of one monomer type
z	numer of end-groups
α, β	extent of reaction, transition probability
α_η, a_η	Kuhn–Mark–Houwink exponent
γ	damping factor
ε	draining exponent
ζ	frictional coefficient
η	viscosity
θ	scattering intensity
λ	number of branches per 1000 monomer units
λ	wavelength
μ	chemical potential
ν	scaling exponent
π	osmotic pressure
ρ	ratio of radius of gyration, and hydrodynamic radius
τ	relaxation time
φ	draining parameter
Ψ	interpenetration parameter
ω	frequency

CHAPTER 1

Introduction

1.1 Overview

Macromolecular branching belongs to nature and is of significant importance for the life functions of organisms. Branching in polymer science has been known since almost the time when first synthesis of linear chains was carried out. No doubt when Staudinger and Schulz carried out the synthesis of styrene they also tried to add divinyl benzene as a second monomer for a co-monomer.[1] The effect was dramatic and rather unexpected. In those days besides the monomer conversion the reaction was also followed by a change in viscosity because the increase of viscosity was a qualitative mean for the increase in molar mass. In the corresponding co-polymerization the increase in the viscosity was significantly lower but after certain time a sudden increase occurred and an non-measurable high value was attained. A highly elastic, gel-like material was obtained whose behaviour greatly resembled that of natural rubber.

The sudden change from a low solution viscosity to an elastic gel remained unclear for some time until, in 1941, Paul J. Flory[2] put forward the mathematically simple expression for the condition of the gelation point:

$$\alpha_c = \frac{1}{f-1} \tag{1.1}$$

He described the extent of reaction by the probability α and found a critical conversion α_c when gelation takes place. f is the number of functional groups per monomer of the same reactivity, and solubility remains in the pre-gel state of $\alpha(f-1) < 1$. Branching takes place only if $f > 2$ and gelation will occur at a connversion beyond the critical point defined by eq. (1.1).

RSC Polymer Chemistry Series No. 16
Hyperbranched Polymers: Macromolecules in between Deterministic Linear Chains and Dendrimer Structures
By Albena Lederer and Walther Burchard

Published by the Royal Society of Chemistry, www.rsc.org

Flory intensely tried to derive the molar mass distributions of such randomly branched polymers. It was Walter H. Stockmayer who, in 1943–1944,[3,4] found the solution and with this distribution he could derive the weight average molar mass expressed in term of the probability of reaction:

$$M_w/M_0 = 1 + \frac{\alpha f}{1 - \alpha(f-1)} = \frac{1+\alpha}{1-\alpha(f-1)}, \quad \beta = \alpha/2 \tag{1.2}$$

with M_0 the molar mass of the monomer. Flory, as well as other researchers, was excited because this equation now gave a mathematical proof for his derivation of the prediction for the critical point of gelation. A lifelong friendship between the two pioneers, Flory and Stockmayer, began here.

The derivation of the so-simple looking equation was a hard challenge for Stockmayer. He was frustrated that he could not find a solution. It was Maria Goeppert Mayer (the later Nobel Prize winner in 1963 and wife and co-author with Joseph E. Mayer) who gave Stockmayer the decisive hint. She observed Stockmayer's attempt and after a while she said, 'Stocky, why don't you try to use our cluster description when we treated the interparticle interactions between colloidal spheres?'[5]

Flory confirmed the Stockmayer' distribution by applying combinatoral mathematics which will be given in Chapter 7 of this book. In the course of the time Flory also derived a molar mass distribution for the polymerization of AB_2 monomers[6] and the corresponding weight average degree of polymerization. No critical conversion for gelation was obtained.

This result surprised even Flory, because he realized that a divergence can be obtained only at full conversion, *i.e.* $\alpha = 1$ or in other words, the branching of *such* AB_2 *samples can never lead to a gel or network.* This discovery was the birth date for hyperbranching.

Flory's interest in this type of structure arose from a book by K. H. Meyer in 1950[7] and from Staudinger.[8] The structure of natural polymers was one of the reasons why Staudinger, with co-workers, turned to chemical synthesis of model polymers in the hope of gaining a deeper understanding of these structures. Amylopectin, the branched component of starch, was one of them.[9,10]

It took more than 20 years before the subject of AB_2 polymers was taken up[11,12] but the decisive change was made by Kim and Webster[13] by coining the name '*hyperbranching*'. These two authors were trying to find a less laborious synthesis for dendrimers, which were known to possess advantageous properties as nano-objects. Although Kim and Webster could not achieve the high accuracy of a perfect structure, so they stated, at least the samples are *hyperbranched*. This now generally used notation is rather unfortunate because also co-polymers of the $A_3 + B_2$ type are often included in this class of polymers although this type of co-polymerization was, for Flory and Stockmayer, the model system for random branching. Indeed gelation can be avoided by external constraints by choosing the appropriate composition of the two monomers.

One may wonder why the research with hyperbranched macromolecules was so much delayed compared to the research on network formation with the remarkable property of rubber elasticity. The answer to the question appears obvious, because in the preparation of an AB_2 monomer the various functional groups should not react when it is synthesized. In order to obtain a polymer, the functional groups has to be activated, by an enzyme, an ionic initiator or other activators, for instance for ester formation. When these activating techniques were elaborated, the method of chemical synthesis for hyperbranched polymers was open. The almost boundless chemical possibilities for modifications made hyperbranched polymers preferred candidates for large-scale applications as the cheap analogue of dendrimers, *e.g.* as reactive components in coatings and resins, and as additives in linear systems for improved rheology and processing. Thermal stability and modulus can be easily modified just by modifying the huge number of functional groups in a single macromolecule.[14] The synthetic procedures for these structures have taken unconventional routes over the years and have led to increasing control over the design of hyperbranched macromolecules, giving rise to new applications, particularly in areas of nanotechnology such as sensorics, diagnostics, catalysis and therapeutics.[15–20]

With the challenging requirements for precise design of materials properties, the need to understand the molecular characteristics of hyperbranched macromolecules becomes urgent. The potential for discovery in this area is enormous taking into account the statistical character and versatile modification possibilities of the dendritic architectures. Thus, academic interest in hyperbranched structures has increased continuously in the last 20 years since their innovation by Kim and Webster.[13] At present more than 700 articles per year report on research activities in hyperbranched polymers and the tendency is continuously growing.[21]

In fact, the theoretical treatment is lagging behind synthetic development, and it is a serious challenge. The difficulties with a suitable theory are not based on the complexity of branching alone, but also the polydispersity, excluded volume interaction and segment over-crowding have to be taken into account. We will show that several of these problems are on the way to being solved. However, the cases of special long range ionic or hydrophobic interactions have not been touched yet. Chemists are presently drawing conclusions more or less intuitively. It is our belief that a sensible theory on the new phenomena caused by the hyperbranching will be indispensable to obtaining a reasonable understanding for future applications of hyperbranched polymers.

1.2 Branched Polymers

In 1935 Staudinger and Schulz observed that the changes in the viscosity of polystyrene solutions depended on the reaction conditions during synthesis.[1] Later, Staudinger and Husemann found a higher number of end groups than expected during their studies on the natural polymers amylopectin and glycogen, which was explained by the existence of branching.[22] Both low viscosity and high number of end groups are typical characteristics of branched polymers. In addition, impeded crystallinity and poor mechanical strength are very common in branched polymers. These properties become more pronounced with branching density and depend on branching topology.

The type of branching can strongly influence polymer properties. Examples include natural polysaccharides, which occur after enzymatic polymerization of glucose, forming different types of branching. Glycogen is formed of glucose monomers and possesses high number of branching points (Figure 1.1). There are up to 50 000 monomer units of glucose on glycogen coupled together. The glucose units are arranged in a tree-like branched structure with a branching point every 8–12 glucose units. Glycogen is an energy supplier for animal cells and is the product of an enzymatic decomposition of starch. Starch is composed of amylose and amylopectin, which are also formed of glucose units, but while amylose possesses linear or very slightly branched chains, in amylopectin branching occurs every 25 units. Starch is the product of photosynthesis in plants. Despite the similar chemical character of the monomer units, the properties of starch and glycogen are different; starch is not soluble in cold water and is semi-crystalline, while glycogen is soluble in water and is not crystalline.

Today the variety of branching in synthetic polymers is manifold and 'branching' becomes a generic term for a huge class of polymer structures. The systematic development in branched polymers started with the discovery of long-chain branched polymers in the 1940s, the synthesis of the first star and graft polymers in the 1960s and the invention of dendrimers in the 1980s and of hyperbranched polymers in the 1990s. Nowadays, the combination of different branched topologies in one macromolecule becomes a common approach to control material properties. Some of these hybrid structures are overviewed in part 1.2.3 of this chapter. Roughly, the

Figure 1.1 Glycogen, amylopectin and amylose are chemically identical polysaccharides but with different branching architecture essential for their physiological functions.

structure of branched polymers could be classified depending on different characteristics and could be illustrated as a multidimensional property.

The *regularity* of branching is significant for calculating global and local properties as well as for further calculation of structural parameters. Branching can be regular or random with regard to either the length of the branches or the frequency of branching points. Figure 1.2 elucidates the differences in stars, comb polymers and dendritic polymers.

The quantification of branching is of immense importance for estimation of structure–property relationships, although, its definition is substantially influenced by the type of branching. Stars can be classified by the functionality of the core possessing only one branching point; comb-like and classical long-chain branched polymers are classified by the *number of*

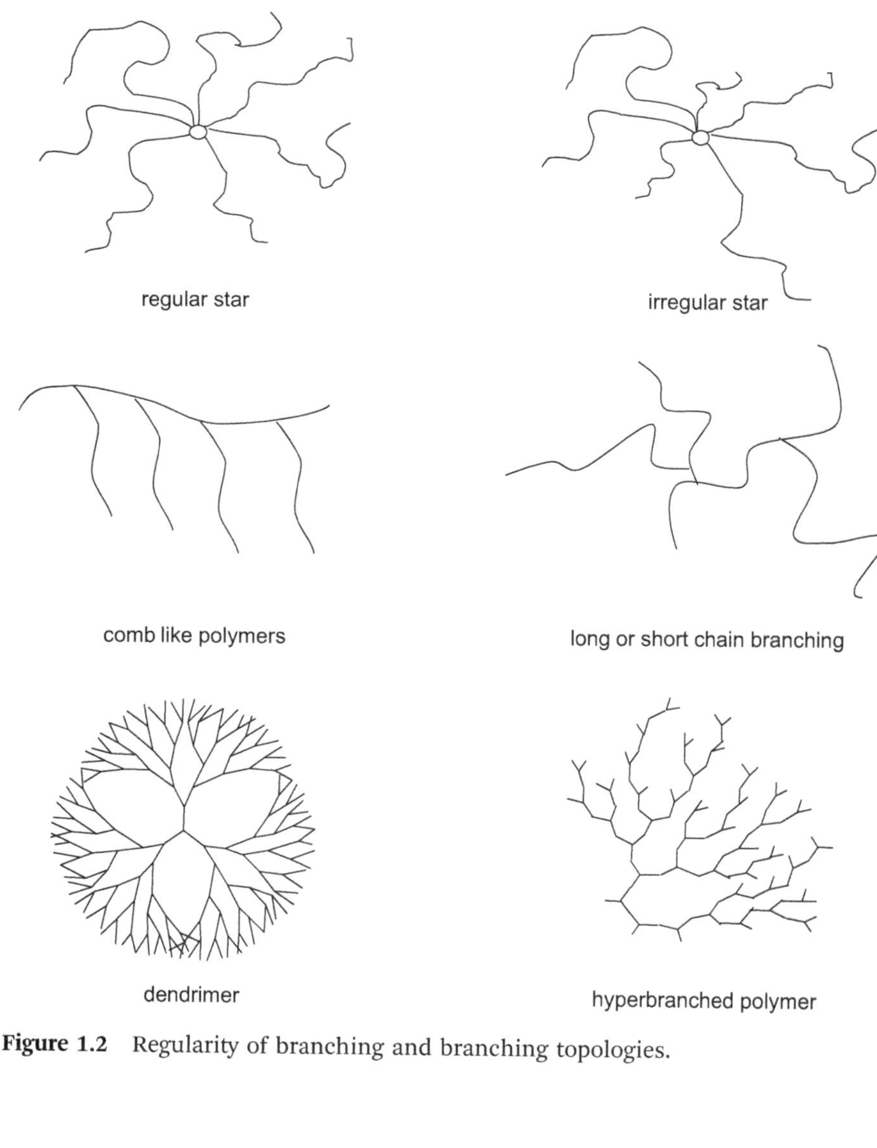

Figure 1.2 Regularity of branching and branching topologies.

branches per molecule or per 1000 C-atoms; dendritic branching is characterized by the relative number of branched units, *e.g. degree of branching*, due to the significantly higher density compared to the above branching types. The calculation and determination of branching degree is described in detail in Chapter 2.

Other characteristics of branched polymers are the *length of branches* and the *functionality of the branching point*. Both are variables which have to be taken into account calculating scaling characteristics of the polymers. The length of the branches also defines their properties in solution and in bulk. Long-chain branched polymers can possess properties typical for entangled macromolecules influencing flow and crystallinity properties. Short-chain branched polymers and dendritic structures are typically non-entangled and non-crystalline objects.

In the next section of this introductory chapter a short historical and structural background of dendritic polymers and their hybrids is given. The correlation to other types of branching is part of this book as far as a systematic comparison of properties is needed for better understanding the particularities of hyperbranched polymers. Additionally, the theoretical considerations developed for long-chain branched or star polymers are critically validated for the case of dendritic polymers.

1.2.1 Dendrimers

The term 'dendritic' comes from the Greek word for tree, *δεντρο*, and is related to the branching character of polymers. The ideal examples for dendritic polymers are dendrimers (Figure 1.3), which are built step-wise and have a branching point at every monomer unit.[23] Due to their perfectly branched nature, aesthetic 'look', complete dimensional control and variety of functional possibilities, dendrimers have fascinated the scientific community for several decades, especially in their role as nano-objects for high-end applications. Organic chemists were attracted by the challenging synthesis,[23] and highly interesting potential applications in the fields of biomaterials and nanotechnology have been found. A drawback of these perfect polymers is their limited industrial relevance because of high manufacturing costs.

The history of dendrimers started in 1978 with the cascade polymers of Buhleier *et al.*,[24] while iterative synthetic concepts were reported in the early 1970s by Lehn[25] and Cram *et al.*[26] In 1982, Aharoni *et al.*[27] studied lysine trees for peptide synthesis. In 1985 the first concepts of dendrimers,[28] or arborols,[23] were reported, giving birth to a new polymer topology class which expanded quickly to a huge variety of chemical structures, modifications and applications.

The special properties of dendritic polymers could easily be explained on the basis of the very well defined structure of dendrimers reflecting complete branching and regularity, which could be achieved in a three-dimensional sense. The formation of dendrimers starts with a core with functionality N_c. The functionality of the branching site N_b depends on the monomer

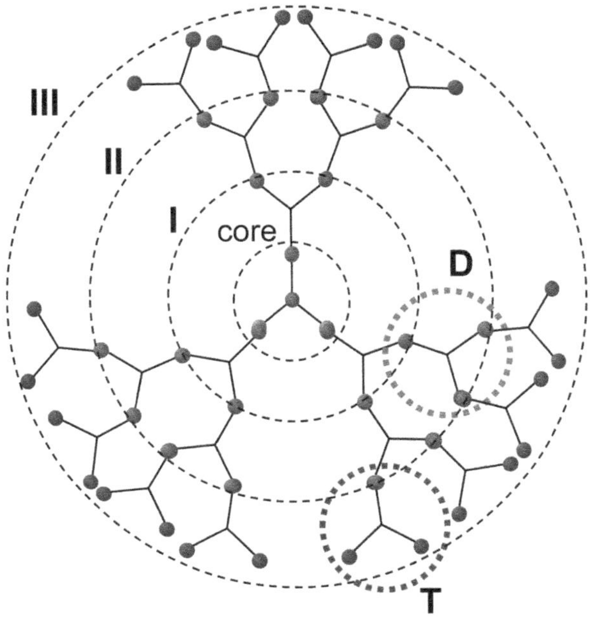

Figure 1.3 Dendrimer in the third generation from AB_2 monomer. Dendrimers possess terminal (*T*) and dendritic units (*D*), and a core.

functionality. Based on these values the development of the dendrimer structure can be simply calculated for every generation *G*. The number of end groups *z* increases according to

$$z = N_c N_b^G \tag{1.3}$$

which means that for the simplest case of a three-functional core and two-functional branching site ($N_c = 3$, $N_b = 2$) the number of end groups increases from 2 in the core to 48 in the fourth generation.

The number of repeating units N_r increases greatly with the generation number even for the simplest case:

$$N_r = N_c \frac{N_b^{G+1} - 1}{N_b - 1} \tag{1.4}$$

and the molar mass *M* accordingly increases exponentially with every generation:

$$M = M_c + N_c \left(N_r \frac{N_b^{G+1} - 1}{N_b - 1} + M_t N_b^{G+1} \right) \tag{1.5}$$

with M_c the mass of the core and M_t, the mass of the terminal units.

The synthesis of dendrimers is well controlled and could proceed according to a divergent or convergent approach, described in Chapter 3.

Dendrimers show very interesting materials properties because of their globular, very compact structure and very high functionality, especially due to their fully defined structure. However, their synthesis is elaborate and very time consuming, which makes their implementation in large-scale applications very complicated and expensive. This fact stimulated scientists and polymer manufacturers to search for alternative routes to produce polymers with similar properties as dendrimers but with easier synthesis.

1.2.2 Hyperbranched Polymers

For applications in which easy and efficient synthesis is more important than a perfectly branched architecture, hyperbranched polymers prepared by only one synthesis step are considered. This was the reason why Kim and Webster, researchers at Du Pont, developed new dendritic polymers that can be synthesized rapidly, in large quantities, for application as rheology modifiers and multifunctional initiators.[13,29] In 1990 they reported on the first *hyperbranched* polymers, the polyphenylenes, which possess irregular structures with linear segments within the dendritic structure and a high polydispersity.

In fact the existence of synthetically produced, highly branched architectures started much earlier, in the 19th century. The first works are not directly available, but they were the basis of later developments on the polymerization of multifunctional monomers ($A_2 + B_3$) based on phthalic anhydride or phthalic acid and glycerol.[30,31] In 1909, Baekeland used the $A_2 + B_3$ reaction between formaldehyde and phenol to produce the first commercialized polymer, a cross-linked material, which exists as an irregularly branched structure just before the network formation.

The theory of polycondensation between multifunctional monomers was extensively treated in the works of Flory in the 1940s and 1950s.[6,32,33] Flory was interested in the statistics of molar mass distribution before gel formation in polymers with trifunctional and tetrafunctional branching points. Before reaching the so-called gel point, *e.g.* the critical conditions for cross-linking, highly branched structures with a certain branching degree are formed. Flory found that, theoretically, the combination of both reaction functionalities in one monomer of type AB_x ($x \geq 2$) should prevent gelation if A reacts only with B (condition 1). The one-pot synthetic route yields dendritic structures that possess highly dense branching with the potential of a branching point at each repeat unit. Simplification of the calculation, which corresponds also to conditions leading to ideal statistics, requires the assumptions that no cyclization or side reactions occur (condition 2) and that the reactivity of B is independent of the degree of polymerization (condition 3).

In the simplest case of an AB_2 monomer this type of reaction leads to the formation of three types of unit (Figure 1.4). If only one of the two

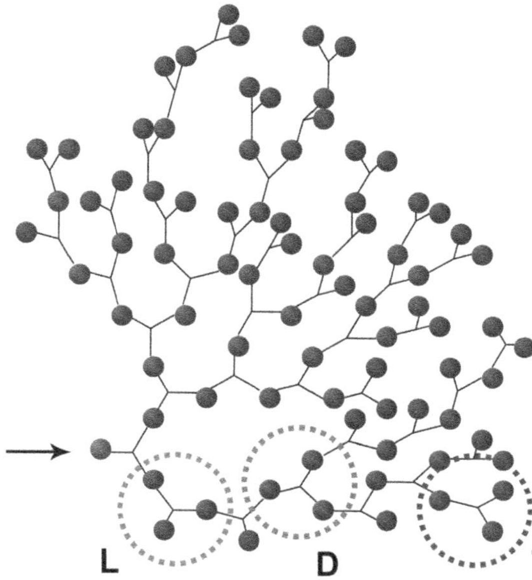

Figure 1.4 Hyperbranched polymer with AB_2 units. Hyperbranched polymers possess terminal (*T*), dendritic units (*D*), linear units (*L*). The point indicated by the arrow belongs to the focal unit, which corresponds to the unreacted A-function.

B-groups in an AB_2 monomer has reacted, a *linear* unit (*L*) is formed, carrying the non-reacted B-group available for further reactions. If both B-groups have reacted a dendritic unit is obtained (*D*); and, finally, if both B-groups no longer have reacted a terminal unit (*T*) is obtained. If the above conditions (1 to 3) for ideal statistics are fulfilled, the frequencies of the three types of polymer units at high conversion are

$$\frac{L}{L+D+T} \approx 0.5, \quad \frac{D}{L+D+T} \approx 0.25, \quad \frac{T}{L+D+T} \approx 0.25$$

$$L+D+T \approx \mathrm{DP}_n \tag{1.6}$$

75% of the polymer units carry B-groups, which corresponds to free B-functionalities approximately equal to the degree of polymerization, $z = \mathrm{DP}_n - 1$.

On the basis of these assumptions Flory calculated the molar mass distribution in a polycondensation reaction leading to statistical branching. A monomer with functionality f would possess in this case one A- and $(f-1)$ B-groups. The probability that a B-group has reacted is p_b and the probability for an A-group to react is p_a. The variable p_b corresponds to the fraction of reacted B-groups while the fraction of reacted A-groups would be $p_a(f-1)$ due to the remaining focal group (cyclization reactions are excluded). In this case the conversion of the reaction could be expressed with p_a and the fraction of branching should correspond to

$$\alpha = \frac{p_a}{f-1} \tag{1.7}$$

where α is the extent of reaction, and with this notation the number averaged and weight averaged degrees of polymerization are using α and p_a, then

$$DP_n = \frac{1}{1-p_a} = \frac{1}{1-\alpha(f-1)} \tag{1.8}$$

$$DP_w = \frac{1-\alpha^2(f-1)}{[1-\alpha(f-1)]^2} \tag{1.9}$$

the distribution of the polymerization degree can be expressed as

$$\frac{DP_w}{DP_n} = \frac{1-\alpha^2(f-1)}{1-\alpha(f-1)} = \frac{1-\alpha p_a}{1-p_a} \tag{1.10}$$

This equation shows that if the conversion of A-groups approximates unity, the molar mass distribution should increase infinitely. This means that at full conversion the molar masses could achieve extremely high values but without network formation. In practice, such molar masses are seldom due to non-ideal statistics which the polycondensation reactions follow.

In the next four decades research activities following from Flory's calculations are not reported. The expected macromolecular characteristics which can be hardly controlled, as irregular branching, very broad polydispersities and amorphous materials with poor mechanical properties were found unprofitable. The understanding of these very complex structures with multidimensionally distributed parameters was limited at that time.

With exception of the activities of Kricheldorf *et al.*, on copolycondensation of AB_2 and AB monomers in 1982,[34] the first intentional development of hyperbranched polymers was started in 1988 by Kim and Webster, who also invented the term *hyperbranched*.

Kim and Webster synthesized hyperbranched polyphenylenes by a catalyzed reaction of multifunctional monomers (Figure 1.5) and observed very good solubility of the polymers in organic solvents even at high molar masses, which was affected by the high number of end groups.

By modification of these end groups to carboxylate end functions the polymers became soluble in water. The glass transition temperature of these polymers was less influenced by the molar mass but was controlled by end-group modification. The hyperbranched polyphenylenes were brittle with rather poor mechanical characteristics but with excellent thermal stability. Their application as materials was put aside but their potential as additives for modification of materials properties was rapidly recognized and invented.

Since this time hyperbranched architectures have attracted increasing attention in the scientific community based on the unique properties in

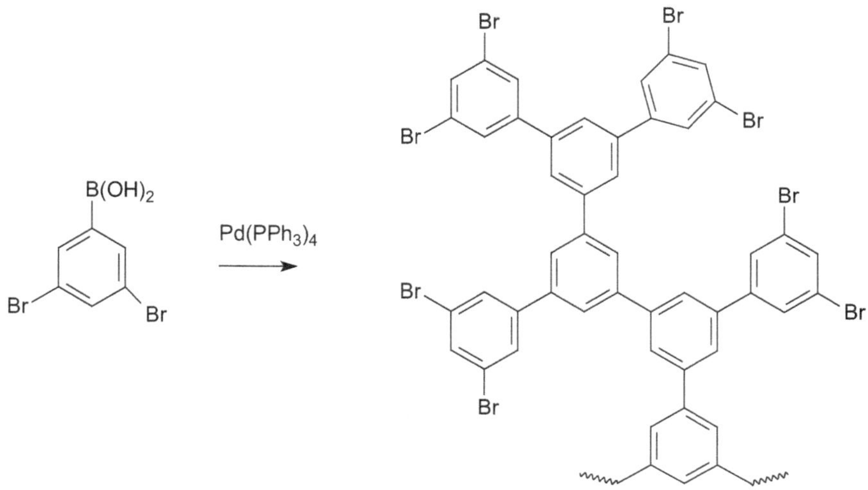

Figure 1.5 Hyperbranched polyphenylene first synthesized by Kim and Webster.[13]

combination with straightforward preparation. The classical synthesis for hyperbranched polymers remains polycondensation, but in the last 20 years great effort has been made to obtain new chemical structures using unconventional synthetic routes, driven by the occurrence of new fields of application.[15,16,35,36] In addition to their well-known use in resins and coating formulations, hyperbranched polymers appeared as additives in linear polymers for rheology modification, or for the improvement of processing, thermal stability, mechanical and surface properties. This huge range of application fields has to be ascribed to the highly branched nature which leads to excellent solubility, low melt and solution viscosities, combined with a high number of reactive end functionalities, which can arbitrarily be greatly modified.[37–40] Furthermore, interesting areas of application have been found in thin films for sensor devices and diagnostics by tuning the confinement of the hyperbranched structures, as pore-forming agents in nanofoams by using their globular shape, in drugs and as catalytic carriers using molecular cavities and functional group geometry.[14–18] Nowadays, there is a number of suppliers who specialize in the development of application-oriented hyperbranched structures and commercial products. Some well-known examples of hyperbranched polymers on the market are the aliphatic polyester Boltorn™,[37] the poly(esteramide) Hybrane™,[38] and the poly(ethylene imine) Lupasol™, as well as a number of hyperbranched polyesters[41] and polyurethanes[42–44] for applications in coatings and resins, and hyperbranched polyglycerols and polyesters for nanotechnology and bioapplications.[45–47]

1.2.3 Dendritic Hybrids

The high functionality and special geometry of branching conceals great potential not only in the classical dendritic structures, dendrimers and hyperbranched polymers. The combination of dendritic and linear chains opens many more variations of polymer architectures and new molecular properties. Such types of combination can be found in dendritic–linear hybrids, which could exist in a different structure but two prominent examples are *dendronized polymers* and hypergraft polymers (Figure 1.6). The concept of dendronized polymers was motivated by the challenging aspects to realize giant macromolecules (DenPols) with high functionality. Along with the perfectly branched dendrimers and hyperbranched polymers, DenPols[48] also belong to the structural family of dendritic macromolecules. They can be described as hybrid macromolecules possessing a linear backbone decorated with dendrons. The dendronization of a linear polymer results in a high concentration of functional groups along the linear backbone and a very dense packing leading to a cylindrical shape.[48–50] In contrast to dendrimers, where the dendrons are coupled to a core and the size and functionality of the molecules are limited, the fabrication of dendronized linear structures enables the realization of diverse nano-objects which vary in their length, branch point frequency, and dendron generation.

The dendronization could be performed instead of perfectly branched dendrons in different generation, also with imperfect, hyperbranched

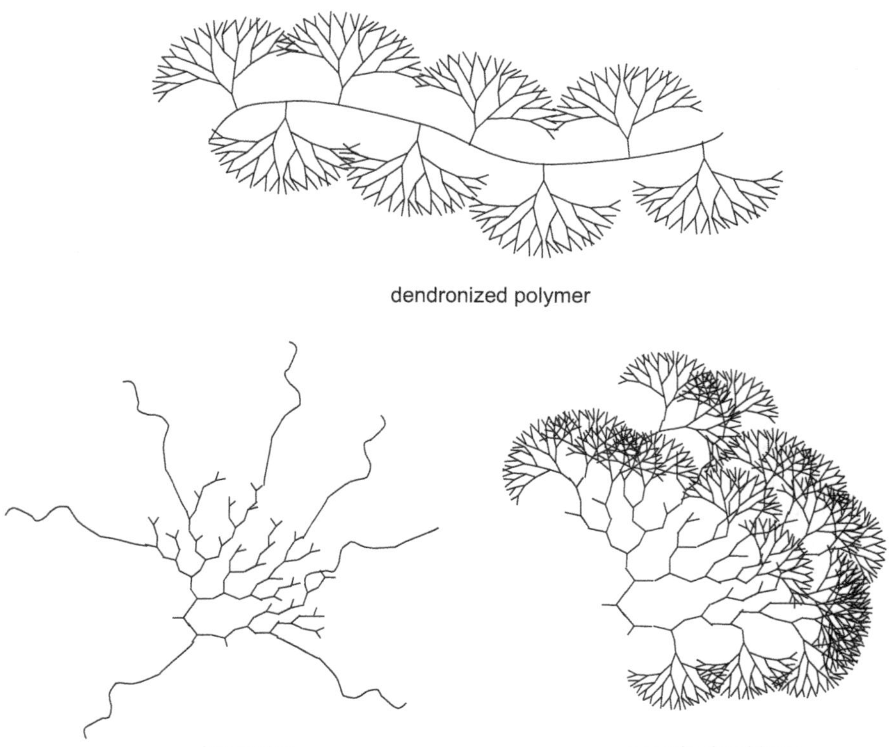

Figure 1.6 Dendritic hybrids.

dendrons leading to linear-hyperbranched graft-co-polymers, or hypergraft polymers.[51–53] This alternative approach was shown recently on polyglycerol using different grafting procedures.[54]

The opposite case to dendronized polymer architecture is dendritic stars and *hyperstars* in which a dendritic core is modified with linear chains (Figure 1.6). In this type of dendritic hybrids the properties of the dendritic molecules usually have to be modified for the purpose of different applications. The globular dendritic core could improve rheology; at the same time the high number of end groups enables high reactivity and long-chain shells could improve solubility or miscibility with the surrounding material. Numerous interesting properties, *e.g.* hydrophilicity, responsive and LCST behaviour, can be adjusted by hybridization with linear arms. The cavities within the molecule give enough space for multivalent interactions and are interesting for controlled drug release.[55]

Hyperstars can be prepared by different approaches, but usually the core first approach is used by grafting from or grafting onto linear polymers to the dendritic core.[56–59] Hyperstars were actively investigated in anionically cured epoxy thermosets,[60] or for their use in biomedical fields as core–multishell nanocarriers.[20,61]

Another dendritic hybrid are the *pseudo-dendrimers*, which possess perfectly branched dendritic units coupled to the end functionalities of a hyperbranched polymer as a core (Figure 1.6). This type of modification leads to complete branching, although an irregular structure. At the same time the number of functional groups increases immensely, leading to properties very similar to those of dendrimers. Lach and Frey, and Caldéron *et al.* first viewed pseudo-dendrimers as polyglycerol architectures and investigated their potential in bioapplications.[52,61] Later, higher generations of pseudo-dendrimers were shown to have a unique similarity to perfectly branched structures due to a special type of molecular density distribution.[62,63] Their particular solution properties and structural characteristics are discussed in detail in Chapters 3 and 5.

1.3 Why Solution Properties?

The characterization studies on dendritic polymers commenced simultaneously with their synthetic development. There are several reviews discussing the characterization of these materials by the groups of Mori *et al.*,[67] McKee *et al.*,[68] McLeish,[69] Teerstra and Gauthier,[70] Jikei and Kakimoto,[71] Burchard,[72] and Voit and Lederer,[66] covering spectroscopic, melt rheology, solution properties of bio- or synthetic dendritic polymers. All these works confirm the particular behaviour of the dendritic structures in bulk and in solution. However, in order to obtain information about the correlation between the chemistry of the polymer backbone and bulk properties, an insight into the parameters of a single molecule is required. Determination of the dimensions of hyperbranched macromolecules and investigation into how are they influenced by molecular geometry, size, branching, flexibility, and environment are the first steps on the way to understanding their material properties and enabling target-oriented molecular design. This might be straightforward for dendrimers, where discrete branching and size are given in one sample, but understanding the multifaceted structure of hyperbranched polymers needs more comprehensive investigation. The physical characterization of natural hyperbranched macromolecules started in the 1960s.[73] When hyperbranching as a result of polycondensation started to develop in the early 1990s, there was already a great deal known about the molecular parameters of dendrimers. Dilute solution properties and scattering measurements gave an insight into excluded volume effects and density gradients of their perfect molecular structure.[64,74–79] In contrast to them, hyperbranched polymers are fairly easy to synthesize but far more challenging when it comes to their characterization. The number of parameters which statistically vary within one sample is highly complex. Their molar masses are usually broadly distributed; chemical heterogeneity and branching distribution cannot be excluded for most samples. On the other hand, possibilities to control the degree of branching and molar mass during the synthetic procedure are extremely limited. The need to improve on this motivated various research groups to develop new synthetic routes for polymerization in the presence of core molecules,[80,81] slow monomer addition,[80,82–84] controlled monomer ratio combinations of $A_2 + B_3$,[85] $AB_x + B_y$[34,68] or inimer + co-monomer.[67]

However, the discrete properties of dendrimers were never achieved by these approaches due to the statistical nature of the hyperbranched reaction, and hence the focus turned towards an enhancement of separation techniques with separation on a molecular level and analysis of monodisperse properties being the targets. Hyphenated size exclusion separations in combination with viscosity and light scattering analysis have been applied,[72,86–89] giving the opportunity to estimate the influence of molar mass on the compactness of hyperbranched molecules. At the same time, huge efforts have been made to simulate the solution behaviour of dendritic structures as an alternative to the limited model polymers for real experiments[90–93] and to correlate them to the degree of branching of statistically branched structures. Surprisingly, the theoretical calculations never fitted the experimental results. The theoreticians found an influence of the molar mass on the molecular compactness in solution resembling dendrimers and changing from a very open to a very compact structure dependent on the molar mass (Figure 1.7).[64] At the same time, the experiments on real samples showed an uninspiringly constant molecular shape common for linear or long-chain branched polymers.[65,67]

The discrepancies of results from different scientific approaches open the question of the difference between simulation and experimental studies and the limited overlap between the broad and multidimensional distribution of hyperbranched samples. When separating highly branched samples according to size one cannot guarantee separation in only one dimension, the molar mass, due to the simultaneous distribution in branching and

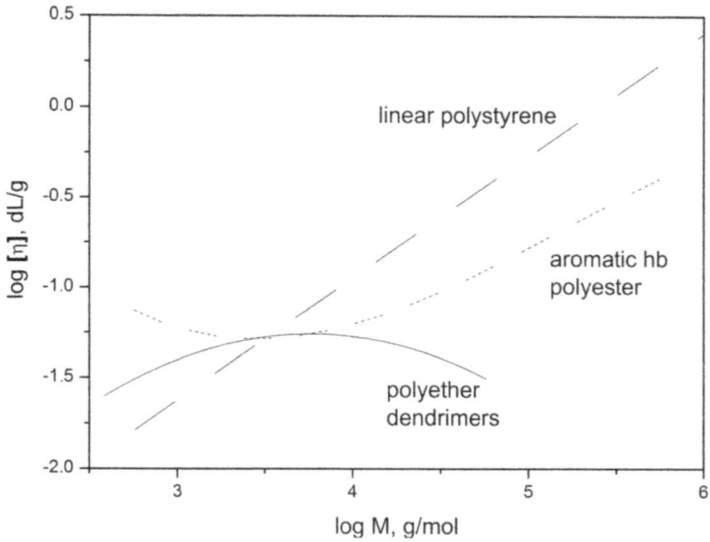

Figure 1.7 Dependence of the intrinsic viscosity on the molar mass for dendrimers and for hyperbranched polymers. The curve shape is schematically adapted from the works of Mourey *et al.*,[64] on polyether dendrimers and Turner *et al.*,[65] on aromatic hyperbranched polyesters.
Reprinted with permission from ref. 66. Copyright © 2009 American Chemical Society.

molecular density. An additional issue is the complexity of the branched structure, which is a combination of linear and branched units leading to a broad diversity of possible isomers.

Complete analysis of the structure–properties relationship could be achieved after understanding single parameters, molar mass, branching degree, chemical structure and the flexibility of the backbone, type and number of the end functionalities. All these tasks should be studied on model polymers, prepared by a combination of accurately chosen synthetic and separation techniques. Detailed characterization of the molecular conformation should show to what extent dendritic polymers can be described by the generalized theories developed for linear polymers in solution. During the last few years great effort was made to systematically develop model polymers with a strict variation of parameters for the validation and development of calculations designed for the special structural needs of hyperbranched polymers. Within this book we believe we offer an exceptional overview on hyperbranched structures showing their particular properties positioned between ideal linear polymers and dendrimers. This overview is based to a great extent on a uniquely systematic study of polyesters, the classical hyperbranched structure, although comparison with other chemical structures and topologies is inevitable and greatly needed for future adaptation of the proposed studies on further statistically branched macromolecular systems.

An understanding of the physics of molecules in solution closely correlates with the chemical background of the macromolecules, and this book aims to build a bridge between chemistry and physics in this respect.

References

1. H. Staudinger and G. V. Schulz, *Ber. Dtsch. Chem. Ges.*, 1935, **68**, 2320.
2. P. J. Flory, *J. Am. Chem. Soc.*, 1941, **63**, 3083.
3. W. H. Stockmayer, *J. Chem. Phys.*, 1943, **11**, 45.
4. W. H. Stockmayer, *J. Chem. Phys.*, 1944, **12**, 125.
5. W. Burchard, *personal communication.*
6. P. J. Flory, *Principles in Polymer Chemistry*, Cornell University Press, Ithaca, New York, 1953.
7. K. Meyer, *Natural and Synthetic High Polymers*, Interscience, New York, 1950.
8. H. Staudinger, *Die Hochmolekularen Organischen Verbindungen – Kautschuk und Cellulose*, Julius Springer, Berlin, 1932.
9. H. Staudinger and W. Heuer, *Ber. Dtsch. Chem. Ges.*, 1934, **67**, 1164.
10. H. Staudinger and E. Husemann, *Ber. Dtsch. Chem. Ges.*, 1935, **68**, 1691.
11. M. Kurata, M. Abe, M. Iwama and M. Matsushima, *Polym. J.*, 1972, **3**, 729–738.
12. W. Burchard, *Macromolecules*, 1972, **5**, 604.
13. Y. H. Kim and O. W. Webster, *J. Am. Chem. Soc.*, 1990, **112**, 4592.
14. B. Voit, *J. Polym. Sci., Part A: Polym. Chem.*, 2005, **43**, 267.
15. C. Gao and D. Yan, *Prog. Polym. Sci.*, 2004, **29**, 183.
16. C. R. Yates and W. Hayes, *Eur. Polym. J.*, 2004, **40**, 1257.
17. K. Inoue, *Prog. Polym. Sci.*, 2000, **25**, 453.
18. M. Seiler, *Chem. Eng. Technol.*, 2002, **25**, 237.
19. K. Olofsson, O. C. J. Andrén and M. Malkoch, *J. Appl. Polym. Sci.*, 2014, **131**, 39876.
20. M. A. Quadir and R. Haag, *J. Controlled Release*, 2012, **161**, 484.
21. www.webofknowledge.de, July 2014.
22. H. Staudinger and E. Husemann, *Justus Liebigs Ann. Chem.*, 1937, **527**, 195.
23. G. R. Newkome, C. N. Moorefield and F. Vögtle, *Dendrimers and Dendrons. Concepts, Syntheses, Applications*, Wiley-VCH, Weinheim, 2001.
24. E. Buhleier, W. Wehner and F. Vögtle, *Synthesis*, 1978, 155.
25. J.-M. Lehn, in *Alkali Metal Complexes with Organic Ligands*, Springer Berlin Heidelberg, 1973, ch. 1, vol. 16, p. 1.
26. R. C. H. D. J. Cram, L. R. Sousa, J. M. Timko, M. Newcomb, P. Moreau, F. de Jong, G. W. Gokel, D. H. Hoffman, L. A. Domeier, S. C. Peacock, K. Madan and L. Kaplan, *Pure Appl. Chem.*, 1975, **43**, 327.
27. S. M. Aharoni, C. R. Crosby and E. K. Walsh, *Macromolecules*, 1982, **15**, 1093.
28. D. A. Tomalia, H. Baker, J. Dewald, M. Hall, G. Kallos, S. Martin, J. Roeck, J. Ryder and P. Smith, *Polym. J.*, 1985, **17**, 117.

29. Y. H. Kim and O. W. Webster, *Macromolecules*, 1992, **25**, 5561.
30. R. H. Kienle, P. A. v. d. Meulen and F. E. Petke, *J. Am. Chem. Soc.*, 1939, **61**, 2258.
31. R. H. Kienle, P. A. v. d. Meulen and F. E. Petke, *J. Am. Chem. Soc.*, 1939, **61**, 2268.
32. P. J. Flory, *J. Am. Chem. Soc.*, 1941, **63**, 3096.
33. P. J. Flory, *J. Am. Chem. Soc.*, 1952, **74**, 2718.
34. H. R. Kricheldorf, Q.-Z. Zang and G. Schwarz, *Polymer*, 1982, **23**, 1821.
35. A. Calmark, C. J. Hawker, A. Hult and M. Malkoch, *Chem. Soc. Rev.*, 2009, **38**, 352.
36. D. D. Yan, C. Gao and H. Frey, *Hyperbranched Polymers: Synthesis, Properties and Applications*, John Wiley & Sons, New York, 2001.
37. A. Hult, M. Johannson and E. Malmström, *Adv. Polym. Sci.*, 1999, **143**, 1.
38. P. Froehling, *J. Polym. Sci., Part A: Polym. Chem.*, 2004, **42**, 3110.
39. B. Voit, *J. Polym. Sci., Part A: Polym. Chem.*, 2000, **38**, 2505.
40. M. Jikei and M. Kakimoto, *J. Polym. Sci., Part A: Polym. Chem.*, 2004, **42**, 1293.
41. J.-F. Stumbé and B. Bruchmann, *Macromol. Rapid Commun.*, 2004, **25**, 921.
42. B. Bruchmann, *Macromol. Mater. Eng.*, 2007, **292**, 981.
43. B. Bruchmann, R. Koeniger and H. Renz, *Macromol. Symp.*, 2002, **187**, 271.
44. B. Bruchmann, H. Malz, H. Renz and H. Schaefer, *Science et Techniques, Double Liaison*, 2004, **541**, 36.
45. R. Haag and M. Radowski, *WIPO Patent Application*, WO/2006/018295.
46. http://www.hyperpolymers.com/prodinf.html.
47. http://www.polymerfactory.com.
48. A. D. Schlüter, in *Topics in Current Chemistry*, ed. A. D. Schlüter, Springer, 2005, vol. 245, p. 151.
49. S. Boye, D. Appelhans, V. Boyko, S. Zschoche, H. Komber, P. Friedel, P. Formanek, A. Janke, B. Voit and A. Lederer, *Biomacromolecules*, 2012, **13**, 4222.
50. B. Zhang, R. Wepf, K. Fischer, M. Schmidt, S. Besse, P. Lindner, B. T. King, R. Sigel, P. Schurtenberger, Y. Talmon, Y. Ding, M. Kröger, A. Halperin and A. D. Schlüter, *Angew. Chem., Int. Ed.*, 2011, **50**, 737.
51. C. Schüll, L. Nuhn, C. Mangold, E. Christ, R. Zentel and H. Frey, *Macromolecules*, 2012, **45**, 5901.
52. C. Lach and H. Frey, *Macromolecules*, 1998, **31**, 2381.
53. C. Schüll and H. Frey, *ACS Macro Lett.*, 2012, **1**, 461.
54. C. Schüll, H. Rabbel, F. Schmid and H. Frey, *Macromolecules*, 2013, **46**, 5823.
55. X. Cao, Z. Li, X. Song, X. Cui, P. Cao, H. Liu, F. Cheng and Y. Chen, *Eur. Polym. J.*, 2008, **44**, 1060.
56. C. Gao and X. Zheng, *Soft Matter*, 2009, **5**, 4788.
57. A. Kowalczuk, A. Vandendriessche, B. Trzebicka, B. Mendrek, U. Szeluga, G. Cholewiński, M. Smet, A. Dworak and W. Dehaen, *J. Polym. Sci., Part A: Polym. Chem.*, 2009, **47**, 1120.

58. A. Kowalczuk-Bleja, B. Trzebicka, H. Komber, B. Voit and A. Dworak, *Polymer*, 2004, **45**, 9.
59. H. Hong, Y. Mai, Y. Zhou, D. Yan and Y. Chen, *J. Polym. Sci., Part A: Polym. Chem.*, 2008, **46**, 668.
60. M. Morell, A. Lederer, X. Ramis, B. Voit and A. Serra, *J. Polym. Sci., Part A: Polym. Chem.*, 2011, **49**, 2395.
61. M. Caldéron, M. A. Quadir, S. K. Sharma and R. Haag, *Adv. Mater.*, 2010, **22**, 190.
62. A. Lederer, W. Burchard, T. Hartmann, J. S. Haataja, N. Houbenov, A. Janke, P. Friedel, R. Schweins and P. Lindner, *Angew. Chem. Int. Ed.*, 2015, DOI: 10.1002/anie.201504059.
63. A. Lederer, T. Hartmann and H. Komber, *Macromol. Rapid Commun.*, 2012, **33**, 1440.
64. T. H. Mourey, S. R. Turner, M. Rubinstein, J. M. J. Fréchet, C. J. Hawker and K. L. Wooley, *Macromolecules*, 1992, **25**, 2401.
65. S. R. Turner, B. I. Voit and T. H. Mourey, *Macromolecules*, 1993, **26**, 46173.
66. B. I. Voit and A. Lederer, *Chem. Rev.*, 2009, **109**, 5924.
67. H. Mori, A. H. E. Müller and P. F. W. Simon, in *Macromolecular Engineering: Precise Synthesis, Materials, Properties, Applications*, ed. K. Matyjaszewski, Y. Gnanou and L. Leibler, Wiley-VCH, Weinheim, 2007, vol. 2, p. 973ff.
68. M. G. McKee, S. Unal, G. L. Wilkes and T. E. Long, *Prog. Polym. Sci.*, 2005, **30**, 507.
69. T. C. B. McLeish, in *Macromolecular Engineering: Precise Synthesis, Materials, Properties, Applications*, ed. K. Matyjaszewski, Y. Gnanou and L. Leibler, Wiley-VCH, Weinheim, 2007, vol. 2, p. 1605ff.
70. S. J. Teerstra and M. Gauthier, *Prog. Polym. Sci.*, 2004, **29**, 277.
71. M. Jikei and M.-A. Kakimoto, in *Progress in Polycondensation*, ed. T. Matsumoto, Research Signpost, Trivandrum, India, 2002, p. 19ff.
72. W. Burchard, in *Branched Polymers II*, ed. J. Roovers, 1999, vol. 143, p. 113.
73. W. Burchard, *Makromol. Chem.*, 1960, **42**, 151.
74. D. A. Tomalia, D. M. Hedstrand and L. R. Wilson, in *Encyclopedia of Polymers Science*, Wiley, New York, 2nd edn, 1990.
75. K. Esumi, *Top. Curr. Chem.*, 2003, **227**, 31.
76. E. M. M. de Brabander, J. Brackman, M. Mure-Mak, H. de Man, M. Hogeweg, J. Keulen, R. Scherrenberg, B. Coussens, Y. Mengerink and S. van der Wal, *Macromol. Symp.*, 1996, **102**, 9.
77. M. Ballauff, *Top. Curr. Chem.*, 2001, **212**, 177.
78. B. J. Bauer, B. Hammouda, J. D. Barnes, R. M. Briber and D. A. Tomalia, *Polym. Mater. Sci. Eng.*, 1994, **74**, 706.
79. D. Potschke, M. Ballauff, P. Lindner, M. Fischer and F. Vogtle, *Macromolecules*, 1999, **32**, 4079.
80. D. Yan, Z. Zhou and A. H. E. Müller, *Macromolecules*, 1999, **32**, 245.
81. W. Radke, G. Litvinenko and A. H. E. Müller, *Macromolecules*, 1998, 31.

82. D. Hölter and H. Frey, *Acta Polym.*, 1997, **48**, 298.
83. R. Hanselmann, D. Hölter and H. Frey, *Macromolecules*, 1998, **31**, 3790.
84. G. Litvinenko and A. H. E. Müller, *Macromolecules*, 2002, **35**, 4577.
85. D. Schmaljohann and B. Voit, *Macromol. Theory Simul.*, 2003, **12**, 679.
86. S. Boye, H. Komber, P. Friedel and A. Lederer, *Polymer*, 2010, **51**, 4110.
87. M. Al Samman, W. Radke, A. Khalyavina and A. Lederer, *Macromolecules*, 2010, **43**, 3215.
88. L. R. Hutchings, *Macromolecules*, 2012, **45**, 5621.
89. W. Radke, *J. Chromatogr. A*, 2014, **1335**, 62.
90. J. Aerts, *Comput. Theor. Polym. Sci.*, 1998, **8**, 49.
91. A. V. Lyulin, D. B. Adolf and G. R. Davies, *Macromolecules*, 2001, **34**, 3783.
92. P. F. Sheridan, D. B. Adolf, A. V. Lyulin, I. Neelov and G. R. Davies, *J. Chem. Phys.*, 2002, **117**, 7802.
93. A. H. Widmann and G. R. Davies, *Comput. Theor. Polym. Sci.*, 1998, **8**, 191.

CHAPTER 2

Degree of Branching

2.1 Definition of Branching

Branching in synthetic polymers was discovered in the 1940s in high pressure polyethylene and was systematically studied and used for modifying the properties of this mass product. With the increasing commercialization of branched polyethylene, the need to understand the structure–properties relationship, *i.e.* the influence of the branching, became essential. The theories and methods for the characterization of branching are therefore generally based on the studies of long-chain branched polymers and model star polymers.[1] In the next decades the theoretical treatment of the influence of branching on macromolecular properties was the focus of the renowned studies by Stockmayer, Zimm, Flory and many other scientists.

The definition of branching depends on the branching topology. In turn, the topology depends mainly on the method of branching formation. Finally, the type of branching defines the technique to be used for its determination. In star polymers, branching occurs only at the central unit and corresponds to the functionality of the core, which is equal to the number of arms. In the case of comb-like polymers we can differentiate between different branching frequencies and branch lengths (Figure 2.1).

A typical example is low density polyethylene (LDPE), which is characterized by long but only a few chain branches (LCB). In contrast, LLDPE possesses more but shorter branches. For such systems it is sensible to define the number of branches per molecule or per 1000 C-atoms, as usually used for polyolefins. For short-chain branched polymers, direct indication *via* spectroscopic methods is possible.[3–6] For long-chain branches, spectroscopy is not applicable because it is limited to chains with up to approximately six C-atoms. For longer chains detection and

RSC Polymer Chemistry Series No. 16
Hyperbranched Polymers: Macromolecules in between Deterministic Linear Chains and Dendrimer Structures
By Albena Lederer and Walther Burchard

Published by the Royal Society of Chemistry, www.rsc.org

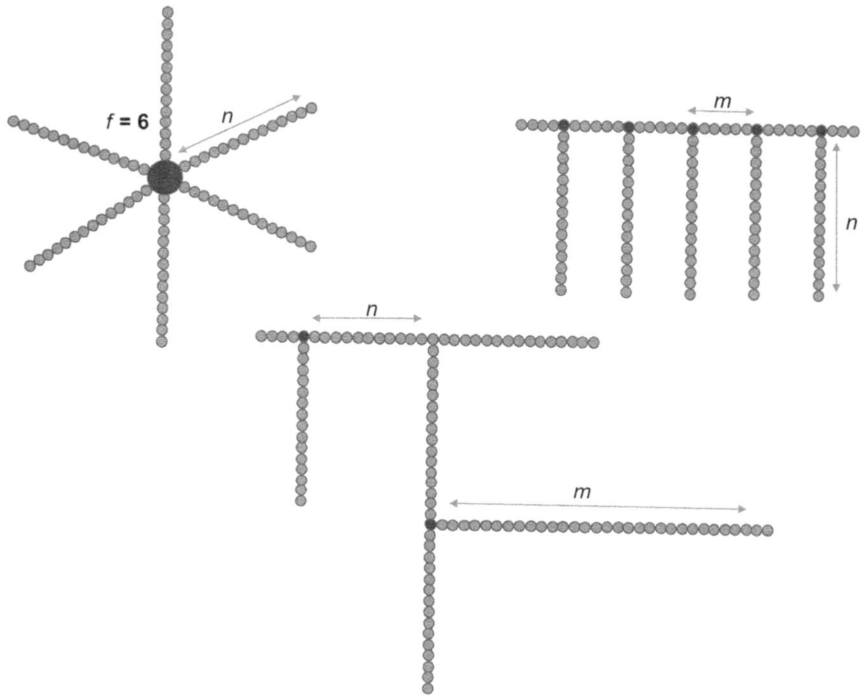

Figure 2.1 Schematic representation of star-like polymer with functionality *f* corresponding to the number of arms. The length of the arms and their distribution is significant for the scaling properties of the stars.[2] Comb-like polymers can have regular or random distribution in the length of the branches as well as in the segments between two branch points. Random branching as in case of long- or short-chain branched polymers is more complex in regard to branching topology.

determination of branching is based on the changes in the size of the molecules and the influence of the branching on their rheology properties in comparison to linear analogues. Figure 2.2 shows the difference between a branched and a linear chain of the *same* chemical structure: the molecular density. The principle is simple: if the molar mass of both topologies is the same, the branched one would occupy a smaller volume and would have higher density.

This important difference lends branched polymers some advantageous properties such as lower melt and solution viscosities. By using branched polymer structures improved polymer processing can be thus achieved. Linear, star or H-shaped polymers have been the subject of theoretical and experimental investigations for several decades and form the fundamentals of the scaling theories of branched polymers.[7–12] Dendritic branching leads to different topology classification and exceptional rheological behaviour,[13,14] which will be discussed in the next chapters. Despite the strong interest in the bulk rheology of branched polymers, only qualitative

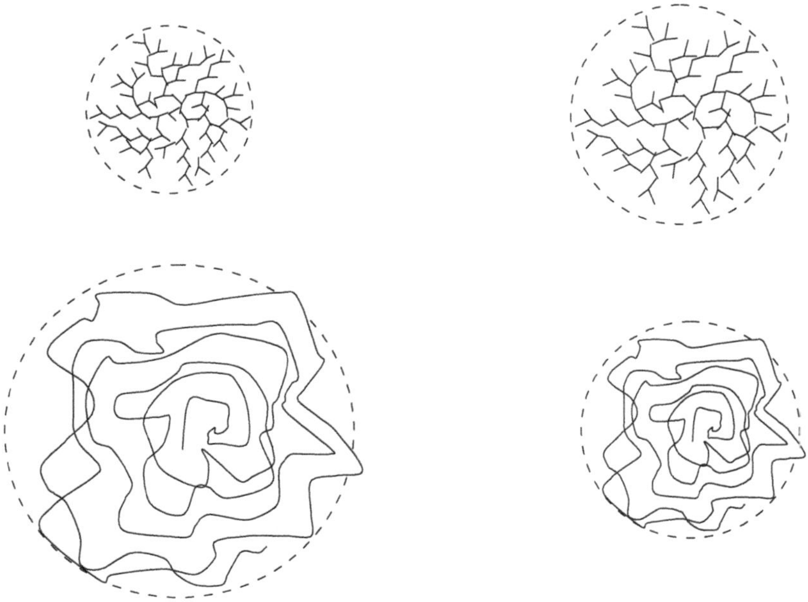

Figure 2.2 Comparison of linear and branched polymer coils.

estimations about their branching density were obtained.[15] Alternatively, solution rheology or dilute solution measurements of the molecular size yields quantitative values for the degree of branching in long-chain branched polymers. These calculations are based on theoretical considerations developed in the works of Zimm and Stockmayer.[16]

2.1.1 Contraction Factors

In order to obtain the frequency of branching points in a polymer sample, the reduction of the molecular size due to branching could be estimated by direct measurement of the radius of gyration (R_g) by scattering techniques leading to a value of the molar mass dependent contraction factor g according to eqn (2.1):

$$g = \frac{R^2_{\mathrm{g,branched}}}{R^2_{\mathrm{g,linear}}} \tag{2.1}$$

The contraction factor is a function of the number of branches per 1000 monomer units (λ) which is different for trifunctional and tetrafunctional branching points and for monodisperse and polydisperse samples.[16]

For monodisperse tri- and tetrafunctional branching, g can be calculated by the following equations:

$$g_3 = \left[\left(1+\frac{n}{7}\right)^{1/2} + \frac{4n}{9\pi}\right]^{-1/2} \tag{2.2}$$

$$g_4 = \left[\left(1+\frac{n}{6}\right)^{1/2} + \frac{4n}{3\pi}\right]^{-1/2} \tag{2.3}$$

For polydisperse samples the contraction factors can be calculated as follows:

$$g_3 = \frac{6}{n}\left\{\frac{1}{2}\left(\frac{2+n}{n}\right)^{1/2} \ln\left[\frac{(2+n)^{1/2}+n^{1/2}}{(2+n)^{1/2}-n^{1/2}}\right] - 1\right\} \tag{2.4}$$

$$g_4 = \frac{\ln(1+n)}{n} \tag{2.5}$$

where n is the number of branches per molecule. The influence of the type of functionality and polydispersity on the contraction could be followed in Figure 2.3.

The number of branches per molecule n can be converted to λ using the molar mass of the polymer M and of the monomer unit M_0:

$$\lambda = 1000n\frac{M_0}{M} \tag{2.6}$$

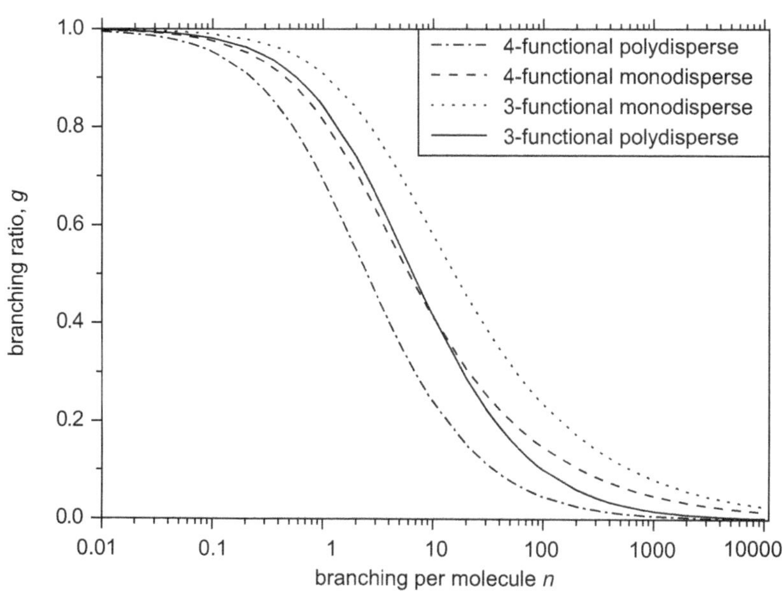

Figure 2.3 Influence of the number of branches per molecule on the contraction depending on branching functionality and distribution.

Due to the fact that the intrinsic viscosity is a function of the size of the molecules, as described by the Flory–Fox equation [eqn (2.7)],[17] the size reduction can be expressed using the ratio of the intrinsic viscosities of the branched and linear polymer [eqn (2.8)]:

$$[\eta] = \Phi\left(\frac{R_g^3}{M}\right) \tag{2.7}$$

$$g' = \frac{[\eta]_{\text{branched}}}{[\eta]_{\text{linear}}} \tag{2.8}$$

The determination of the intrinsic viscosity is significantly less elaborative than the measurement of radius of gyration. Thus a simple relationship between g and g' would be quite useful for calculation the number of branches according eqn (2.1)–(2.7). Therefore theoretical as well as experimental efforts were made to find a correlation between the two contraction factors.[17–19] The relationship between g and g' is the drainage exponent ε introduced by Kurata and Fukatsu,[20] which theoretically was predicted to lie between 0.5 and 1.5 depending on the type of branching[21,22] and is discussed in detail in Chapter 5.

Direct application of the branching calculations of Stockmayer and Zimm to the dendritic polymer class is not possible due to their significantly higher branching density. This feature leads to particular properties in conjunction with molecular shape and high functionality.

2.1.2 Degree of Branching

Theoretical considerations on hyperbranched chain formation were put forward in the 1950s by Paul Flory[23] in the course of his studies on gelation. At this time, practical experience with those materials was rather limited and the beginning of systematic experimental studies started in the 1980s. Flory predicted the formation of statistically branched macromolecules after self-condensation of multifunctional monomers leading to very broad molar mass distributions, non-entangled and non-crystalline polymers, properties unprofitable at that time and not considered to be of further scientific or application interest. However, Flory's theory supported the first synthetic steps in the classical hyperbranched chemistry based on AB_n monomers and predicted general physical properties, which helped to understand and optimize characterization procedures. According to these studies, if an AB_n monomer with B-functionality is capable of reacting only with A and the reactivity of the B-functionalities is constant, this would lead to a step growth reaction. In contrast to other multifunctional monomers, this reaction proceeds without the danger of cross-linking. The introduced repeating units in a hyperbranched molecule are dendritic when all B-functionalities are converted; linear if one B-function is converted; and terminal if no B-function is converted (Figure 2.4).

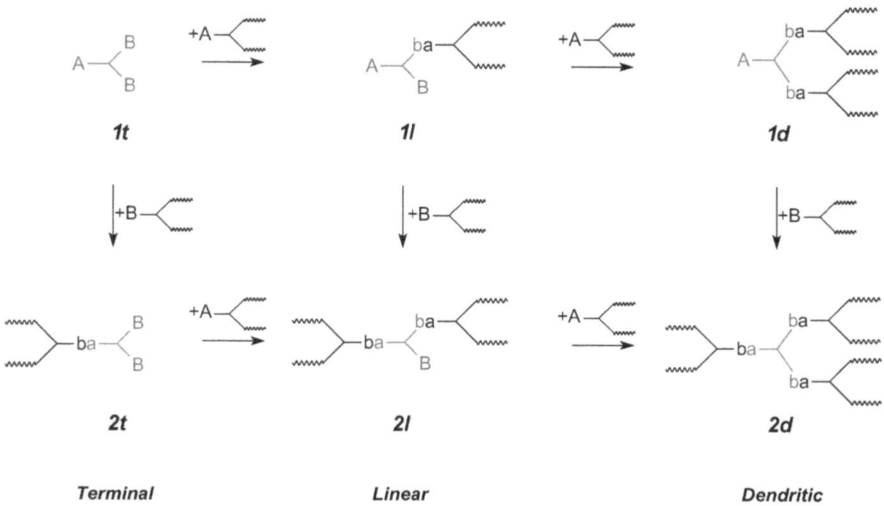

Figure 2.4 Structural units characterizing a hyperbranched polymer after self-condensation of an AB_2 monomer. Representation of possible terminal (T), linear (L) and dendritic (D) structural units. The first row of structures corresponds to unit with unreacted A-group, which is the *focal unit*.

Ideally, if the complete statistics is fulfilled, the fraction of the linear units should be the highest one with 50%, and the fraction of terminal units should be with 25% equal to the fraction of the dendritic units. In reality, the number of the different types of repeating units deviates from the ideal values due to the formation of by-products or cyclization during the polymerization process. This also leads to deviations from the ideally 50% degree of branching (DB), as discussed below. The estimation of the degree of branching based on the amount of the different types of repeat units – terminal, dendritic and linear (T, D and L) – was first proposed in the 1990s by Kim and Webster[24] and Hawker *et al.*:[25]

$$DB_{\text{Fréchet}} = \frac{T+D}{T+D+L} \tag{2.9}$$

In the extreme case of perfect dendritic branching, *i.e.* for dendrimers, $DB = 1$ (or 100%), because dendrimers do not possess linear units. The other extreme would be a completely linear chain, which should correspond to $DB = 0$. For hyperbranched polymers, we should take into account that $T = D + 1$. According to eqn (2.9), even fully linear structures would possess a $DB > 0$. This definition can indeed accurately describe the degree of branching at products with higher polymerization degrees, where the number of dendritic units approximates the number of the terminal units, but in the low molar mass region the amount of terminal units surpasses the degree of branching values. In fact, the higher degrees of polymerization could be well described by this equation, and since the number of dendritic units is equal to the number of terminal units, as predicted by Flory for ideal statistics, for high molar mass polymers the equation could be modified to[26]

$$\mathrm{DB} = \frac{2T}{2T + L} \tag{2.10}$$

In analogy to eqn (2.10) another relationship was suggested by Hölter *et al.*[27] and Yan *et al.*,[28] which has been found to be more suitable for the calculation of degree of branching in a broader sense, and successfully reflects the degree of branching for low and for high molar masses [eqn (2.11)]:

$$\mathrm{DB}_{\mathrm{Frey}} = \frac{2D}{2D + L} \tag{2.11}$$

Studies correlating the degree of branching with the degree of polymerization show that at lower degrees of polymerization the values of the degree of branching determined by eqn (2.9) and (2.11) deviate strongly from each other (Figure 2.5), although the degrees of branching calculated by both equations become closer at high degrees of polymerization. This fact was demonstrated in experimental studies on the kinetics of the polycondensation reaction of hyperbranched aliphatic–aromatic polyesters (Figure 2.5).[29,30]

Usually, the degree of branching is calculated from the number of the different polymer units using eqn (2.9)–(2.11), but another suitable way is by using the conversion of the functional groups. Taking into account the

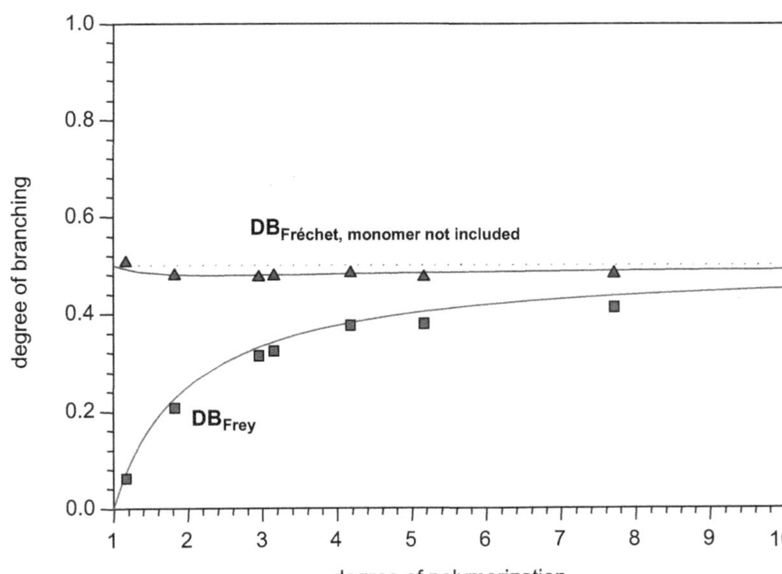

Figure 2.5 Dependence of the degree of branching on the polymerization degree as shown by Schmaljohann *et al.*[30]

A-functionalities in an AB_x-type polymer, only one group is expected to remain free in the case of full conversion, which means that

$$p_a = 1 - \frac{1}{DP} \tag{2.12}$$

$$DB = \frac{p_a}{2} \tag{2.13}$$

with p_a the conversion of the A-groups and DP the degree of polymerization. A prerequisite for the validity of this relationship is the validity of equal reactivity between the functional B-groups, *e.g.* fulfilled ideal statistics,[31] which is not always the case.[30,32]

Another definition of branching is the density of branching, *e.g.* the average number of branches in hyperbranched polymers out of AB_x monomers.[27] The average number of branches describes the deviation from linearity for the non-terminal units and depends on the number of functional B-groups x. As a result, in the case of ideal statistics and full conversion, for $x = 2$ the theoretical average number of branches is 0.333 and increases with increasing x.

2.1.3 Correlation Between the Degree of Branching and Contraction

So far definition of branching degree in dendritic chemistry does not correlate to the number of branching per 1000 monomer units, λ, well known in the analysis of long-chain branched and star polymers, which was discussed above. In order to obtain a general theory of branched polymers in solution such a correlation would be significant. For this reason, the concept of branch point fraction (FB)[33] has been introduced as

$$FB = \frac{D}{2D + L} = \frac{1}{2} DB = \frac{1}{1000} \lambda \tag{2.14}$$

Eqn (2.14) is an attempt to assign dendritic polymers to the polymer branch theory of Zimm and Stockmayer outlined above. Nevertheless, the specific dendritic branching topology with the potential for every monomer to form a branching point does not allow for direct comparison of dendritic branching to long-chain branched or star-branched polymers, for which physical properties are to a large extent governed by macromolecular entanglements.

This holds not only for the classical AB_n polycondensation described by Flory. Even hyperbranched polymers, products of synthetic approaches extended to $A_2 + B_m$ monomer combinations,[34–37] self-condensing vinyl polymerization,[38] ring-opening polymerization and proton transfer polymerization[33,39–41] as well as very complex self-assembled dendritic structures or branched hybrids[42–50] possess broad structural distributions and irregular branched structures which cannot fit into the physical theories of long-chain branching. Additionally, the common relationships that govern the molecular parameters of linear and long-chain branched polymers cannot be directly applied to dendritic polymers and, therefore, the characterization of these structures stands out as a great challenge. However, the immense interest in applications and the need to control and improve the material properties call for a clear understanding of their structure–property relationship. Therefore enormous effort has been made to fully characterize the structures of hyperbranched polymers, often driving polymer characterization techniques to their limits.

2.2 Determination of the Degree of Branching

2.2.1 Analysis by Spectroscopic Approaches

The calculation of the degree of branching is based on the spectroscopic determination of the structural units and their ratio using eqn (2.9)–(2.11). NMR measurements can be performed as shown in Figure 2.6 on a spectrum

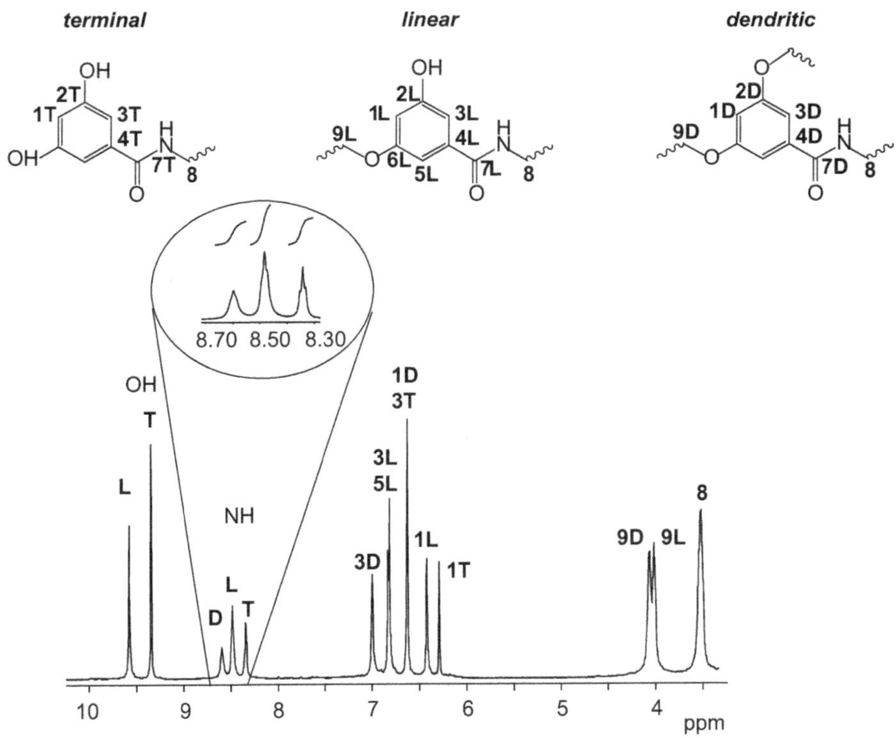

Figure 2.6 ^{1}H NMR spectrum used for the determination of the amount of *T*, *L* and *D* units in hyperbranched poly(etheramide).
Reprinted with permission from ref. 51. Copyright © 1999 Wiley-VCH Verlag GmbH & Co. KGaA, Weinheim.

of a hyperbranched poly(etheramide),[51] a product of the ring-opening polymerization of 2-oxazoline as an AB_2 monomer. The definition and assignment of the different structural units in this case is straightforward. Using the ratios of the intensities of the different signals in eqn (2.11), the degree of branching can be calculated.

Highly branched structures can be obtained after reaction of $A_2 + B_y$ monomers too, which possess several advantages, *e.g.* the combination of easily accessible monomers. Branching very similar to hyperbranching can be achieved during the process of cross-linking just before the critical conversion (gel point) is reached. This approach draws on the classical network formation, where A_2 and B_2 monomers are combined with B_y as a cross-linker, which is well known and has high technical relevance.[52] The gelation theory developed for this system can be applied also for $A_2 + B_y$ systems, postulating that the critical molar ratio of A- and B-groups at which gelation is reached corresponds to the full conversion of the groups, which are in a minority. Before reaching full conversion, highly branched polymers are obtained, which are still soluble. There are many studies showing the potential of this approach for hyperbranched polymers and tuning the synthetic conditions for control over molar mass, degree of branching and gel point.[53–56]

These molecular parameters depend on many factors, including reaction time and ratio of the monomers. Still, the typically broadly distributed molecular properties in terms of molar mass and branching topology and degree are the result. High molar masses are reached close to the gel point. It was found that reasonable molar masses can be achieved without gelation at a functionality ratio A : B of 2 : 3 leading to functionality conversion of A below 87%.[52,57] A functionality ratio of A : B of 1 : 2 results in systems with properties very similar to the AB_2 systems, in which the same functionality ratio is given. The danger of cross-linking in these systems is strongly reduced; however, lower molar masses can be produced. At an A : B ratio of 1 : 1 a degree of branching above 50% can easily be obtained.[57,58]

The definition of the terminal (T), linear (L) and dendritic (D) units after reaction of $A_2 + B_y$ is rather complicated compared to AB_2 poly-condensation. For determination of the degree of branching of co-polymers of type $A_2 + B_3$ different structural units have to be considered, as shown in Figure 2.7. In this case the dendritic groups are of type b_3 with three reacted functionalities, the Bb_2 groups would correspond to the linear units with two reacted functionalities and the B_2b group are the terminal units with one reacted B:[59]

$$\mathrm{DB} = \frac{2D}{2D + L} = \frac{2(\mathrm{b_3})}{2(\mathrm{b_3}) + (\mathrm{Bb_2})} \tag{2.15}$$

The interpretation of the NMR spectrum for calculation of degree of branching is not trivial and for an $A_2 + B_3$ reaction a number of possible

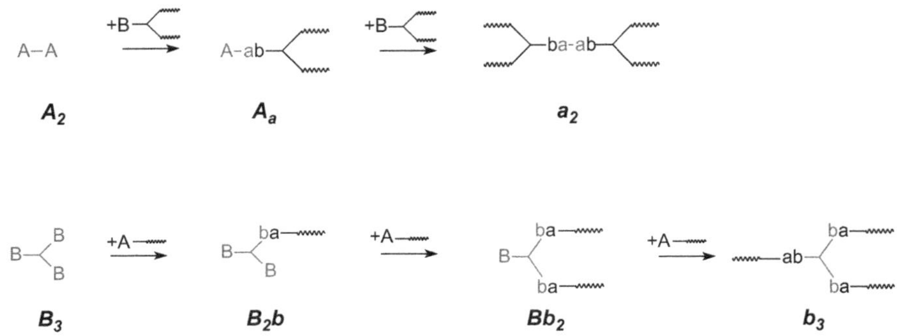

Figure 2.7 Possible reactions of the functional groups in an $A_2 + B_3$ systems and the resulting structural units.

structural combinations have to be considered.[59–63] Significant support to the interpretation of the NMR spectra is possible using the information from model compounds.[60,62,63]

Figure 2.8 shows an example for the ^{13}C NMR spectrum of hyperbranched polyphenylene synthesized from $A_2 + B_3$ in different ratios.[59] The different structural units were recognized only on the basis of model substances synthesized for this purpose.

The formation of different structural units is strongly influenced by the ratio of A_2 and B_3 monomers, so is the degree of branching, molar mass and gel formation. An important disadvantage of $A_2 + B_3$ co-polymers is their affinity to cross-link, leading to gel structures. Gel formation can be well controlled not only by the reaction time but especially by tuning the ratio between the two monomers as described above. Attempts to shift or even remove the gel point lead to the development of approaches with monomers having higher reactivity of one A or B function, *e.g.* AA' or $B'B_y$ monomers.[64–67] One typical example is hyperbranched poly(urea–urethane)[68,69] in which the monomers AA' and $B'B_2$ form an intermediate $A(A' - B')B_2$, which strongly resembles an AB_2 monomer and behaves as predicted by Flory's theory (Figure 2.9). The hyperbranched structure contains both urea and urethane groups and an NMR structural analysis verifies a degree of branching higher than 60%.[62]

2.2.2 Influence of the Synthetic Approach on the Degree of Branching

In the case of $A_2 + B_y$ polycondensation it is rather easy to reach a degree of branching higher than 50%, while in AB_x systems 50% is usually the highest value. In reality, dependence between the degree of branching and the degree of polymerization exists. Figure 2.5 shows an increasing degree of branching with the degree of polymerization, which contradicts the development of the degree of branching during a reaction in the ideal case.

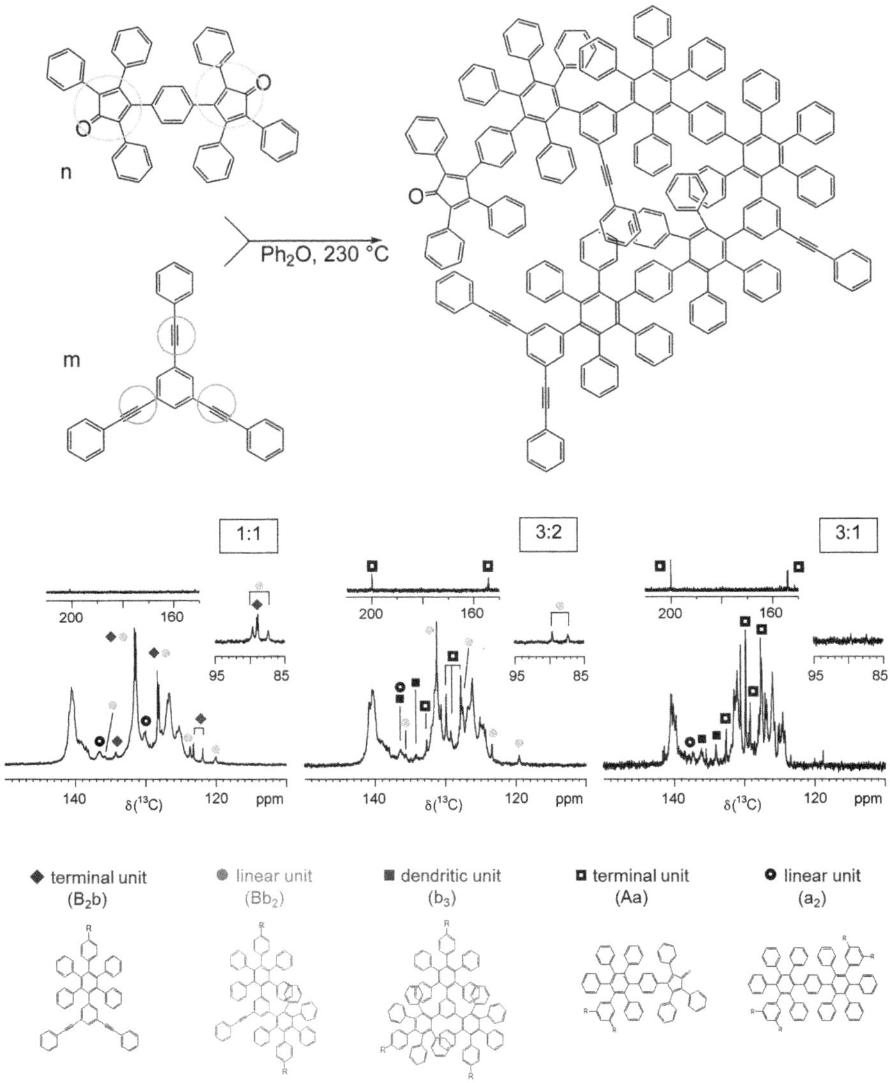

Figure 2.8 ^{13}C NMR spectra and different structural units of polyphenylenes produced by different ratios of $A_2 : B_3$.
Reprinted with permission from ref. 59. Copyright © 2006 Wiley-VCH Verlag GmbH & Co. KGaA, Weinheim.

According to the theory, the degree of branching remains constant and independent of the degree of polymerization. However, even in statistical polycondensation a dependence of the degree of branching on the reaction conversion has been observed. Using Frey's equation [eqn (2.11)], the kinetics of the formation of a hyperbranched aliphatic–aromatic polyester was investigated. Detailed interpretation of ^{13}C and ^{1}H NMR measurements show a strong increase in the degree of branching from DB = 0.06 to 0.40

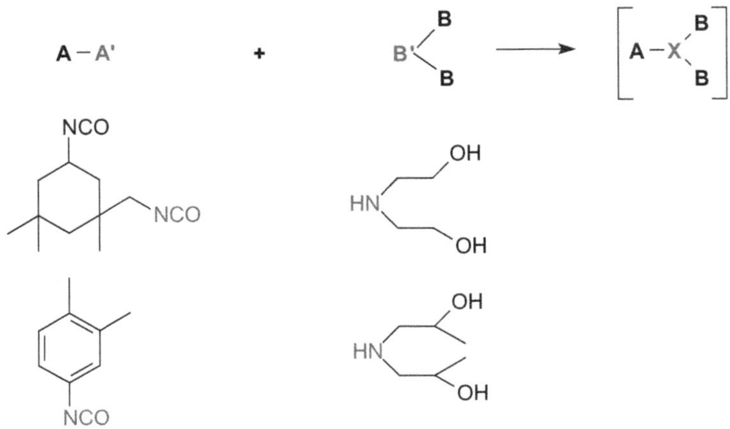

Figure 2.9 Intermediate formation out of AA′ and B′B_2 monomers, in which A′ and B′ react preferably with each other and an example of such monomers leading to hyperbranched poly(urea–urethanes).[62]

during the early stage of the polymerization (DP = 1–8). At high conversion of the A-group the expected degree of branching of DB = 0.5 can be reached, corresponding to ideal statistics.

A *degree of branching higher than* 50% corresponds to an increased number of functional groups and molecular density, approaching those of dendrimers. This was the goal in a number of investigations discussing different strategies. Enhanced degree of branching can be reached within certain limits by performing the reaction in the presence of a core molecule or by slow monomer addition during the reaction.[70,71] Special monomer design which enables tandem reactions was also used for this purpose.[72] All these techniques lead to a degree of branching higher than 50% but lower than complete branching of 100% in which only dendritic and terminal units in the dendritic structure exist. A 100% degree of branching was achieved using labile linear units, which decompose to the educts or react immediately with a terminal unit to a dendritic unit.[73–77] In this way, in the branched structure, no linear units can appear. Combinations of different monomer functionalities or post-synthetic modification of the functional groups with *e.g.* AB_2 units[26,35,78] are one direct way to reduce the linear units (see Chapter 3).

Tuning of the degree of branching within the formation of hyperbranching is certainly limited also in the case of $A_2 + B_3$ reaction. The sensitivity of the degree of branching in this case is controlled additionally by the ratio of the monomers, as discussed above. The sequence of the monomer addition is, however, a further parameter, which could lead to a tuned degree of branching and it was experimentally observed and theoretically treated.[57,79,80] The simulations by Schmaljohann and Voit[57] showed that the sequence of monomer addition at a constant 1 : 1 ratio of $A_2 : B_3$ influences strongly DB. Applying simultaneous stepwise addition of A_2 and B_3 a

DB = 0.65 could be achieved, while the stepwise addition of B_3 to A_2 solution leads to a DB = 0.91.

A *degree of branching below* 50% in hyperbranched polymers can be achieved by the co-polycondensation of multifunctional AB_x with bifunctional AB monomers.[81–85] In a similar manner, the combination of an inimer and a co-monomer in self-condensing vinyl polymerization (SCVP) could lead to a reduced branching as well.[38] The calculation of the degree of branching in co-polymer systems was investigated in detail by Frey and Hölter taking into account the conversions of the different functionalities.[86] For a co-condensation of AB_2 + AB, the degree of branching can be described simply by the following equation:

$$\mathrm{DB} = \frac{2D}{2D + L_{\mathrm{AB_2}} + L_{\mathrm{AB}}} \tag{2.16}$$

Another way to calculate the degree of branching is by using the conversion of the functional groups which is described in more detail, using a particular example, in Section 3.2.

By controlling the ratio between AB_2 and bifunctional AB monomers the degree of branching can be adjusted, as shown in the work of Schallausky *et al.*[83] Figure 3.7 in Chapter 3 shows the ratio $r_p = [AB]/[AB_2]$ of the monomer units in the formed hyperbranched polymer as indicated by ^{13}C NMR and the ratio $r_m = [AB]_0/[AB_2]_0$ in the feed. Full incorporation of the monomer into the hyperbranched polymer leads to a tuned reduction of the degree of branching. Unfortunately, in this case a linear product from the AB monomers was not produced due to cyclization reactions in the early stage of the reaction.

Successful control over the degree of branching from linear (0%) up to hyperbranched (50%) was achieved in homopolymers of the AB_2 monomer. For this purpose protection of the one B-group leading to an ABB′ monomer must be performed in order to obtain a bifunctional diluting agent of a same chemical structure. After polycondensation of AB_2 and ABB′, a deprotection results in the desired polymer (see Chapter 3).[87,88]

2.2.3 Side Reactions and the Degree of Branching

The statistical polycondensation process can be strongly influenced by additional parameters, which lead to deviation from the theoretically calculated 50% degree of branching. The equal reactivity of the different functional groups is responsible for reaching high degrees of branching but even a low extent of side reactions could strongly influence these high values. Typical products of side reactions are intramolecular ring formation or cross-linking when A does not react selectively enough with B. The cyclization is a well-known side reaction in AB_x and $A_2 + B_y$ polymerization systems and is discussed in a number of reports.[61,89–98] Kricheldorf[99] performed extensive investigations on the formation of these structures, establishing a theory which explains the deviations of the hyperbranched structures from the cascade theory of Flory. Ring formation is a complex analytical problem, since it cannot be identified by the trivial spectroscopic and chromatographic characterization methods used in the routine characterization of synthetic materials, although cyclic structures could strongly influence the number of functional groups and eventually the material properties. A systematic analysis *via* combination of techniques could be helpful for analysing cyclic structures. MALDI-TOF-MS is a powerful technique giving information about different structural formations and repeating units in complex polymer systems. It could lead to quantitative information about cycle formation in hyperbranched molecules, as was done in the work of Jaumann *et al.*[98] on polyalkoxysiloxanes, calculating an exact number of one cycle per seven silicon atoms. The question of the distribution and extent of cycles in a hyperbranched sample, however, cannot be given by this technique. A combination with NMR[89,97] measurements and SEC separation of the broadly distributed samples can lead to a comprehensive data set about the distribution of cycles within a complex mixture of hyperbranched macromolecules.

A further possibility for detecting cycles is SEC-UV, in the case that the UV absorbance of the formed rings deviates from the main structure. Simon and Müller[96] used the special UV absorbance of β-ketoesters to indicate their formation. β-Ketoesters are the product of the back-biting reaction during the self-condensing group transfer polymerization of methacrylate in the low molar mass region. However, quantitative analysis of the extent of cycles by this technique is not common.

The cyclization behaviour of the well-known Boltron-type aliphatic polyester[100] was a subject of several studies. It was analysed extensively, *e.g.* by applying polymer degradation techniques[101] or kinetic model calculations,[94] and the strong affinity to ring formation has been repeatedly confirmed. It was found that *intra*molecular bonding and *inter*molecular side reactions with the same chemical character could lead to cross-linking or cycles. Komber *et al.*[89] investigated these processes and verified etherification as a side reaction during the polycondensation of 2,2-bis(hydroxymethyl)propionic acid. On the one hand, self-condensation leads to ether or ester bond formation on an intramolecular level. On the other hand, cross-linking through ether formation between the molecules takes place. Both possible reactions are shown in Figure 2.10. The indication of the latter process is based on the detection of more than one focal group per molecule, which can be shown by NMR. At very high polymerization degrees this side reaction could even lead to gelation.

High-resolution ^{13}C NMR shows up to 12% ether bonds after hydrolysis of the hyperbranched polyester. Using kinetics analysis it was shown that ether formation takes place during the whole polymerization process and not only at high conversion of the A-groups. Depending on the incorporation of core molecule this content can be controlled to a certain extent, but values up to 50% ether formation were found as characterized by MALDI-TOF.[97]

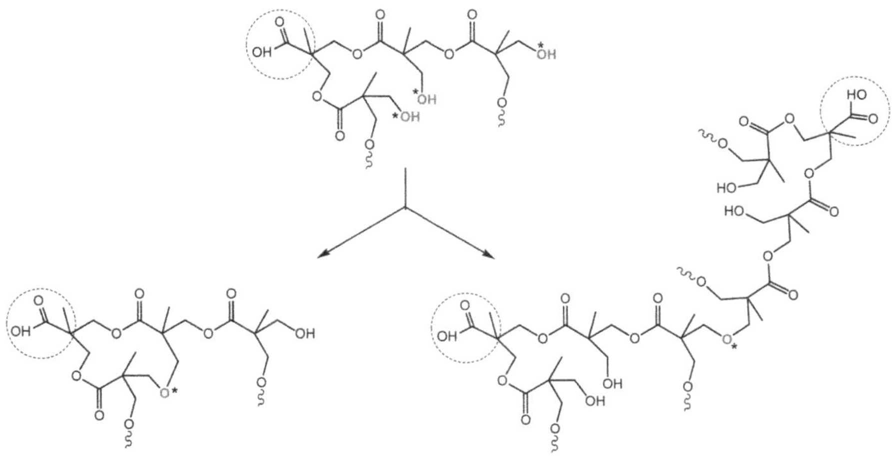

Figure 2.10 Etherification as a side reaction in aliphatic hyperbranched polyester. Symbols indicated by an asterisk are reacting functional groups. Intramolecular etherification leads to cycles. Intermolecular etherification leads to molecules with more than one focal unit (marked with a circle).[89]

References

1. J. R. Schaefgen and P. J. Flory, *J. Am. Chem. Soc.*, 1948, **70**, 2709.
2. W. Burchard, in *Branched Polymers II*, ed. J. Roovers, 1999, vol. 143, p. 113.
3. K. Klimke, M. Parkinson, C. Piel, W. Kaminsky, H. W. Spiess and M. Wilhelm, *Macromol. Chem. Phys.*, 2006, **207**(4), 382.
4. M. Pollard, K. Klimke, R. Graf, H. W. Spiess, M. Wilhelm, O. Sperber, C. Piel and W. Kaminsky, *Macromolecules*, 2004, **37**, 813.
5. W. Kaminsky, in *Handbook of Polymer Synthesis*, Plastics Engineering, New York, 2nd edn, 2005, vol. 70, p. 1ff.
6. L. Wild, T. Ryle and D. Knobeloch, *Polym. Prepr. (Am. Chem. Soc., Div. Polym. Chem.)*, 1982, **23**, 133.
7. W. W. Graessley and J. Roovers, *Macromolecules*, 1979, **12**, 5.
8. P. G. DeGennes, *J. Phys., Lett.*, 1975, **36**.
9. T. C. B. McLeish and S. T. Milner, *Adv. Polym. Sci.*, 1999, **143**, 195.
10. T. Pakula, D. Vlassopoulos, G. Fytas and J. Roovers, *Macromolecules*, 1998, **31**, 8931.
11. J. Roovers, *Macromol. Symp.*, 1997, **121**, 89.
12. D. Vlassopoulos, T. Pakula, G. Fytas, J. Roovers, K. Karatasos and N. Hadjichtistidis, *Europhys. Lett.*, 1997, **39**, 617.
13. P. J. Farrington, C. J. Hawker, J. M. J. Fréchet and M. E. Mackay, *Macromolecules*, 1998, **31**, 5043.
14. I. Sendijarevic and A. J. McHugh, *Macromolecules*, 2000, **33**, 590.
15. D. Auhl, J. Stange, H. Münstedt, B. Krause, D. Voigt, A. Lederer, U. Lappan and K. Lunkwitz, *Macromolecules*, 2004, **37**, 9465.

16. B. H. Zimm and W. H. Stockmayer, *J. Chem. Phys.*, 1949, **17**, 1301.
17. P. J. Flory and T. G. J. Fox, *J. Am. Chem. Soc.*, 1951, **73**, 1904.
18. A. Lederer, W. Burchard, A. Khalyavina, P. Lindner and R. Schweins, *Angew. Chem., Int. Ed.*, 2013, **52**, 4659.
19. A. Lederer, W. Burchard, T. Hartmann, J. S. Haataja, N. Houbenov, A. Janke, P. Friedel, R. Schweins and P. Lindner, *Angew. Chem. Int. Ed.*, 2015, DOI: 10.1002/anie.201504059.
20. M. Kurata and M. Fukatsu, *J. Chem. Phys.*, 1964, **41**, 2934.
21. W. H. Stockmayer and M. Fixman, *Ann. N. Y. Acad. Sci.*, 1953, **57**, 334.
22. B. H. Zimm and R. W. Kilb, *J. Polym. Sci.*, 1959, **37**, 19.
23. P. J. Flory, *Principles in Polymer Chemistry*, Cornell University Press, Ithaca, New York, 1953.
24. Y. H. Kim and O. W. Webster, *Macromolecules*, 1992, **25**, 5561.
25. C. J. Hawker, R. Lee and J. M. J. Fréchet, *J. Am. Chem. Soc.*, 1991, **113**, 4583.
26. C. J. Hawker and F. Chu, *Macromolecules*, 1996, **29**, 4370.
27. D. Hölter, A. Burgath and H. Frey, *Acta Polym.*, 1997, **48**, 30.
28. D. Yan, A. H. E. Müller and K. Matyjaszewski, *Macromolecules*, 1997, **30**, 7024.
29. D. Schmaljohann, H. Komber, J. G. Barratt, D. Appelhans and B. I. Voit, *Macromolecules*, 2003, **36**, 97.
30. D. Schmaljohann, H. Komber and B. Voit, *Acta Polym.*, 1999, **50**, 196.
31. P. J. Flory, *J. Am. Chem. Soc.*, 1941, **63**, 3096.
32. C. Cheng, K. L. Wooley and E. Khoshdel, *J. Polym. Sci., Part A: Polym. Chem.*, 2005, **43**, 4754.
33. H. Mori, A. H. E. Müller and P. F. W. Simon, in *Macromolecular Engineering: Precise Synthesis, Materials, Properties, Applications.*, ed. K. Matyjaszewski, Y. Gnanou and L. Leibler, Wiley-VCH, Weinheim, 2007, vol. 2, p. 973ff.
34. C. Gao and D. Yan, *Prog. Polym. Sci.*, 2004, **29**, 183.
35. M. Jikei and M. Kakimoto, *J. Polym. Sci., Part A: Polym. Chem.*, 2004, **42**, 1293.
36. M. Jikei, S.-H. Chon, M. Kakimoto, S. Kawauchi, T. Imase and J. Watanabe, *Macromolecules*, 1999, **32**, 2061.
37. T. Emrick, H.-T. Chang and J. M. J. Fréchet, *Macromolecules*, 1999, **32**, 6380.
38. J. M. J. Fréchet, H. Henmi, I. Gitsov, S. Aoshima, M. R. Leduc and R. B. Grubbs, *Science*, 1995, **269**, 1080.
39. P. Kubisa, *J. Polym. Sci., Part A: Polym. Chem.*, 2003, **41**, 457.
40. A. Sunder, R. Hanselmann, H. Frey and R. Mülhaupt, *Macromolecules*, 1999, **32**, 4240.
41. M. Suzuki, A. Ii and T. Saegusa, *Macromolecules*, 1992, **25**, 7071.
42. S. Xu, Y. Luo and R. Haag, *Macromol. Biosci.*, 2007, **7**, 968.
43. A. D. Schlüter, in *Topics in Current Chemistry*, ed. A. D. Schlüter, Springer, 2005, vol. 245, p. 151.

44. A. D. Tomalia, D. M. Hedstrand and M. S. Ferrito, *Macromolecules*, 1991, **24**, 1435.
45. R. Freudenberg, W. Claussen and A. D. Schlüter, *Polymer*, 1994, **35**, 4496.
46. H. Frauenrath, *Prog. Polym. Sci.*, 2005, **30**, 325.
47. M. Gauthier and M. Möller, *Macromolecules*, 1991, **24**, 4548.
48. M. Gauthier, *J. Polym. Sci., Part A: Polym. Chem.*, 2007, **45**, 3803.
49. C. Liu, G. Wang, Y. Zhang and J. Huang, *J. Appl. Polym. Sci.*, 2008, **108**, 777.
50. A. R. Brenner and B. Voit, *Macromol. Chem. Phys.*, 1996, **197**, 2673.
51. T. Huber, F. Böhme, H. Komber, J. Kronek, J. Luston, D. Voigt and B. Voit, *Macromol. Chem. Phys.*, 1999, **200**, 126.
52. P. J. Flory, *J. Am. Chem. Soc.*, 1952, **74**, 2718.
53. T. Emrick, H.-T. Chang and J. M. J. Fréchet, *Macromolecules*, 1999, **32**, 6380.
54. M. Jikei, S.-H. Chon, M.-a. Kakimoto, S. Kawauchi, T. Imase and J. Watanebe, *Macromolecules*, 1999, **32**, 2061.
55. S. Russo, A. Boulares, A. Da Rin, A. Mariani and M. E. Cosulich, *Macromol. Symp.*, 1999, **143**, 309.
56. S. M. Aharoni, *Polym. Adv. Technol.*, 1995, **6**, 373.
57. D. Schmaljohann and B. Voit, *Macromol. Theory Simul.*, 2003, **12**, 679.
58. A. Reisch, H. Komber and B. Voit, *Macromolecules*, 2007, **40**, 6846.
59. H. Komber, K. Stumpe and B. Voit, *Macromol. Chem. Phys.*, 2006, **207**, 1814.
60. A. R. Fornof, T. E. Glass and T. E. Long, *Macromol. Chem. Phys.*, 2006, **207**, 1197.
61. S. Unal, C. Oguz, E. Yilgor, M. Gallivan, T. E. Long and I. Yilgor, *Polymer*, 2005, **46**, 4533.
62. M. Abdelrehim, H. Komber, J. Langenwalter, B. Voit and B. Bruchmann, *J. Polym. Sci., Part A: Polym. Chem.*, 2004, **42**, 3062.
63. K. Stumpe, H. Komber and B. I. Voit, *Macromol. Chem. Phys.*, 2006, **207**, 1825.
64. S. P. Rannard and N. J. Davis, *J. Am. Chem. Soc.*, 2000, **122**, 11729.
65. S.-W. Zhu and W.-F. Shi, *Polym. Int.*, 2002, **51**, 223.
66. C. Gao, D. Yan and W. Tang, *Macromol. Chem. Phys.*, 2001, **202**, 2623.
67. C. Gao and D. Yan, *Macromolecules*, 2003, **36**, 613.
68. B. Bruchmann, R. Koeniger and H. Renz, *Macromol. Symp.*, 2002, **187**, 271.
69. B. Bruchmann and W. Schrepp, *e-Polym.*, 2003, **014**.
70. A. Mock, A. Burgath, R. Hanselmann and H. Frey, *Macromolecules*, 2001, **34**, 7692.
71. D. Hölter and H. Frey, *Acta Polym.*, 1997, **48**, 298.
72. J. Wang and M. Johnson, *Polym. Int.*, 2009, **58**, 1234.
73. Y. Fu, A. Vandendriessche, W. Dehaen and M. Smet, *Macromolecules*, 2006, **39**, 5183.
74. W. Sinananwanich and M. Ueda, *J. Polym. Sci., Part A: Polym. Chem.*, 2008, **46**, 2689.

75. M. Smet, E. Schacht and W. De Haen, *Angew. Chem., Int. Ed.*, 2002, **41**, 4547.
76. G. Maier, C. Zech, B. Voit and H. Komber, *Macromol. Chem. Phys.*, 1998, **199**, 2655.
77. B. Voit, H. Komber and A. Lederer, in *Synthesis of polymers*, ed. A. D. Schlüter, C. J. Hawker and J. Sakamoto, Wiley-VCH, Weihnheim, 2012.
78. A. Lederer, T. Hartmann and H. Komber, *Macromol. Rapid Commun.*, 2012, **33**, 1440.
79. S. Unal, Q. Lin, T. H. Mourey and T. E. Long, *Macromolecules*, 2005, **38**, 3246.
80. S. Unal, C. Oguz, E. Yilgor, M. Gallivan, T. E. Long and I. Yilgor, *Polymer*, 2005, **46**, 4533.
81. J.-B. Baek and L.-S. Tan, *Polymer*, 2003, **44**, 3451.
82. M. Jikei, K. Fujii and M.-a. Kakimoto, *Macromol. Symp.*, 2003, **199**, 223.
83. F. Schallausky, M. Erber, H. Komber and A. Lederer, *Macromol. Chem. Phys.*, 2008, **209**, 2331.
84. M. G. McKee, S. Unal, G. L. Wilkes and T. E. Long, *Prog. Polym. Sci.*, 2005, **30**, 507.
85. H. R. Kricheldorf, Q.-Z. Zang and G. Schwarz, *Polymer*, 1982, **23**, 1821.
86. H. Frey and D. Hölter, *Acta Polym.*, 1999, **50**, 67.
87. K. L. Wooley, J. M. J. Fréchet and C. J. Hawker, *Polymer*, 1994, **35**, 4489.
88. A. Khalyavina, F. Schallausky, H. Komber, M. Al Samman, W. Radke and A. Lederer, *Macromolecules*, 2010, **43**, 3268.
89. H. Komber, A. Ziemer and B. Voit, *Macromolecules*, 2002, **35**, 3514.
90. A. Burgath, A. Sunder and H. Frey, *Macromol. Chem. Phys.*, 2000, **201**, 782.
91. C. A. Martínez and A. S. Hay, *J. Polym. Sci., Part A: Polym. Chem.*, 1997, **35**, 2015.
92. V. Percec, P. Chu and M. Kawasumi, *Macromolecules*, 1994, **27**, 4441.
93. J. K. Gooden, M. L. Gross, A. Mueller, A. D. Stefanescu and K. L. Wooley, *J. Am. Chem. Soc.*, 1998, **120**, 10180.
94. K. Dušek, J. Šomvársky, M. Smrčková, W. J. Simonsick Jr and L. Wilczek, *Polym. Bull.*, 1999, **42**, 489.
95. D. Parker and W. J. Feast, *Macromolecules*, 2001, **34**, 2048.
96. P. F. W. Simon and A. H. E. Müller, *Macromol. Symp.*, 2006, **240**, 83.
97. L. Chikh, M. Tessier and A. Fradet, *Polymer*, 2007, **48**, 1884.
98. M. Jaumann, E. A. Rebrov, V. V. Kazakova, A. M. Muzafarov, W. Goedel and M. Möller, *Macromol. Chem. Phys.*, 2003, **204**, 1014.
99. H. R. Kricheldorf, *Macromol. Rapid Commun.*, 2007, **28**, 1839.
100. A. Hult, M. Johannson and E. Malmström, *Adv. Polym. Sci.*, 1999, **143**, 1.
101. P. Kamborius and C. J. Hawker, *J. Chem. Soc., Perkin Trans. 1*, 1993, 2717.

CHAPTER 3

Control over the Conformation of Dendritic Polymers

3.1 Introduction

Dendrimers and hyperbranched polymers are prepared by entirely different synthetic approaches, reflected in their structural properties. Dendrimers are synthesized *via* multi-step reactions leading to the formation of one generation at every step. This procedure leads to well-defined macromolecules with a theoretical molar mass distribution of 1, symmetrical branches and a degree of branching of 100% corresponding to the formation of dendritic and terminal units (Figure 3.1). This synthesis is very elaborative and heterogeneity is to be expected only in the case of defects, which could have a genealogical character.

As a dendritic polymer class made by the 'one-pot' approach, the hyperbranched polymers possess one clear advantage: the ease of their synthesis. Generally, the hyperbranching is obtained by the self-condensation of AB_x $(x \geq 2)$ monomers. In Figure 3.2 the polycondensation of an AB_2 monomer resulting in a hyperbranched polymer is shown.

In the ideal case, the bonding and branching formation during the reaction are controlled by statistics. This fact does not allow for tuning narrowly distributed molecular parameters in hyperbranched polymers as this is the case in the step-wise synthesis of dendrimers. The influence of this statistical character on the structure–properties relationship is, however, essential with respect to the high application potential of hyperbranched structures.

Systematical structural characterization is possible only if samples with variation in only one parameter are available. However, commonly synthesized hyperbranched structures possess at least two distributions: in

RSC Polymer Chemistry Series No. 16
Hyperbranched Polymers: Macromolecules in between Deterministic Linear Chains and Dendrimer Structures
By Albena Lederer and Walther Burchard

Published by the Royal Society of Chemistry, www.rsc.org

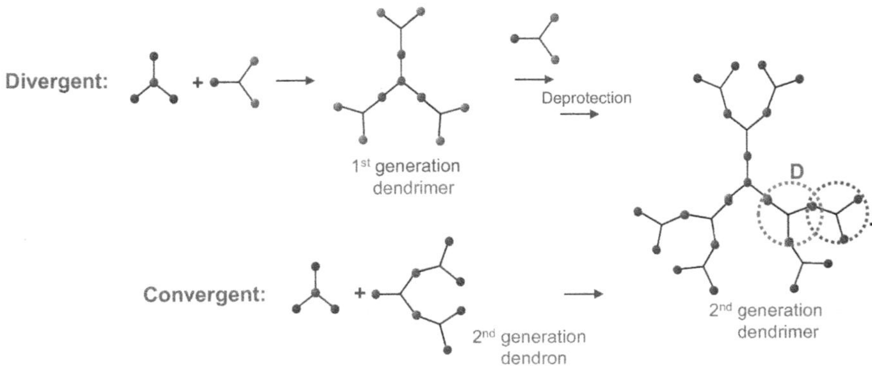

Figure 3.1 Scheme of different approaches for 2nd generation dendrimer synthesis. Dendrimers consist of dendritic (*D*) and terminal (*T*) units only.

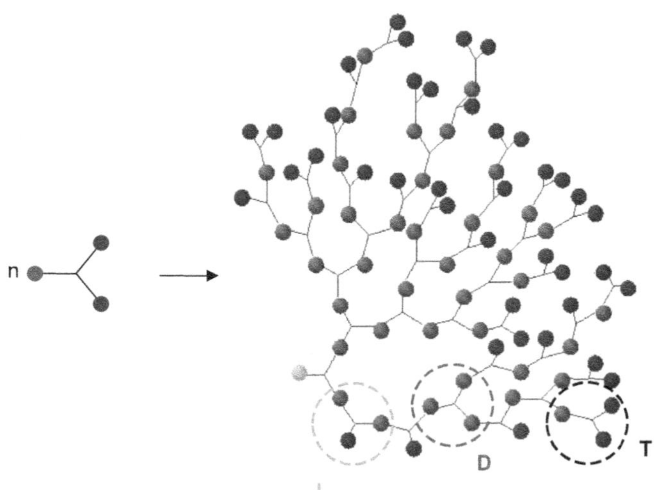

Figure 3.2 Schematic representation of the polycondensation of an AB_2 monomer to a hyperbranched polymer within a one-pot reaction. The light grey point is the free A-functionality, the black point is the free B-functionality and dark grey is the reacted A with B. (See also Figure 1.4.)

molar mass and in branching or chemical character. The *broad molar mass distribution* is not only experimentally observed but it was also theoretically predicted by Flory[1] within his calculations on polycondensation reactions of AB_x ($x \geq 2$) monomers. In contrast to linear polycondensation or living polymerization, the molar mass dependence of the degree of polymerization degree increases approximately with $M^{1/2}$ (Figure 3.3).

The existence of *branching distribution* was observed experimentally. On the basis of polyesters Schmaljohann *et al.* showed that the degree of branching could strongly change with the degree of polymerization especially at lower degrees of polymerization (Figure 2.5).[2]

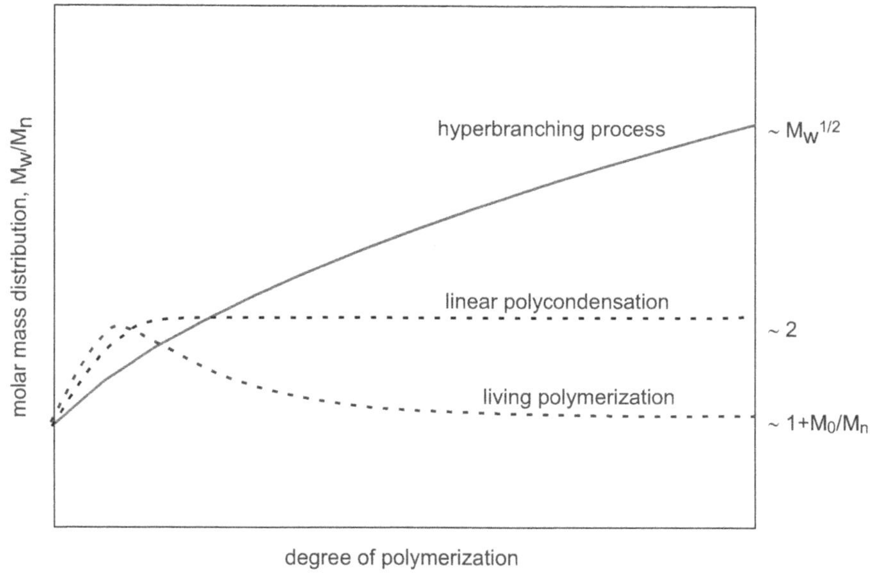

Figure 3.3 Schematic representation of the dependence of the molar mass on the degree of polymerization for different types of reaction.

Finally, the chemical character distribution could be also a non-discrete value, as is usual situation in macromolecular chemistry. More specifically, for multifunctional polymers the *number and type of end groups* plays an essential role in defining their solution or bulk properties. The reason for that is their high number which in AB_2-type polymers is theoretically nearly equal to the degree of polymerization, *e.g.* $N = DP - 1$.

Understanding the influence of the different parameters is only possible if these parameters are systematically varied in model polymers. In order to fulfil the requirements for model polymers with systematically fixed structural parameters, a portfolio of samples with a controlled variation in degree of branching, molar mass and chemical structure is needed.

Some examples of well-known synthetic routes for hyperbranched polymers are discussed below. In this book hyperbranched poly(etheramide)s,[3] polyesters[4–6] and polyurethanes[7] prepared by well-understood synthesis from commercially available monomers form the synthetic background for the discussion on the physical properties in the next chapters. An important aspect regarding the chemical character of these systems is also their analogy to some commercially relevant polymer classes, which could facilitate direct adoption of the proposed interpretation to real applications. Poly(etheramide)s are chosen due to their high application potential in blends with linear polymers for improving their processing and mechanical properties.[8,9] Poly(urea–urethane)s are known as cross-linkers in coating formulations.[10,11] Together with Boltorn™-type aliphatic polyesters with well-known application fields,[12–15] aliphatic–aromatic polyesters with high commercialization potential in coatings,[16,17] nano-composite formulations[18–20] and in thin films for protein adsorption[21] are also polymers described in this chapter.

3.2 Tuning the Degree of Branching

In order to achieve a full picture of the influence of the degree of branching on the solution properties, controlled variation of this parameter at fixed chemical structure and molar mass should be performed. Certain control over the degree of branching (DB) could be achieved by using the slow monomer addition approach or core molecule,[22,23] but leading to a limited portfolio of degrees of branching. Other methods are based on the combination of bifunctional with multifunctional monomers and this approach seems rather auspicious for this purpose.[4,5,24–26] Some other approaches are discussed also in Chapter 2. Quite successfully, the degree of branching in polymers based on AB_2 monomers could be tailored by adding different amounts of bifunctional monomer to the reaction mixture (Figure 3.4). The major challenge here is to keep the type of chemical structure constant at every degree of branching. For this purpose two approaches could be used: dilution of the branching degree (1) with well-accessible AB monomer of similar chemical character (the AB_2+AB approach),[5,27,28] and (2) with mono-protected ABB* monomer, with A and B being active functions, which is far more challenging (AB_2+ABB^*).[4]

3.2.1 The AB_2+AB Approach

As an example for the AB_2+AB approach here the co-polycondensation of 4,4-bis(4′-hydroxyphenyl)valeric acid as the AB_2 monomer and 3-(4-hydroxyphenyl)propionic acid as the AB monomer will be described (Figure 3.5).[5]

The self-condensation of the AB_2 monomer in solution using a condensing agent has already been successfully applied to linear polyesters[29] and to hyperbranched poly(triazene-ester)s[30] and is used in the case of

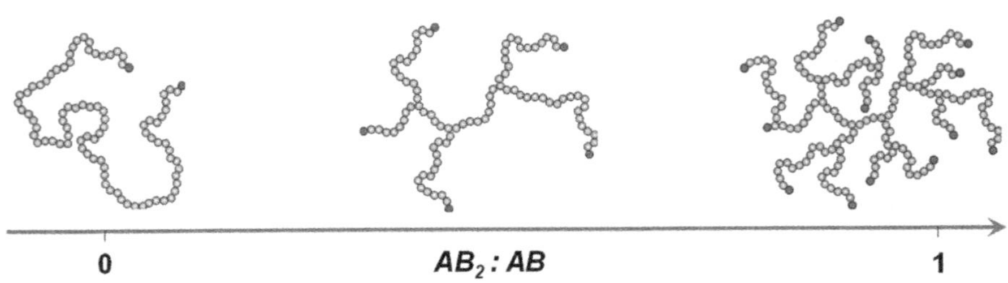

Figure 3.4 Dilution of branching in hyperbranched polymers leads in the limit to linear polymers with DB = 0 and reduced number of end groups.

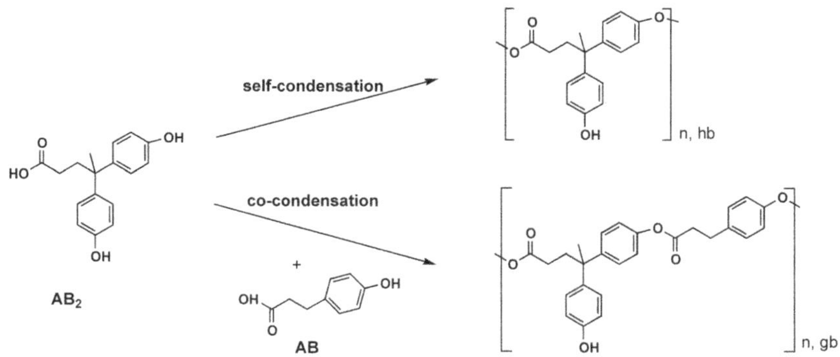

1

2

Figure 3.5 Synthesis of the differently branched $AB_2 + AB$ polyesters.

4,4-bis(4′-hydroxyphenyl)valeric acid. The absolute molar masses of approx. 40 000 g mol^{-1} are determined using size exclusion chromatography (SEC) coupled to a multi-angle light scattering detection (MALS, see also Chapter 4). Compared to products from melt polycondensation[2] lower polydispersities were indicated, M_w/M_n of 1.5–2. This polydispersity is even significantly narrower than those theoretically expected for a statistical polycondensation of AB_2 monomers at high degrees of polymerization.[1] This fact is a result of the fractionation processes during the cleaning procedures, by which the low molar mass samples are partially lost. This is in any case advantageous for obtaining well-defined polymers with low molar mass distributions but systematic variation in degree of branching.

The calculation of the degree of branching can be performed according to eqn (2.9) and (2.11) using the intensities of the NMR signals corresponding to terminal (*T*), linear (*L*) and dendritic (*D*) units. The degree of branching of the end product calculated by these equations is 50%, which corresponds to a statistical type of branching. The reaction procedure in solution can be applied to the co-polymerization of $AB_2 + AB$ monomers, as shown on Figure 3.5. Simply by variation of the initial monomer ratio $r_m = AB/AB_2$ in the feed, variation of the degree of branching is achieved using this approach. The quality of the reaction products is confirmed by ^1H and ^{13}C NMR. Figure 3.6 depicts the ^{13}C NMR spectrum of the co-polymer with 37% degree of branching. Several signals from the incorporated AB_2 units show the characteristic splitting, corresponding to *T*, *L* and *D* units. However, distinguishing the type of *B* units originating from AB_2 or AB monomer is not possible. For the incorporated AB monomer two signal groups are found, caused by linear units with reacted A- and B-groups (C10 to C16) and by terminal units if only the carboxylic group has reacted (C10′ to C16′).

For calculation of the AB/AB_2 ratio the ^1H NMR spectra are used. For this purpose the signal intensities of the methylene proton signals (H11, H12)

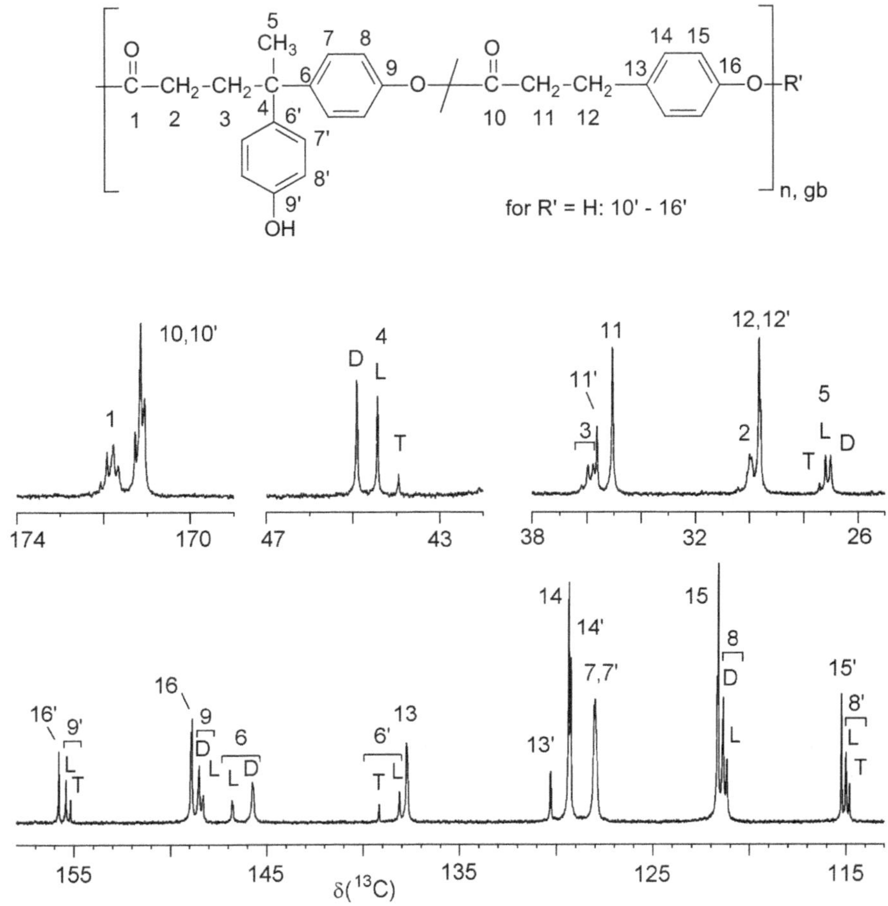

Figure 3.6 ^{13}C NMR spectrum of gradually branched (gb) co-polyester with DB = 37% (solvent: dimethyl sulfoxide (DMSO)-d_6).

of the AB unit and the methyl signals (H5) of the AB_2 unit are needed. Very good agreement between the theoretically (r_m) and experimentally (r_p) found values is obtained for the gradually branched polyesters (Figure 3.7). A similar result was obtained by Frey and Hölter, who observed that the co-polymer composition equals the monomer mixture at higher degrees of polymerization.[28] Nevertheless, by increasing the amount of AB units in the structure above 75%, an abrupt change in the reproducibility of the initial ratio and the solubility of the polymer is observed, which is a result of a cyclization reaction at lower conversions and limits the reaction to very low molar masses. Due to this fact the degree of

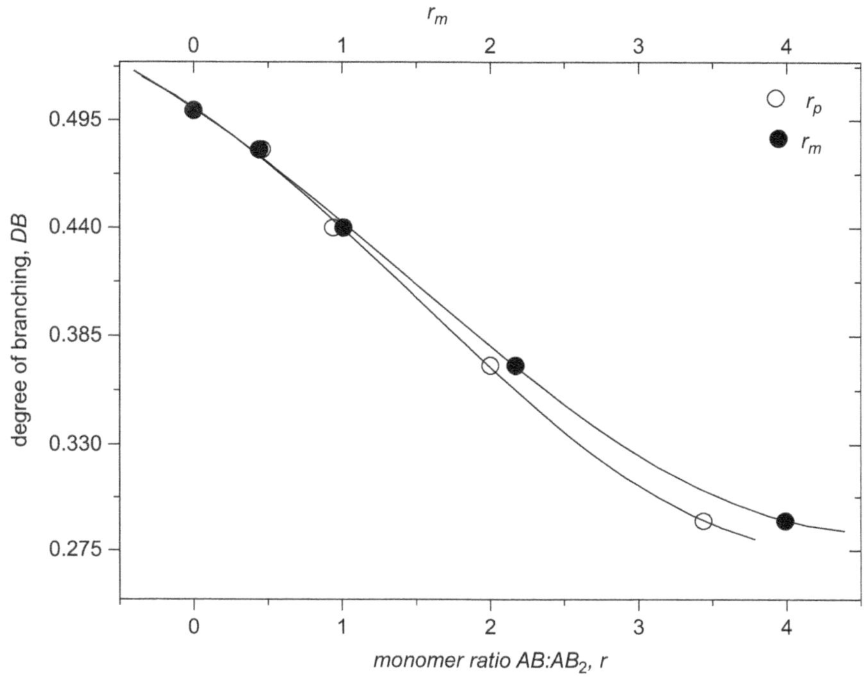

Figure 3.7 Degree of branching DB *versus* r_p and r_m ratio. Adapted with permission from ref. 31. Copyright © 2009 American Chemical Society.

branching obtained by this approach is limited and in this particular case DB ≥ 37%.

Determination of the degree of branching of the co-polyester is more complex than for the homopolyester, where eqn (2.9) and (2.11) can simply be applied. In their work Frey and Hölter postulated[28] the following relation between the degree of branching and r_p for co-polyesters:

$$\mathrm{DB}_{\mathrm{Frey}} = 2p_{\mathrm{A}}\frac{r_{\mathrm{p}}+1}{(r_{\mathrm{p}}+2)^2} = \frac{\overline{\mathrm{DP}_n}-1}{\overline{\mathrm{DP}_n}\left(2+\dfrac{r_{\mathrm{p}}^2}{2(r_{\mathrm{p}}+1)}\right)}. \quad (3.1)$$

For this purpose the determination of degree of polymerization $\overline{\mathrm{DP}_n}$ is required, which is possible by ^{1}H NMR on the basis of the focal unit (Figure 3.2). In case the focal unit is not available due to, for example, cyclization reactions preferably taking place in solution, the determination of $\overline{\mathrm{DP}_n}$ can be performed using the absolute number-average molar mass determined by SEC-MALS. Using the M_n values from the SEC measurements the $\overline{\mathrm{DP}}_n$ for the homopolyester and the co-polyesters can be determined by the following equations:

$$\overline{\mathrm{DP}}_n = \frac{\overline{M}_n - M_{\mathrm{H_2O}}}{M_{\mathrm{monomer}} - M_{\mathrm{H_2O}}} \tag{3.2}$$

$$\overline{\mathrm{DP}}_n = \frac{(1 + r_\mathrm{p})(\overline{M}_n - M_{\mathrm{H_2O}})}{r_\mathrm{p} M_{\mathrm{AB}} + M_{\mathrm{AB_2}} - (1 + r_\mathrm{p}) M_{\mathrm{H_2O}}} \tag{3.3}$$

More accurate is certainly the determination of the degree of branching calculated from by ^{13}C NMR spectra using eqn (2.16), which could be used for a verification of the results from eqn (3.1). In this equation, the dendritic units formed by AB_2 monomers and the total number of linear units originating from AB and from AB_2 monomers are taken into account.[28] The results of these calculations are shown in Figure 3.7. The degree of branching decreases with the increasing amount of AB monomers and the r_m- and r_p-value, as expected.

The combination of two different types of monomers in order to achieve branching variation at the same chemical structure has its limitations from the synthetic point of view as it results in polymers with limited degree of branching portfolio. The formation of linear polymers from AB monomers is limited due to competitive cyclization at very low degrees of polymerization, leading to oligomeric cyclic structures. The influence of the degree of branching on the number of functional groups is most important (Figure 3.8).

The variation of only one structural parameter in this case cannot be achieved since with the increasing degree of branching the number of functional groups also increases, which leads to overlapping effects in solution and bulk properties as discussed in the next chapters. In order to obtain a more accurate variation of the structural parameters, the rather

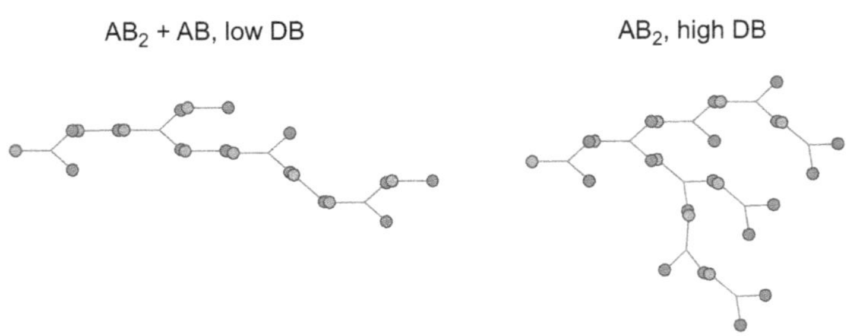

Figure 3.8 The co-polymerization of AB_2 and AB monomers leads to a reduced number of functional groups at the same degree of polymerization, which depends on the number of AB units and the degree of branching, respectively.

demanding $AB_2 + ABB^*$ should be applied, which leads to homopolymers and is introduced below.

3.2.2 The $AB_2 + ABB^*$ Approach

The basic idea behind this approach is the design of linear, gradually branched and hyperbranched polymers from identical monomers. The bifunctional monomer playing the role of branching dilution agent has to be prepared from the same AB_2 monomer used for the self-condensation of the hyperbranched sample. Therefore protection of one of the B-groups should lead to a monomer with only one B- and one A-reactive group. The principle of this approach is described in Figure 3.9.

The synthesis of linear and hyperbranched structures of similar chemical origin by polycondensation has been used in several works,[5,32–35] but only a few use the same monomer. The elaborative procedure of mono-protection was presented by Wooley *et al.*[34] who compared linear hyperbranched polymers with dendrimers. While in this case the polycondensation reaction leads to relatively low molar masses, the polycondensation in a solution of 4,4-bis(4′-hydroxyphenyl)valeric acid as the AB_2 monomer leads to comparable and sufficiently high molar masses. The protection of the AB_2 monomer is carried out by *t*-butyldimethylsilylchloride (TBDMS chloride). The ABB* monomer can be prepared according to Wooley *et al.*[34] with additional separation of the mono-protected product by flash chromatography.

As shown in Figures 3.10 and 3.11 the polycondensation of ABB* by the DPTS/DCC[29] procedure gives a linear polymer. The polycondensation of the ABB* monomer leads to high molar masses (up to 100 000 g mol^{-1}) in contrast to an AB monomer as discussed above. However, the cyclization in the lower molar mass region cannot be excluded completely. SEC and MALDI-TOF-MS measurements show that even in the case of rather stiff monomers, a small number of cyclic structures with a molar mass of up to 6000 g mol^{-1} are obtained (Figure 3.12). After deprotection with trifluoroacetic acid (TFA)[36] the OH-terminated linear polymer can be obtained.

The synthesis of gradually branched polymers employs the co-polymerization of the ABB* and AB_2 monomers in different ratios. According to eqn (3.6), derived by Frey and Hölter,[28] and Mock *et al.*,[37] the degree

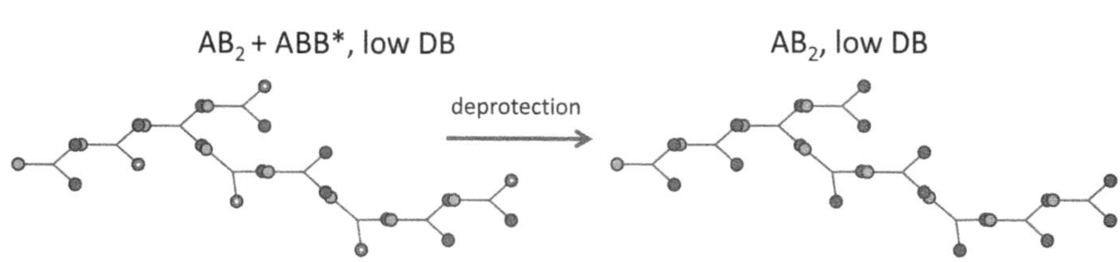

Figure 3.9 The co-polymerization of AB_2 and ABB* monomers leads after deprotection to one functional group per monomer unit, on average, independent of the degree of branching.

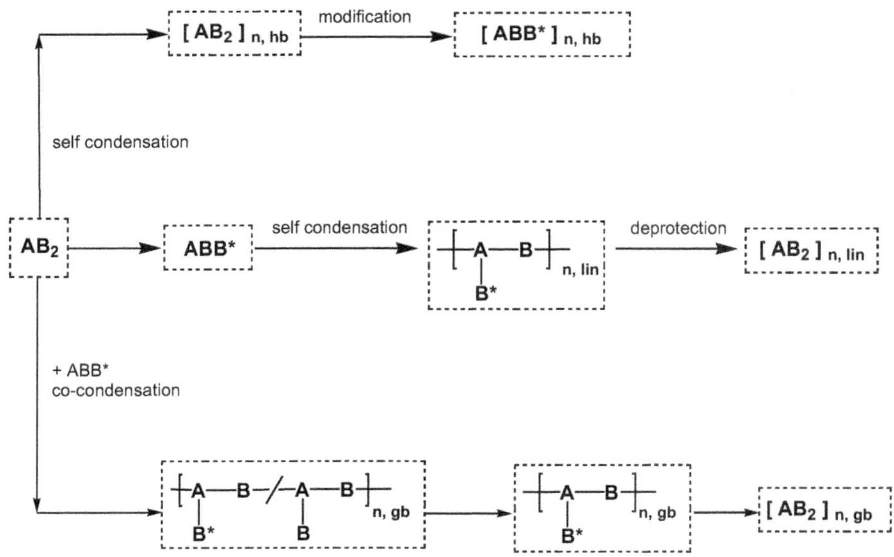

Figure 3.10 Schematic representation of the $AB_2 + ABB^*$ approach. Adapted with permission from ref. 4. Copyright © 2010, American Chemical Society.

of branching is a function of a given molar fraction of the bifunctional co-monomer:

$$X_{ABB^*} = \frac{ABB^*}{AB_2 + ABB^*} \tag{3.4}$$

where X_{ABB^*} can be expressed as a function of r_m:

$$X_{ABB^*} = \frac{r_m}{1 + r_m} \tag{3.5}$$

Assuming full conversion and equal reactivity of the monomers, the following equation for estimation of the theoretically expected degree of branching can be applied:

$$DB_{theor} = 2\frac{1 - X_{ABB^*}}{(2 - X_{ABB^*})^2} \tag{3.6}$$

The theoretically expected degrees of branching are in good agreement with the bifunctional co-monomer molar fractions used. For SY-protected samples (TBDMS protected, see Figure 3.11) a full portfolio of degrees of branching between 0% and 50% with comparable molar mass of approx. 50 000 g mol^{-1} can be obtained. After deprotection, the OH-terminated samples differ in their molar mass, due to the additional deprotection and purification procedures. A fully linear OH-terminated structure is not achieved due to minor branching formation during the deprotection step.[4]

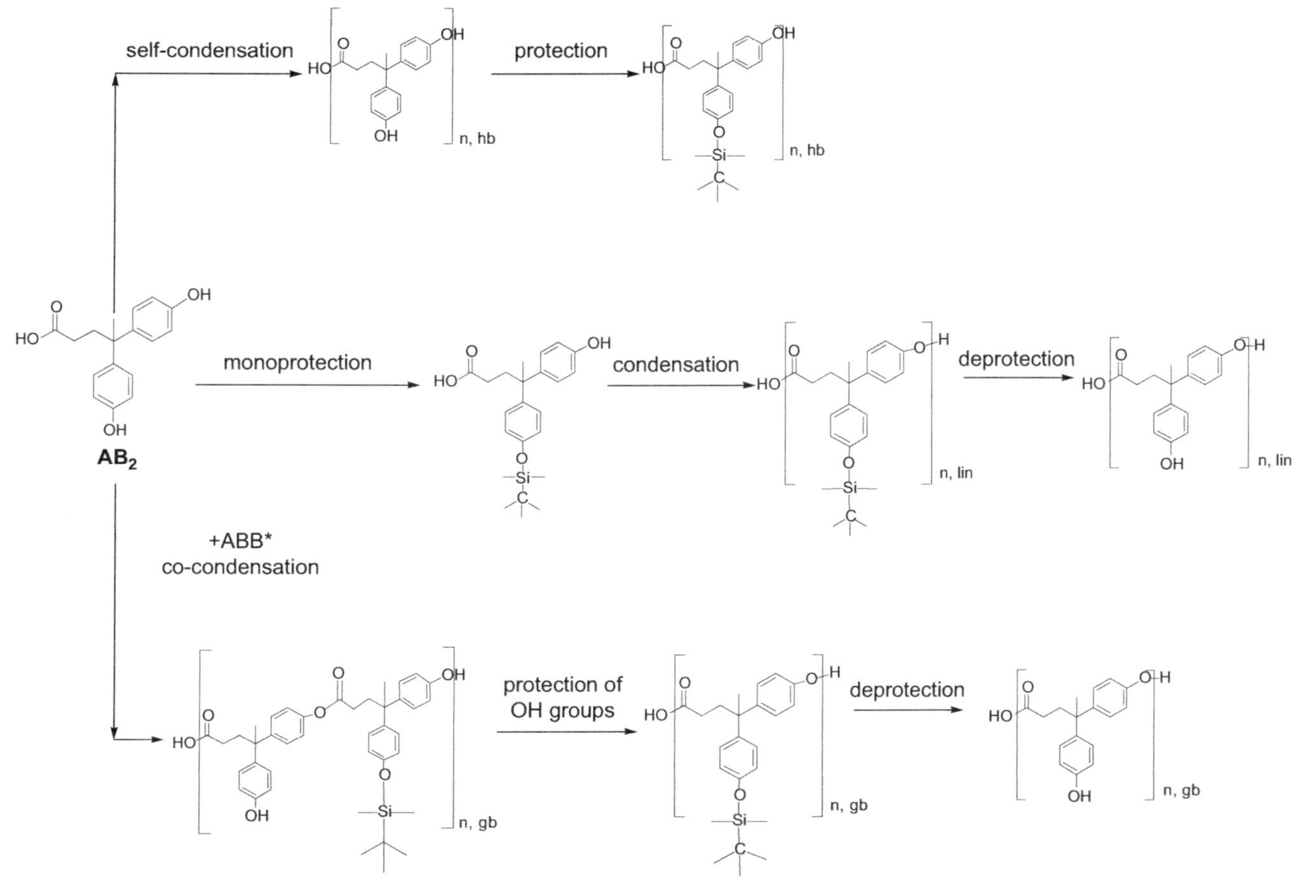

Figure 3.11 Chemical reactions in the $AB_2 + ABB^*$ approach. Adapted with permission from ref. 4. Copyright © 2010, American Chemical Society.

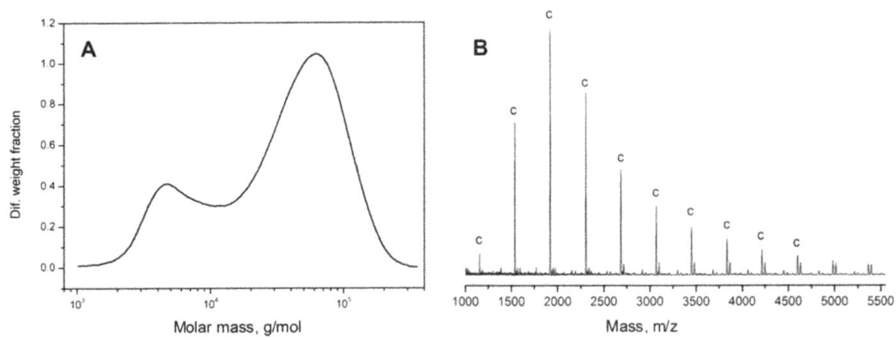

Figure 3.12 Differential weight distribution of the molar mass determined by SEC-MALS-RI chromatogram (A) and MALDI-TOF-MS mass spectrum with peaks corresponding to cyclic structures (**c**) (B) of linear polyester. Adapted with permission from ref. 4. Copyright © 2010, American Chemical Society.

The experimental evaluation of the degree of branching (DB_{exp}) can be performed by NMR according to well-known equations.[5] Figure 3.13 shows regions in the ^{13}C NMR spectra of the linear polymer with a 0% degree of branching containing only *L* units and of two branched samples with different degrees of branching. The lower content of *T* and *D* units of the gradually branched sample with 22% degree of branching compared to the hyperbranched sample with 50% degree of branching can be clearly identified. The experimental degree of branching (DB_{exp}) was determined according to eqn (2.11) based on the intensities of C_4, C_6/C_6 and C_9/C_9 signals which are characteristic for terminal, linear and dendritic units (Figure 3.13).

Full modification of the gradually branched as well as of the hyperbranched structures with TBDMS leads to a series of SY polyesters with variation of the degree of branching as shown in Figure 3.14 depending on the amount of ABB* co-monomer in the reaction mixture.

Complete covering of the entire degree of branching between 0% and 50% is achieved by the AB_2/ABB* approach at completely constant chemical structure and, importantly, identical number of end groups per monomer unit (see Figure 3.9). In this way, at comparable molar masses, the degree of branching remains the only variable parameter for the further structural elucidation of these architectures.

3.2.3 Pseudo-dendrimers by AB_2 Modification

The maximum degree of branching obtained for hyperbranched polymers is 50%. This value is predicted for ideal conditions and for equal reactivity of the two B-groups in the AB_2 units.[1] The increase of the degree of branching up to 100% is not possible using ideal statistics. For this purpose, the linear units have to be excluded and the branched structures have to contain only

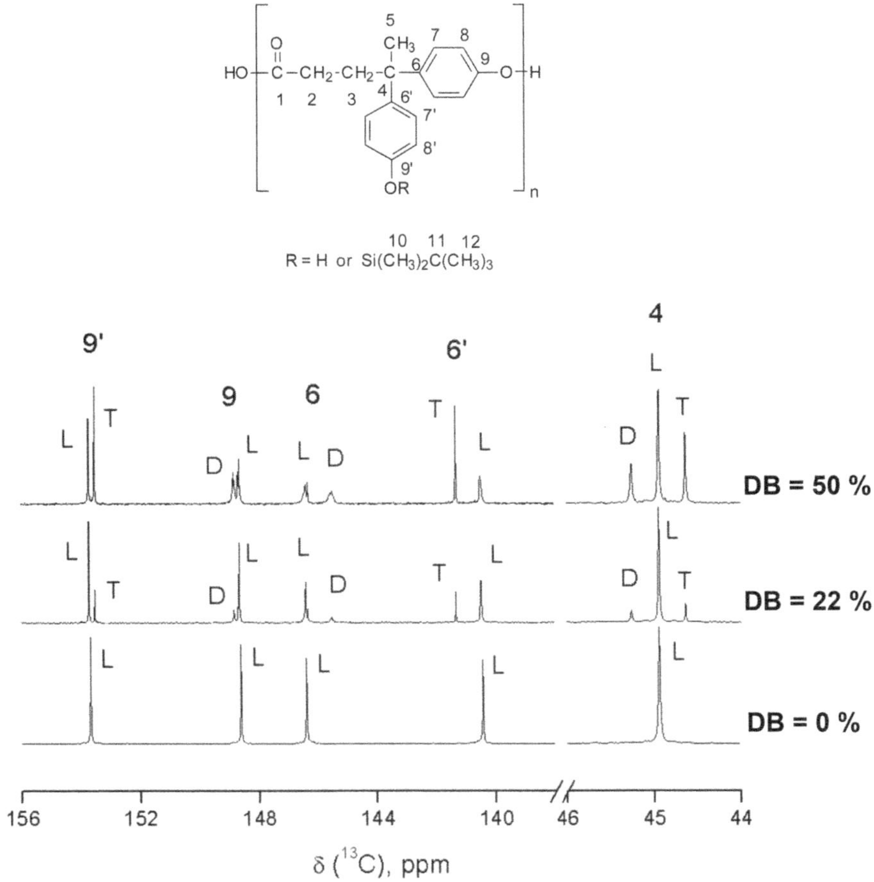

Figure 3.13 Numbering of carbon atoms for the assignment of ^{13}C NMR and ^{13}C NMR spectra of the differently branched, SY-terminated polymers (solvent: $CDCl_3$-d_1).
Adapted with permission from ref. 4. Copyright © 2010, American Chemical Society.

dendritic and terminal units as in the case of dendrimers (see Figure 3.1). Such a type of branching is interesting for special applications in which a higher segmental density is needed combined with a larger number of functional groups. At present, *e.g.* for drug delivery, perfectly structured dendrimers are preferred,[38] but complicated synthesis and work-up are major drawbacks. Compared to the perfectly constructed dendrimers, hyperbranched polymers possess a lower degree of branching, fewer functional groups, and lower segmental density, but they are easier to prepare. Evidently, it would be beneficial to obtain a high segmental density and a large number of functional groups in hyperbranched polymers, close to those of dendrimers. An increased number of functional groups and the degree of branching were aimed for by developing different strategies, *e.g.* by

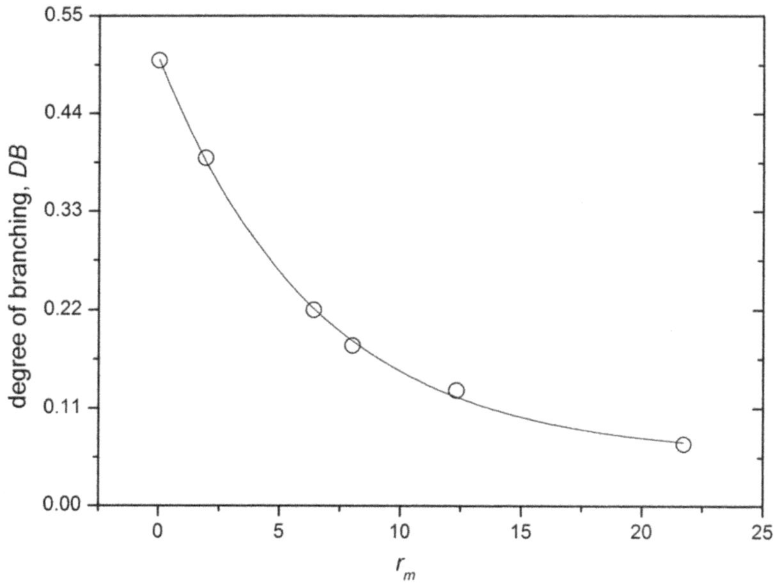

Figure 3.14 Degree of branching dependency of adjusted r_m for SY-terminated gradually branched polyesters.

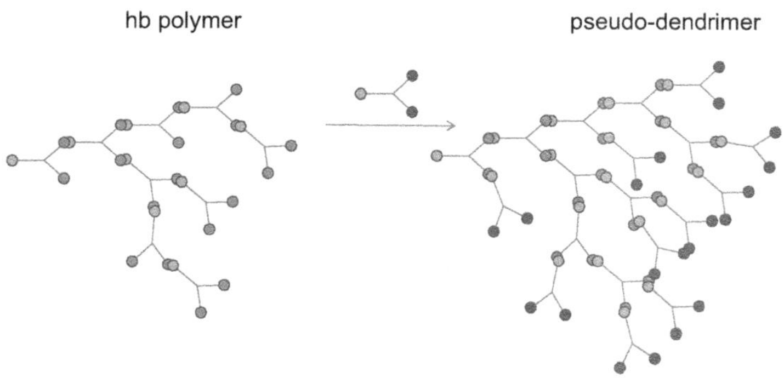

Figure 3.15 Schematic representation of the modification of an AB_2 type hyperbranched polymer by AB^*_2 protected monomers to obtain pseudo-dendrimer of the first generation.

using dendrons,[39,40] introducing core molecules,[22] or designing special architectures suitable for tandem reactions.[41] These strategies led to branching degrees less than 84%; a 100% degree of branching can be reached either with special monomer design[42–45] or after complete post-synthetic modification of the functional groups. The latter approach was first introduced by Frey *et al.* for polycarbosilanes.[46] Haag *et al.* expanded this strategy to polyglycerols[47,48] and introduced the term *pseudo-dendrimers* (Figure 3.15). Similarly to that work, Yamanaka *et al.* reached complete modification of hyperbranched polyimides.[49]

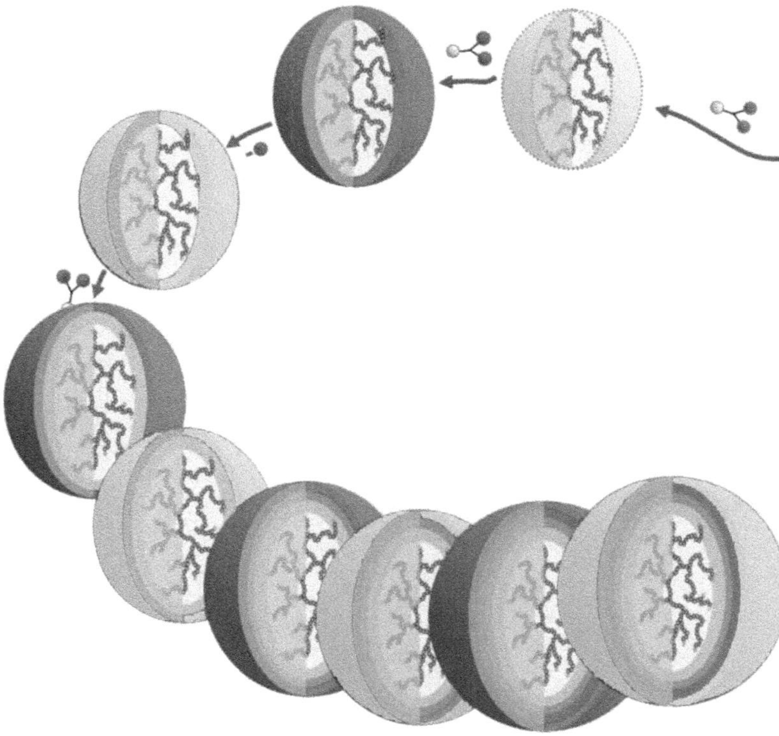

Figure 3.16 Scheme of the principle of a step-wise modification leading to four generations of pseudo-dendrimers. Protection and deprotection procedures lead to pseudo-dendrimers with different end groups – OH- and SY-terminated.
Reprinted with permission from ref. 50. Copyright © 2012, WILEY-VCH Verlag GmbH & Co. KGaA, Weinheim.

These modifications remained limited to the preparation of pseudo-dendritic structures with only one generation which actually corresponds to a derivatization of the free end-groups. The question of how higher generation pseudo-dendrimers would influence the structure density and functionality remained open for more than a decade. Higher generations can be achieved by repeating the modification procedure in a step-wise manner similar to the divergent dendrimer approach (Figure 3.16).

Post-modification of the B-terminated hyperbranched polymer can be performed with an AB^*_2 monomer in which the B-groups are inactivated by protection with a *tert*-butyldimethylsilyl (TBDMS) group.[4] This reaction turns both linear and terminal units into dendritic units resulting in a pseudo-dendrimer with 100% degree of branching according to the definition given by Frey [eqn (2.11)]. Figure 3.17 shows the first synthetic steps leading to the first-generation pseudo-dendrimer with silyl-protected B-groups, *e.g.* TBDMS end groups (SY-G1) and the OH-terminated pseudo-dendrimer (OH-G1) after deprotection of the B-groups. The next three generations are synthesized

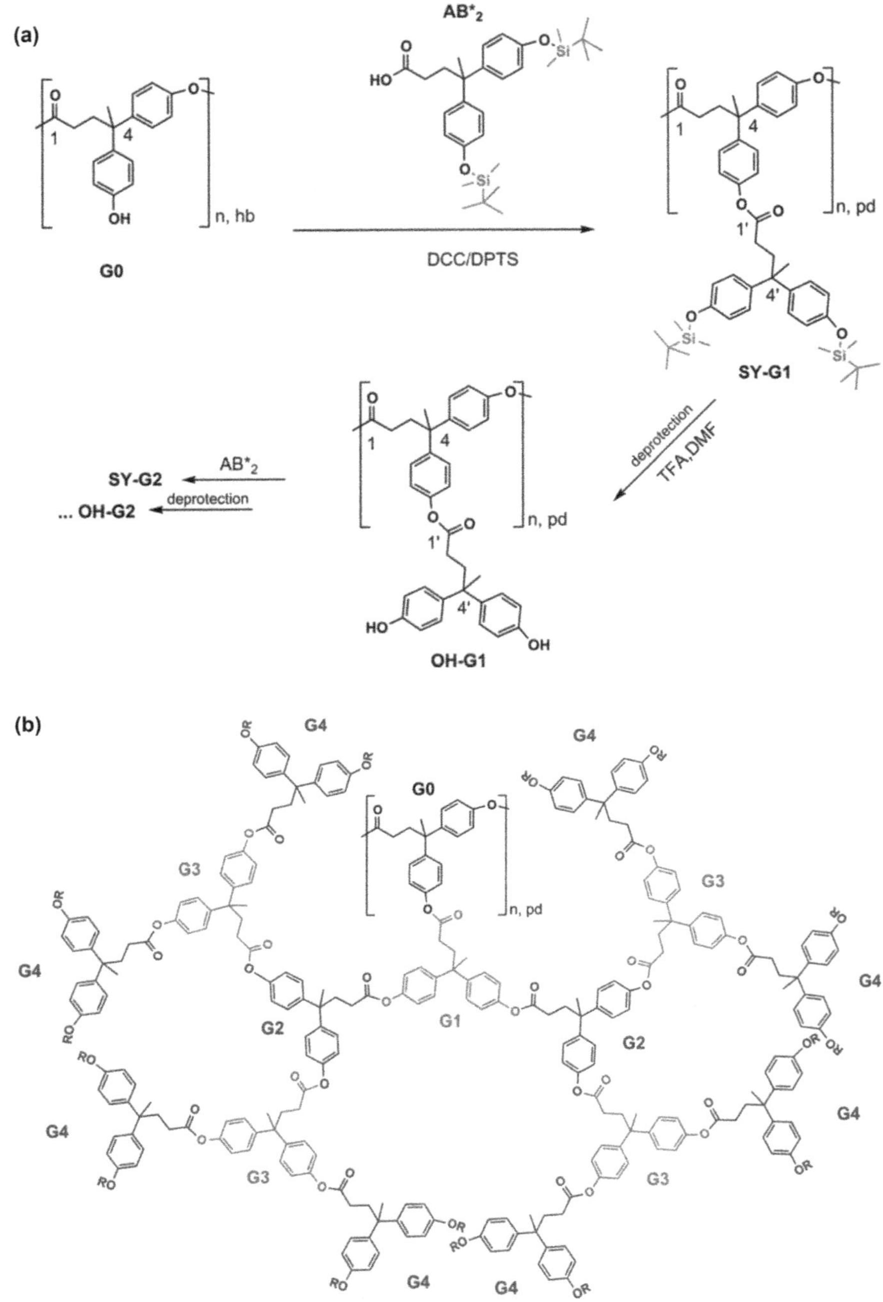

Figure 3.17 (a) Postmodification of hyperbranched polyester leads to first generation protected pseudo-dendrimer (SY-G1) and deprotection gives OH-G1 (see SI). (b) Scheme of the 4th generation (G4) pseudo-dendrimer (R = -H or -TBDMS).
Reprinted with permission from ref. 50. Copyright © 2012, WILEY-VCH Verlag GmbH & Co. KGaA, Weinheim.

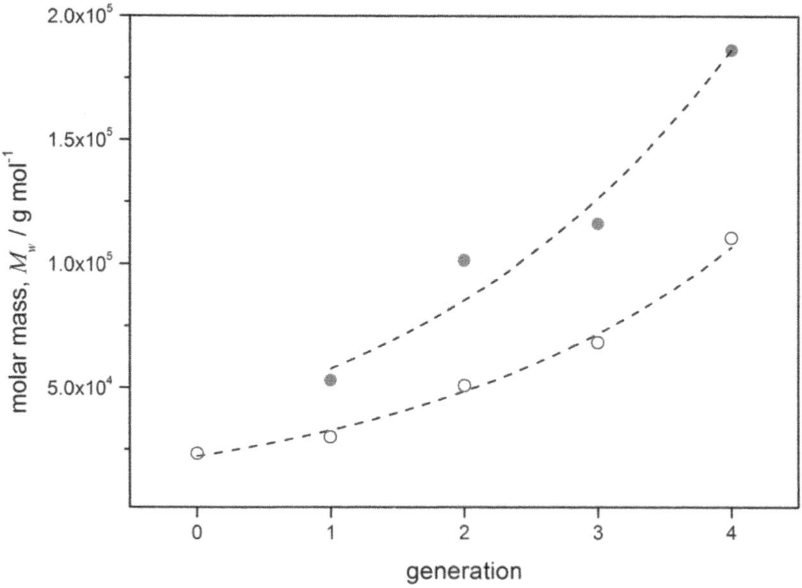

Figure 3.18 Molar mass development of series of pseudo-dendrimers with SY end-groups (open symbols) and OH end-groups (filled symbols).

repeating this procedure, *i.e.* modification of the OH groups with the AB^*_2 monomer followed by deprotection. With this approach four generations of every pseudo-dendrimer type, SY-terminated and OH-terminated, are synthesized (Figure 3.17).

The molar masses of both series increase strongly with the generation number as shown in Figure 3.18 reaching values from 20 000 up to 180 000 g mol^{-1}. The expected molar mass increase with post-modification and decrease after deprotection is accompanied by molar mass distributions M_w/M_n between 1.6 and 2.0.

^{13}C NMR investigation of the products obtained from the different synthetic steps confirms full modification of the OH end-groups into SY-G1 approaching 100% degree of branching (Figure 3.19). The deprotection proceeds completely, yielding OH-G1.

A small fraction of linear groups could be formed during the reaction, which influences the degree of branching. Therefore after the nearly 100% degree of branching in the first generation, slightly changing values are detected in the next generations. In the following three generations the degree of branching decreases slightly as shown in Figure 3.20. Although even in the fourth generation deprotected pseudo-dendrimer the degree of branching achieved is not less than 84%. More likely this decrease is a result of steric hindrance, which is amplified by the generation number and does not allow for full modification of internal OH end-groups because of lower accessibility at higher generations. This steric hindrance is a result of the very high segmental density achieved after the first generation, as discussed in Chapter 5.

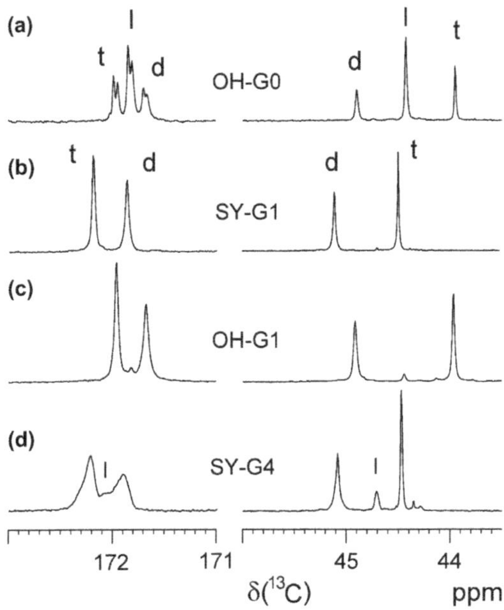

Figure 3.19 Signal region of the carbonyl carbons C1 and C1′ (left) and of the quarternary carbons C4 and C4′ (right) from the ^{13}C NMR spectra of hyperbranched core polymer OH-G0 (a) and pseudo-dendrimers SY-G1 (b), OH-G1 (c) and SY-G4 (d) showing signals of terminal (t), linear (l) and dendritic (d) units (solvent: (a, c) DMSO-d_6, (b, d) $CDCl_3$). Reprinted with permission from ref. 50. Copyright © 2012, WILEY-VCH Verlag GmbH & Co. KGaA, Weinheim.

The multi-step synthesis of perfect dendrimers leads to 100% degree of branching, defined branched shells and a number of repeating units which increases exponentially with the generation number. Only terminal units at the highest generation carry the functional groups. In contrast to dendrimers, the functional groups in hyperbranched polymers are no longer positioned at the outskirt area but they are also found within the macromolecule. Figure 3.21 clarifies the somewhat complex looking structural differences between a hyperbranched polymer, dendrimer and pseudo-dendrimer.

3.3 Control over Molar Mass

Molar mass dependencies of structural parameters are essential for calculation of scaling properties and contraction of branched polymers. Therefore the variation of molar mass at fixed chemical structure and degree of branching is a basic prerequisite for extraction of these parameters. As mentioned at the beginning of this chapter, an important difficulty is to obtain this combination of properties just by changing the reaction time, especially at low conversions, because such an approach leads mostly to

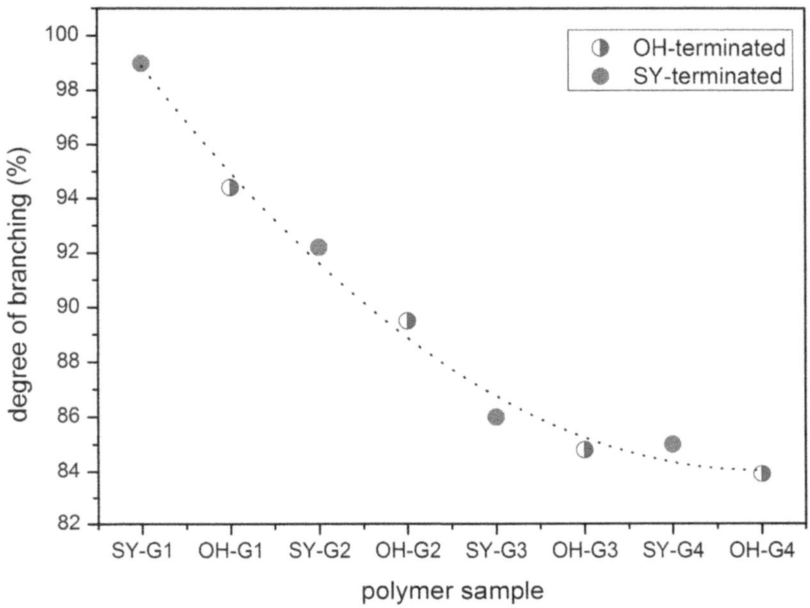

Figure 3.20 Degree of branching of the pseudo-dendrimers determined from the number of dendritic and linear units as indicated by ^{13}C NMR and calculated according to eqn (2.11).
Reprinted with permission from ref. 50. Copyright © 2012, WILEY-VCH Verlag GmbH & Co. KGaA, Weinheim.

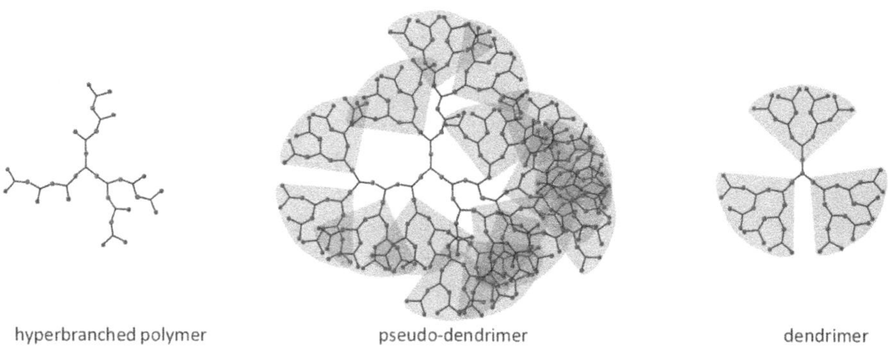

Figure 3.21 Schematic representation of the different structure in hyperbranched polymers, pseudo-dendrimers and dendrimers of three generations. The scheme represents the structures in a two-dimensional manner and is not eligible for the real structure in space. Discussion on the molecular density of pseudo-dendrimers and dendrimers is part of Chapter 5.

simultaneous variation in the degree of branching.[2] Therefore, significant effort has to be made regarding the preparation of fractions from broadly distributed molar masses. The challenge is to obtain sufficiently large amounts in the fractions in order to facilitate further analytical investigations and, hence, preparative fractionation should be taken into consideration. One well-known technique is preparative SEC. However, due to the fact that in one hyperbranched sample, different molecular densities (or degrees of branching) could possess the same hydrodynamic radius and are eluted at the same elution volume, this technique is not always appropriate for this purpose. In order to obtain separation into discrete molar mass fractions without the influence of the volume of the molecule, separation similar to the well-known Baker–Williams fractionation could be applied, which employs both temperature and solvent gradient.[51] The interplay between temperature and solvent quality is not effective in the case of hyperbranched polymers, which stand out with their very good solubility at ambient conditions as a result of their high number of functional groups, relatively low molar masses and small sizes. Therefore, a solvent gradient between well-miscible non-solvent and good solvent could be used, which elutes a thin film of the broadly distributed hyperbranched polymer prepared on a supporting material.[3,6,7] The quality of the solvent can be determined by *e.g.* calculation of the Schulz–Blaschke and Huggins constants.[52] It has to be taken into account that the fractionation according to solubility is governed by two effects: (1) the molar mass and (2) the degree of branching, or number of end groups, which is usually coupled to the degree of branching. The result of the fractionation can be proven by SEC-MALS as well as by spectroscopic methods.

An example of successful fractionation according to molar mass is a hyperbranched poly(etheramide) (PEA), prepared by a ring opening addition reaction (Figure 3.22), which is an extensively studied, broadly distributed hyperbranched polymer.[3,53,54] Figure 3.23 allows a comparison between the polydispersity of the hyperbranched sample directly after one-pot synthesis and after its fractionation. In all fractionations performed on hyperbranched polymers, independently of their chemical origin, whether polyesters, poly(etheramide)s or polyurethanes, it was observed that solubility has been governed entirely by the molar mass. Whether this influence of molar mass is exclusive or not is a question which could be answered by analysis of the degree of branching and chemical structure of the single fractions.

^{1}H NMR studies on the hyperbranched fractions of PEA enable chemical identification and calculation of the degree of branching according to eqn (2.9) and (2.11). Table 3.1 summarizes the calculated degrees of branching, which are evidently distinctive for statistical branching character. By this fractionation we obtain a large portfolio of molar masses. Frey's calculation approach [eqn (2.11)] is considered suitable for this case, due to the fact that it also allows calculation of the degree of branching in hyperbranched molecules with a lower degree of polymerization. The relevance of the different calculation approaches of the degree of branching are discussed in Chapter 2.

A slight increase in the degree of branching with the molar mass is observed. Nevertheless, this marginal effect does not exert an influence on

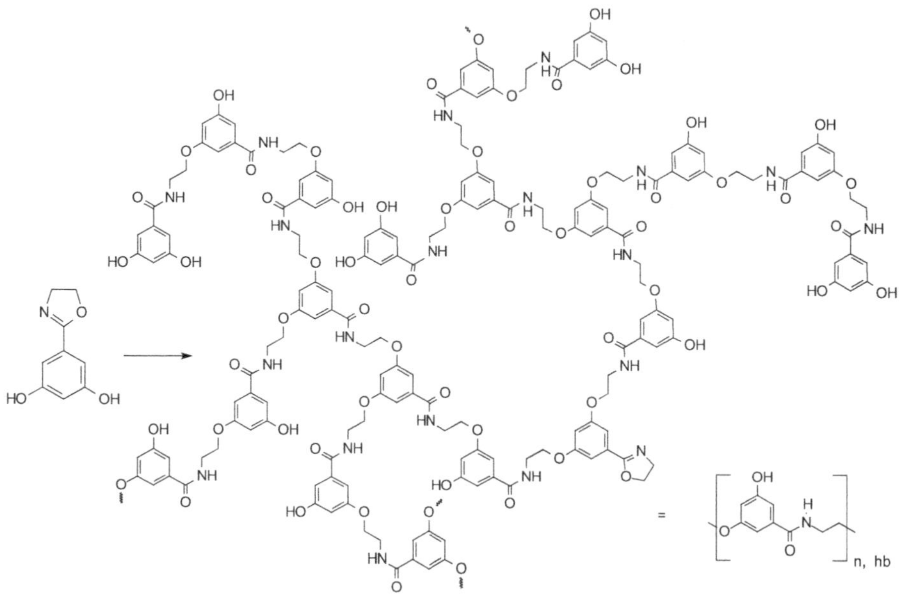

Figure 3.22 Scheme of a hyperbranched poly(etheramide) prepared *via* a nucleophilic ring-opening of the AB_2 monomer 2-(3,5-dihydroxyphenyl)-1,3-oxazolin.
Reprinted with permission from ref. 53. Copyright © 1999, WILEY-VCH Verlag GmbH & Co. KGaA, Weinheim.

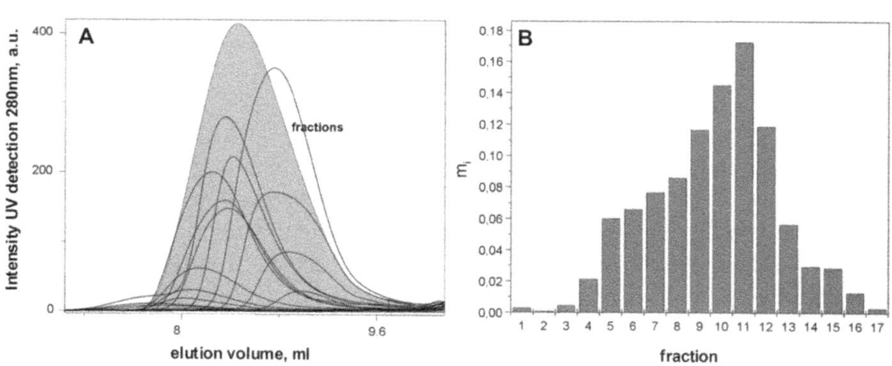

Figure 3.23 (A) SEC-RI elution curves of the starting hyperbranched poly(etheramide) (grey) and different fractions after elution fractionation; (B) relative yield distribution of the different fractions.
Reprinted with permission from ref. 3. Copyright © 2002, Elsevier.

the solution behaviour of the hyperbranched molecules. However, stronger decrease in the degree of branching at lower degrees of polymerization (DP) has been found after fractionation of an aliphatic–aromatic polyester.[33] This is in agreement with previous kinetics studies on the one-pot melt

Table 3.1 Molar masses of fractions of poly(etheramide) determined by SEC-MALS and degree of branching determined by ^{1}H NMR.

M_w (g mol^{-1})	M_w/M_n	$DB_{Fréchet}$ (%)	DB_{Frey} (%)
63 000	1.1	54	45
73 500	1.1	54	47
128 900	1.04	55	51
428 800	—	56	52
724 300	—	56	52

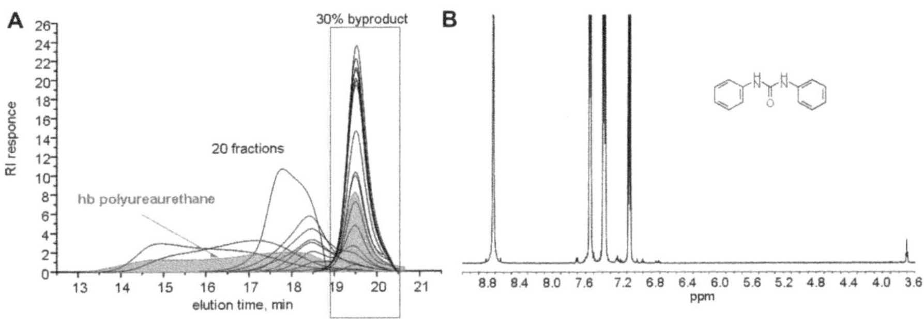

Figure 3.24 (A) SEC-RI elution curves of the starting poly(urea–urethane) and of the fractions; (B) ^{1}H NMR analysis of the fractions containing the by-product diphenylurea.
Reprinted with permission from ref. 7. Copyright © 2006, De Gruyter, Berlin.

polycondensation of this type of aliphatic–aromatic polyester performed by Schmaljohann *et al.*[2] Hence, in these samples, co-existence of two different distributions was proven: molar mass and degree of branching. It should be taken into account that, in contrast to the one-pot synthesis of hyperbranched PEA,[53] self-condensation of the aliphatic–aromatic polyesters leads to relatively low DP and, due to this fact, the distribution of the degree of branching is strongly shifted to lower degrees of branching.

Apart from the degree of branching and molar mass, chemical inhomogeneities as a result of side reactions commonly accompany polymers synthesized by one-pot reactions. On the one hand, the fractionation process is quite strongly influenced by the different chemical nature of by-products, while on the other hand it gives the possibility to separate and identify them. One example of chemically controlled elution fractionation is hyperbranched poly(urea–urethane), which is a product of a very complex $AA^{*} + B_2B^{*}$ reaction of 2,4-toluene diisocyanate (TDI) and diethanolamine (DEA).[55] The fractionation of this sample leads to a bimodal distribution. Analysis of the fractions by SEC with refractive index detection clearly shows that the last peak of the distribution belongs to a low molar mass substance, identified by ^{1}H NMR and MALDI-TOF as the by-product diphenylurea (Figure 3.24). The elution fractionation enables complete separation of the by-product. Further fractionation of the pure hyperbranched

Table 3.2 Molar masses and hydrodynamic radii of fractions from broadly distributed sylilated polyesters with DB = 0% and DB = 50%. Molar mass distributions, M_w/M_n, between 1.1 and 1.5.

Linear DB = 0%		Hyperbranched DB = 50%	
M_w,g mol^{-1}	R_h,nm	M_w,g mol^{-1}	R_h,nm
13 600	4.4	17 100	4.65
29 800	7.15	29 400	5.5
44 600	9.15	47 700	6.5
53 400	9.9	84 000	8.1
64 000	10.75	188 000	11.1
96 700	12.95		

poly(urea–urethane) results in the molar mass dependent separation of fractions for comprehensive solution analysis.[7]

The elution fractionation for preparing fractions of different molar masses is not limited to one degree of branching. Fractionations of a series of polyesters with different degrees of branching, described in Chapter 5, led to a sample set with a *complex, two-dimensional variation of properties: of degree of branching, and molar mass.* These well-defined samples enable full investigation of physical properties with developing of statistical type of branching degree, which is exceptional. In this way it was possible for a first time to follow the contraction parameters and to prove drainage exponents in dependence of the degree of branching. Table 3.2 shows an example of a fractionated linear polymer compared to a hyperbranched polymer (for details see Section 5.5).

Generally, it can be concluded that preparative elution fractionation with a solvent gradient for a certain, broadly distributed hyperbranched polymer is mainly controlled by its molar mass. Strong dependence on the degree of branching could be observed only as far as it is coupled to the degree of polymerization. However, the number and type of end groups play an important role in defining the solubility behaviour of the hyperbranched samples. Because the number of end groups is constant at a constant degree of branching, the separation might not be manipulated according to the molar mass of one sample. However, by converting the polarity of the end groups, the elution performance is drastically changed. An example is OH-terminated aromatic–aliphatic polyester, which is not soluble in *n*-hexane. As soon as the OH-groups are modified with benzoyl chloride, they become less polar and are soluble in *n*-hexane. Hence, the elution conditions are completely changed. The influence of the end functionalities become even more complex with the selective modification of linear and terminal end-functionalities, which will be described in the next section.

3.4 Variation of End-group Functionality

A particular characteristic of dendritic polymers is their huge number of functional groups. That is what makes their fields of application so wide and so versatile. Only by changing the type of the end groups can mechanical, thermal,[56] as well as solution properties[47] be drastically influenced and fine-tuned.[12,31,39,57,58] A very large number of polar functional groups leads to strong tendency of H-bond formation with neighbouring molecules, solvent, water or interactive surfaces.[59,60] The elucidation of the influence of end-group polarity on the structural characteristics of hyperbranched polymers is possible by systematically varying the polarity at the same degree of branching and molar mass.

Variation of functionality at the same degree of branching works quite well in the course of protection/deprotection reactions, as described in Section 3.2.2 and shown in Figure 3.11 for aromatic–aliphatic polyesters synthesized by the $AB_2 + ABB^*$ approach. Step-wise protection and deprotection of the functional groups leads to polymers with an identical number of end groups per monomer unit and which are fully polar (OH-terminated) or fully non-polar (TBDMS-capped). The full modification of the 50% branched homopolymer produced by self-condensation of the AB_2 monomer could be targeted at different polarity.[4] In Figure 3.25 end-group modification by benzoylchloride[33] and TBDMS-Cl is shown.

Modification of functional groups of hyperbranched polymers has been repeatedly reported and especially in the case of OH-terminal groups it is a straightforward procedure with high yields.[61] More complicated is the selective modification of the functional groups appearing in terminal and linear repeat units with opposite polarity. Such investigations enable deeper

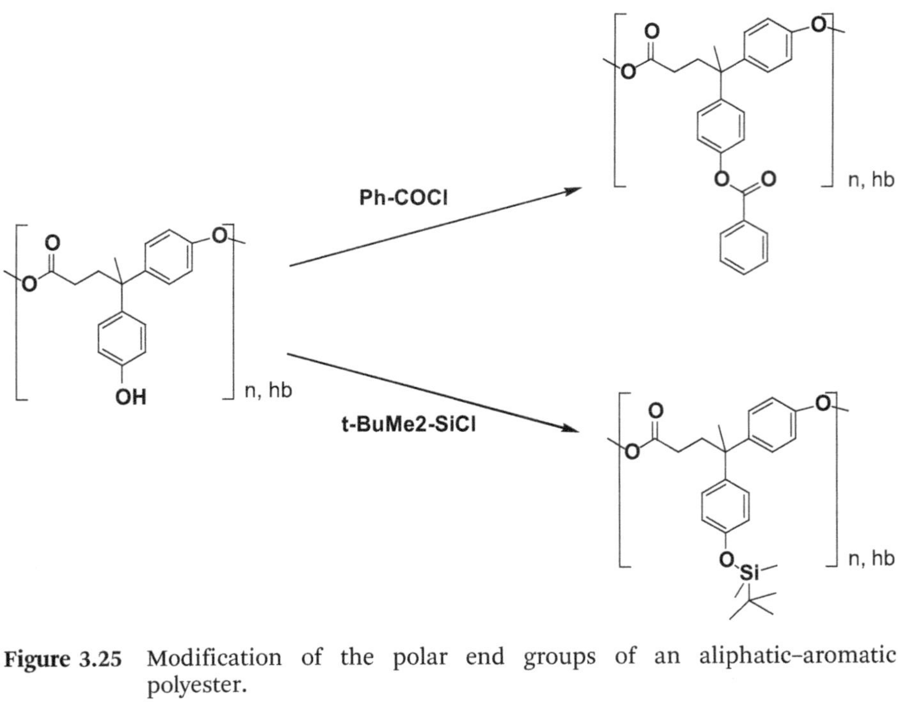

Figure 3.25 Modification of the polar end groups of an aliphatic–aromatic polyester.

understanding about the role of the linear and the terminal units in shaping properties of hyperbranched structures in solution.

Selective modification of linear and terminal end groups in statistically branched polymers is not a trivial task due to the fact that they possess equal reactivity, especially in the case of ideal statistics.[1] The only way for successful modification is indirectly, by protecting terminal types of end groups, subsequently modifying the linear type and finally deprotecting the terminal type, as shown in Figure 3.26. Comprehensive work on the selective modification of hyperbranched polyglycerols has been carried out by Haag *et al.*[47] showing the strong influence of end-group type on solubility and thermal properties.

The synthesis of the Boltorn™-type aliphatic polyester (Figure 3.27) is carried out according to the well-known self-condensation in the melt leading to products with molecular properties as described in the work of Komber *et al.*[62] The degree of branching determined by ^{1}H NMR spectroscopy is found to be 44%, respectively, corresponding to non-ideal statistics during the reaction.[62] Low amounts of ether bonds are typical for these polymers, which were found not to significantly influence the solution properties and reactivity. For the selective modification the unique geometry of the terminal end groups in this chemical structure is used. Protection with dimethoxypropane is confined to the diol character of the *terminal* units leading to their modification, while the *linear* units remain unchanged (Figure 3.27). ^{1}H NMR spectra show complete protection of the terminal units by disappearance of the terminal OH signals (4.58 ppm) and the presence of new signals corresponding to the formed acetonides of the protection group (1.24, 1.34, 3.59 and 3.98 ppm) (Figure 3.28). In the next step the *linear* units react with acetyl chloride leading to a hb polymer only with non-polar functionalities. The successful modification can be proven by the appearence of the acetyl group signal at 1.98 ppm in the NMR spectra in Figure 3.28. In the last reaction step, the deprotection of the *terminal* units, 99% recovery of the terminal OH-groups without changes in the non-polar linear units was achieved.

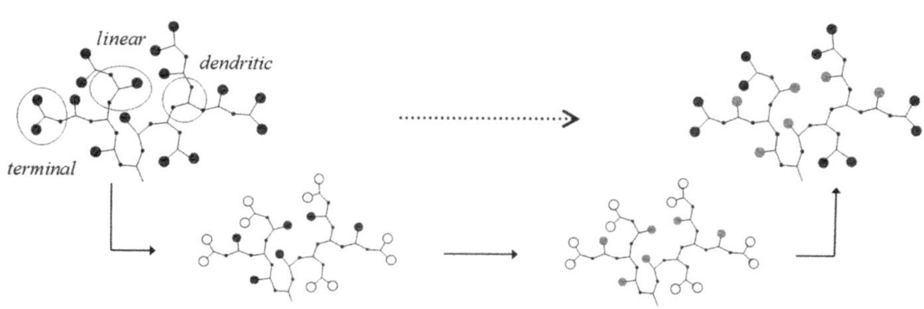

Figure 3.26 Synthetic pathway for selective modification of linear end groups of the hyperbranched polyester.
Reprinted with permission from ref. 6. Copyright © 2010, Elsevier.

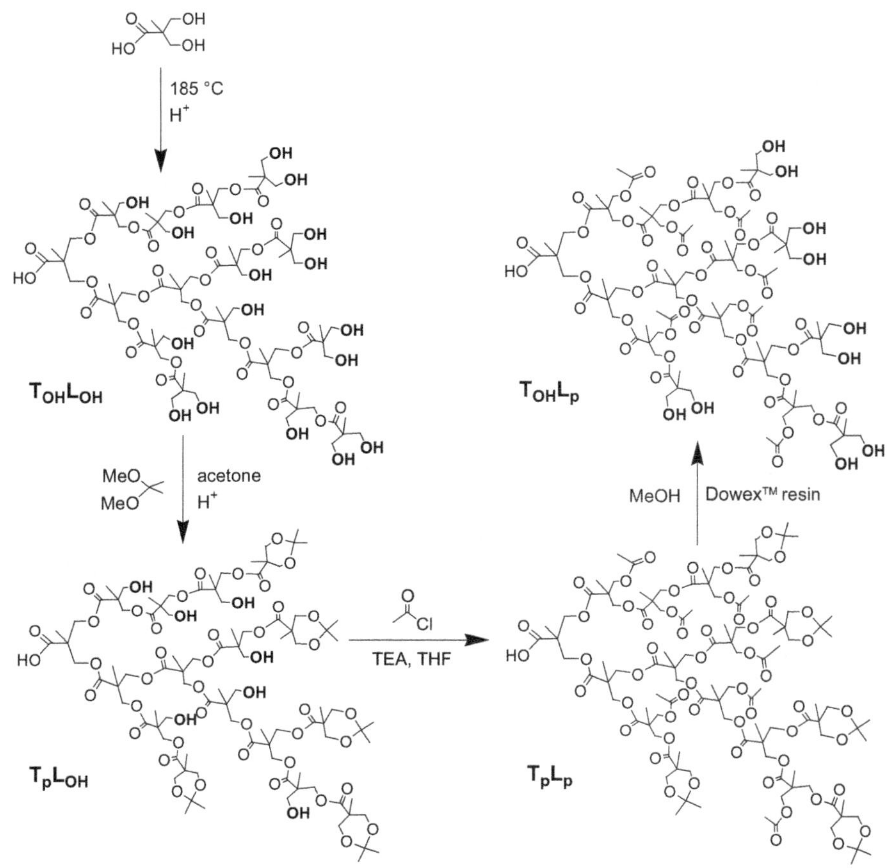

Figure 3.27 Synthetic steps for selective modification of linear and terminal end groups in a hyperbranched aliphatic polyester.
Adapted with permission from ref. 6. Copyright © 2010, Elsevier.

An indication of changed solution properties with the modification is the change in aggregation behaviour of the aliphatic polyesters reflected in the molar mass distribution. Molar masses of hyperbranched polymers can be determined *via* SEC-MALS. At weight average molar mass of approx. 30 000 g mol^{-1} and very broad distribution, different light scattering signals are obtained (Figure 3.29).

In addition to the main MALS signal of the original OH-terminated polyester in Figure 3.29, a strong shoulder at lower elution volumes is clearly visible corresponding to species in very low concentration, not detected by the RI detector. This shoulder could be observed also for the partially OH-modified samples (**T_pL_{OH}** and **$T_{OH}L_p$**) but not for the fully modified *one* (**T_pL_p**). This shoulder corresponds to aggregates due to hydrogen bonding between the polar end groups, investigated in detail for Boltorn™-type

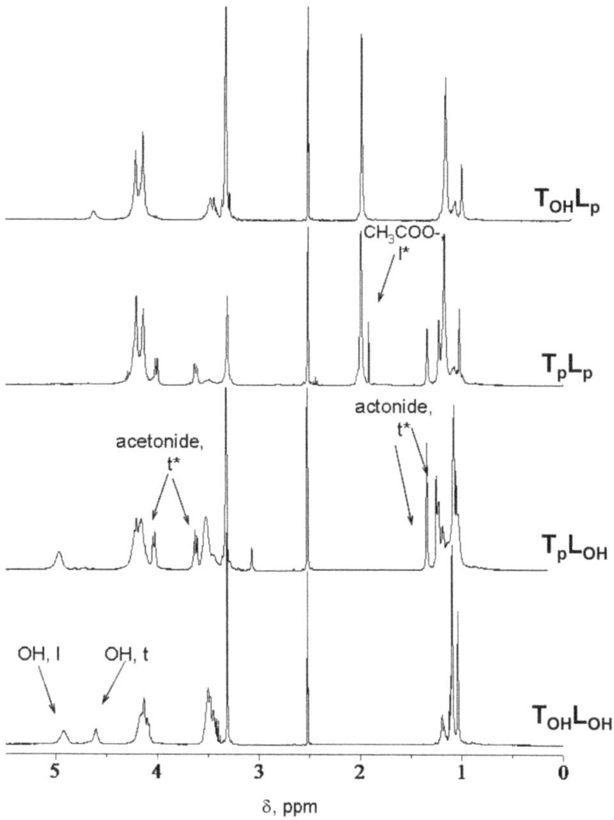

Figure 3.28 ^{1}H NMR spectra in DMSO-d_6 of selectively protected hyperbranched polyesters.
Adapted with permission from ref. 6. Copyright © 2010, Elsevier.

polymers by Žagar and Grdadolnik.[59] The disappearance of this shoulder in the MALS response of the completely non-polar polymer supports this fact.

The influence of the type of end groups on dilute solution properties of these polymers as well as their elution behaviour can be proven on the basis of scaling and elution properties. For this purpose, parameters that are dependent on molar mass are needed, which is coupled to *another set of two-dimensional properties variation – in end-group type and molar mass.* Elution fractionation of all types of end-group variation was performed leading to interesting behaviour of the intrinsic viscosity (Chapter 5) and elution properties (Chapter 4).

The purpose of the selective modification is to exclude the influence of the linear units and thus, to tune the solution and interaction properties. Hence, by systematically changing the solvent content it is possible to separate the hyperbranched polymers by their degree of branching.

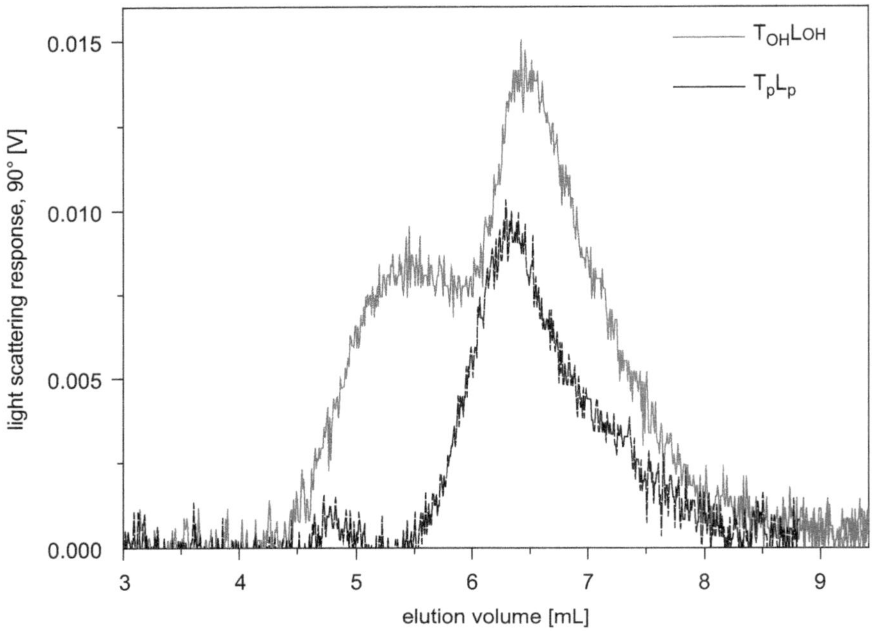

Figure 3.29 MALS-chromatograms of OH terminated hyperbranched aliphatic polyester (**$T_{OH}L_{OH}$**) and its fully modified product (**T_pL_p**). Reprinted with permission from ref. 6. Copyright © 2010, Elsevier.

This hyperbranched polymer series is an interesting candidate for investigating chromatographic separation according to degree of branching, although, due to the specific type of chemical system which could be well-treated by special protection procedures, the approach discussed would most probably remain an exception. While such separation is still not proven, successful separation of statistically branched polymers according to degree of branching was recently performed using liquid chromatography under critical conditions. This approach is discussed in detail in Section 4.3.

References

1. P. J. Flory, *Principles in Polymer Chemistry*, Cornell University Press, Ithaca, New York, 1953.
2. D. Schmaljohann, H. Komber and B. Voit, *Acta Polym.*, 1999, **50**, 196.
3. A. Lederer, D. Voigt, C. Clausnitzer and B. Voit, *J. Chromatogr. A*, 2002, **976**, 171.
4. A. Khalyavina, F. Schallausky, H. Komber, M. Al Samman, W. Radke and A. Lederer, *Macromolecules*, 2010, **43**, 3268.
5. F. Schallausky, M. Erber, H. Komber and A. Lederer, *Macromol. Chem. Phys.*, 2008, **209**, 2331.
6. S. Boye, H. Komber, P. Friedel and A. Lederer, *Polymer*, 2010, **51**, 4110.

7. A. Lederer, M. Abd Elrehim, F. Schallausky, D. Voigt and B. Voit, *e-Polym.*, 2006, **039**, 1.
8. C. Clausnitzer, PhD Thesis, TU Dresden, 2002.
9. F. Böhme, C. Clausnitzer, F. Gruber, S. Grutke, T. Huber, P. Pötschke and B. Voit, *High Perform. Polym.*, 2001, **13**, 21.
10. B. Bruchmann and W. Schrepp, *e-Polym.*, 2003, **014**.
11. B. Bruchmann, R. Koeniger and H. Renz, *Macromol. Symp.*, 2002, **187**, 271.
12. A. Hult, M. Johannson and E. Malmström, *Adv. Polym. Sci.*, 1999, **143**, 1.
13. C. C. Lee, J. A. MacKay, J. M. J. Frechet and F. C. Szoka, *Nat. Biotechnol.*, 2005, **23**, 1517.
14. R. Vestberg, M. Malkoch, M. Kade, P. Wu, V. V. Fokin, K. B. Sharpless, E. Drockenmuller and C. J. Hawker, *J. Polym. Sci., Part A: Polym. Chem.*, 2007, **45**, 2835.
15. P. Antoni, M. Malkoch, G. Vamvounis, D. Nystrom, A. Nystrom, M. Lindgren and A. Hult, *J. Mater. Chem.*, 2008, **18**, 2545.
16. M. Sangermano, A. Priola, G. Malucelli, R. Bongiovanni, A. Quaglia, B. Voit and A. Ziemer, *Macromol. Mater. Eng.*, 2004, **289**, 442.
17. D. Schmaljohann, B. I. Voit, J. F. G. A. Jansen, P. Hendriks and T. J. A. Loontjens, *Macromol. Mater. Eng.*, 2000, **275**, 31.
18. M. Sangermano, A. Di Gianni, R. Bongiovanni, A. Priola, B. Voit and D. Appelhans, *Macromol. Mater. Eng.*, 2005, **290**, 721.
19. A. Di Gianni, S. Trabelsi, G. Rizza, M. Sangermano, H. Althues, S. Kaskel and B. Voit, *Macromol. Mater. Eng.*, 2007, **208**, 76.
20. A. Di Gianni, M. Sangermano, G. Malucelli and B. Voit, *Macromol. Mater. Eng.*, 2006, **8**, 1004.
21. S. Reichelt, K.-J. Eichhorn, D. Aulich, K. Hinrichs, N. Jain, D. Appelhans and B. Voit, *Colloids Surf., B*, 2009, **69**, 169.
22. D. Hölter and H. Frey, *Acta Polym.*, 1997, **48**, 298.
23. R. Hanselmann, D. Hölter and H. Frey, *Macromolecules*, 1998, **31**, 3790.
24. S. Unal, C. Oguz, E. Yilgor, M. Gallivan, T. E. Long and I. Yilgor, *Polymer*, 2005, **46**, 4533.
25. H. Mori and A. H. E. Müller, *Prog. Polym. Sci.*, 2003, **28**, 1403.
26. H. Mori, A. H. E. Müller and P. F. W. Simon, in *Macromolecular Engineering: Precise Synthesis, Materials, Properties, Applications*, ed. K. Matyjaszewski, Y. Gnanou and L. Leibler, Wiley-VCH, Weinheim, 2007, vol. 2, p. 973ff.
27. H. R. Kricheldorf, Q.-Z. Zang and G. Schwarz, *Polymer*, 1982, **23**, 1821.
28. H. Frey and D. Hölter, *Acta Polym.*, 1999, **50**, 67.
29. J. S. Moore and S. I. Stupp, *Macromolecules*, 1990, **23**, 65.
30. M. Eigner, H. Komber and B. Voit, *Macromol. Chem. Phys.*, 2001, **202**, 245.
31. B. I. Voit and A. Lederer, *Chem. Rev.*, 2009, **109**, 5924.
32. A. Mock, A. Burgath, R. Hanselmann and H. Frey, *Macromolecules*, 2001, **34**, 7692.
33. F. Schallausky, PhD, Technische Universität Dresden, 2007.

34. K. L. Wooley, J. M. J. Fréchet and C. J. Hawker, *Polymer*, 1994, **35**, 4489.
35. E. De Luca and R. W. Richards, *J. Polym. Sci., Part B: Polym. Phys.*, 2003, **41**, 1339.
36. P. G. M. Wuts and T. W. Greene, *Greene's protective groups in organic synthesis*, Wiley, New Jersey, 2007.
37. A. Mock, A. Burgath, R. Hanselmann and H. Frey, *Macromolecules*, 2001, **34**, 7692.
38. U. Boas and P. M. H. Heegaard, *Chem. Soc. Rev.*, 2004, **33**, 43.
39. M. Jikei and M. Kakimoto, *J. Polym. Sci., Part A: Polym. Chem.*, 2004, **42**, 1293.
40. C. J. Hawker and F. Chu, *Macromolecules*, 1996, **29**, 4370.
41. J. Wang and M. Johnson, *Polym. Int.*, 2009, **58**, 1234.
42. G. Maier, C. Zech, B. Voit and H. Komber, *Macromol. Chem. Phys.*, 1998, **199**, 2655.
43. M. Smet, E. Schacht and W. De Haen, *Angew. Chem., Int. Ed.*, 2002, **41**, 4547.
44. W. Sinananwanich and M. Ueda, *J. Polym. Sci., Part A: Polym. Chem.*, 2008, **46**, 2689.
45. W. Huang, L. Su and Z. Bo, *J. Am. Chem. Soc.*, 2009, **131**, 10348.
46. C. Lach and H. Frey, *Macromolecules*, 1998, **31**, 2381.
47. R. Haag, J.-F. Stumbé, A. Sunder, H. Frey and A. Hebel, *Macromolecules*, 2000, **33**, 8158.
48. R. Haag, A. Sunder and J.-F. Stumbé, *J. Am. Chem. Soc.*, 2000, **122**, 2954.
49. K. Yamanaka, M. Jikei and M. Kakimoto, *Macromolecules*, 2001, **34**, 3910.
50. A. Lederer, T. Hartmann and H. Komber, *Macromol. Rapid Commun.*, 2012, **33**, 1440.
51. P. S. Francis, R. C. Cooke and J. H. Elliott, *J. Polym. Sci.*, 1958, **31**, 453.
52. K.-F. Arndt and G. Müller, *Polymercharakterisierung*, Carl Hanser Verlag, München Wien, 1996.
53. T. Huber, F. Böhme, H. Komber, J. Kronek, J. Luston, D. Voigt and B. Voit, *Macromol. Chem. Phys.*, 1999, **200**, 126.
54. T. Huber, P. Pötschke, G. Pompe, R. Hässler, B. Voit, S. Grutke and F. Gruber, *Macromol. Mater. Eng.*, 2000, **280/281**, 33.
55. M. Abdelerehim, H. Komber, J. Langenwalter, B. Voit and B. Bruchmann, *J. Polym. Sci., Part A: Polym. Chem.*, 2004, **42**, 3062.
56. B. Voit, D. Beyerlein, K.-J. Eichhorn, K. Grundke, D. Schmaljohann and T. Loontjens, *Chem. Eng. Technol.*, 2002, **25**, 7.
57. B. Voit, *J. Polym. Sci., Part A: Polym. Chem.*, 2000, **38**, 2505.
58. P. Froehling, *J. Polym. Sci., Part A: Polym. Chem.*, 2004, **42**, 3110.
59. E. Žagar and J. Grdadolnik, *J. Mol. Struct.*, 2003, **658**(3), 143.
60. Y. Mikhaylova, G. Adam, L. Häussler, K.-J. Eichhorn and B. Voit, *J. Mol. Struct.*, 2006, **788**(1–3), 80.
61. B. Voit and A. Lederer, *Chem. Rev.*, 2009, **109**(11), 5924.
62. H. Komber, A. Ziemer and B. Voit, *Macromolecules*, 2002, **35**, 3514.

CHAPTER 4

Separation and Determination of Molar Mass

4.1 Determination of Absolute Molar Mass

The investigation of dilute solution properties of polymers is based on their molar mass and, therefore, the exact determination of this parameter is crucial. However, one has to note that the determination of the absolute molar mass of hyperbranched polymers is not easy. Due to their compact, branched structure, their overall molecular density is increased compared to linear analogues in a good solvent (Figure 2.2). Therefore, the widely used technique size exclusion chromatography (SEC), using simple detection and calculation procedures, cannot be applied to the separation and determination of the molar mass distribution of hyperbranched polymers. The usual detection in combination with SEC is differential refractive index (RI) or UV detection. These detectors give an intensity signal which corresponds to the frequency of the sample components at a certain elution volume. Subsequent calibration using a linear polymer standard with known molar mass gives the corresponding molar mass to the measured elution volume. This information enables calculation of the molar mass values, weight average, M_w and number average molar mass, M_n as well as the molar mass distribution, M_w/M_n, of the measured sample. Hence, SEC works on the basis of separation not according to molar mass but to volume of the eluted species, although, a branched and a linear sample with same molar mass and chemical structure would elute at different elution volumes. *Vice versa*, a more compact branched sample with higher molar mass could co-elute with a linear chain of the same volume but lower molar mass (Figure 4.1).

In 1976 Benoit's group introduced the universal calibration approach for the determination of molar mass.[1] In this approach, the apparent molecular

RSC Polymer Chemistry Series No. 16
Hyperbranched Polymers: Macromolecules in between Deterministic Linear Chains and Dendrimer Structures
By Albena Lederer and Walther Burchard

Published by the Royal Society of Chemistry, www.rsc.org

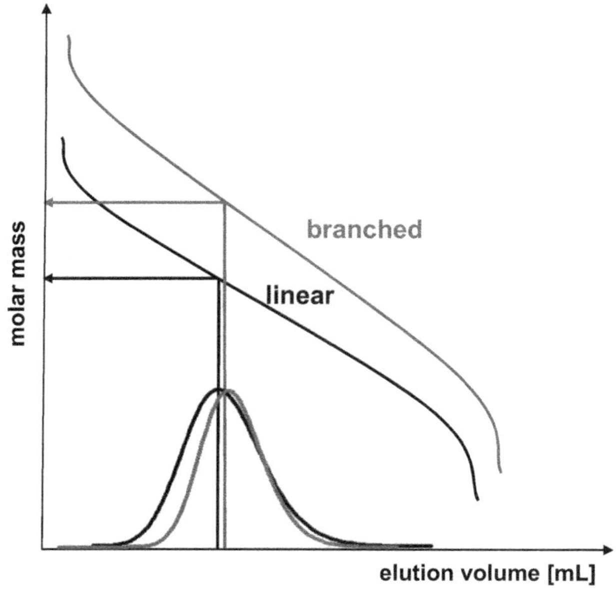

Figure 4.1 Co-elution of branched and linear sample of the same hydrodynamic volume but different molar mass. Branched polymers are more compact than linear and possess at the same volume higher molar mass.

density effect is corrected by multiplying the molar mass with the intrinsic viscosity. Several investigations on polymers with different chemical structure confirmed the universality of this approach (Figure 4.2).

Calibration of the elution behaviour with linear polymer standards, even if the intrinsic viscosity is taken into account, remains only a relative method for molar mass determination. Absolute molar mass determination could be performed by static light scattering as shown in detail in Chapter 6. A combination between SEC and a static light scattering detector provides the absolute molar mass distribution.

The intensity measured by the static light scattering detector is the Rayleigh scattering intensity. Eqn 4.1 shows the dependence of the scattering intensity from the molar mass:

$$R_\theta = \frac{4\pi n_0^2 \left(\frac{dn}{dc}\right)^2}{\lambda_0^4 N_L} cM \tag{4.1}$$

In this equation the only variables are the concentration and the molar mass. The first term contains the refractive index of the solvent, wavelength of the laser source, the Loschmidt's constant and the refractive index increment of the polymer solution. The information on the concentration c is obtained from the refractive index (RI) detector coupled to the light scattering detector. In this way, direct information on the molar mass can be

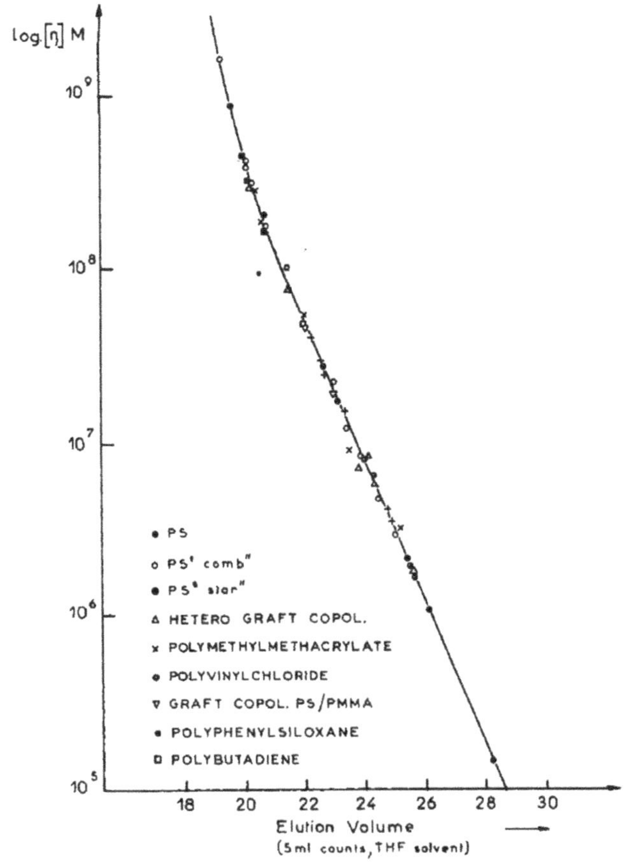

Figure 4.2 Universal calibration according to Grubisic *et al.*[1] Later on it was found that the universal calibration has its limits if the density of the molecules increases significantly as it is the case for highly branched polymers. Details of these relationships are discussed in Chapter 5.
Reprinted with permission from ref. 1. Copyright © 1967 John Wiley & Sons, Inc.

obtained. Detection can be performed under different angles but for small molecules it can be independent of the scattering angle. The situation becomes different for larger molecules for which the scattering intensity depends on the scattering angle. The scattering angle dependency correlates with the size of the molecules and it is useful for further calculation of additional parameters as the radius of gyration or the second virial coefficient. The interpretation of this phenomenon is described in detail in Chapter 6.

If the scattering intensity can be detected under different angles simultaneously, the detector used is called a multi-angle light scattering detector (MALS). The intensity detected by MALS is directly proportional to the molar mass and the interpretation of a SEC-MALS-RI chromatogram results in a plot of the molar mass elution volume, as shown in Figure 4.3.

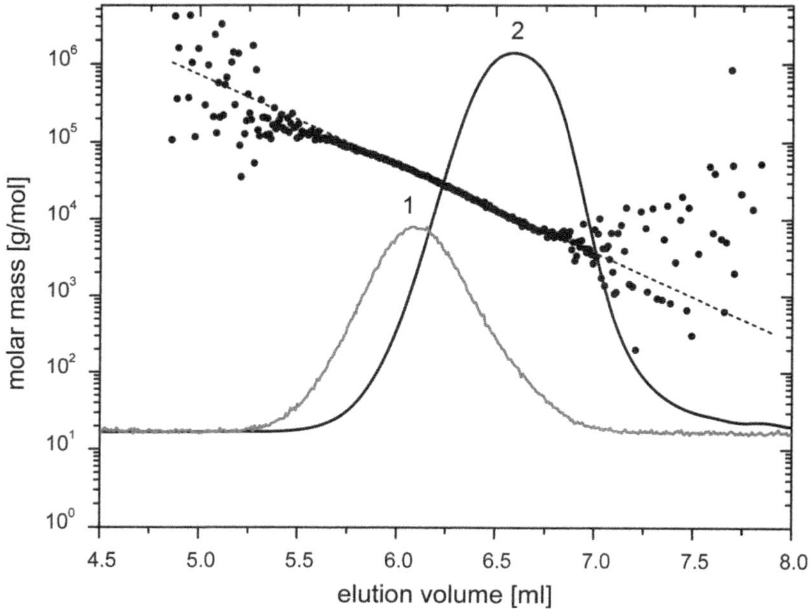

Figure 4.3 SEC Chromatograms from MALS detection at 90° (curve 1) and RI detector (curve 2) of hyperbranched aromatic polyester in *N,N*-dimethylacetamide with LiCl.[2] Molar mass distribution (black points) is calculated from the RI and MALS data using eqn 4.1.

Comprehensive comparison between different interpretations of the SEC separation of hyperbranched poly(etheramide) using polystyrene and poly(ethylene oxide) as calibration standards, as well as the method of universal calibration was performed in order to find the best adequate method for processing the SEC data.[3] The comparison was made on the basis of the molar masses obtained by SEC coupled to RI and MALS detectors, the combination of which also gives absolute molar mass values for complex macromolecular architectures. The deviations between the values obtained by these three methods are extremely large, as shown in Figure 4.4. Even the data from the universal calibration did not get close to the real molar mass values, although the intrinsic viscosity is incorporated in this calculation, taking into account its influence on the molecular volume, according to the Flory–Fox equation [eqn (2.7)]. It has to be noted that the molar masses calculated *via* SEC-MALS do not follow the same linear behaviour as calculated by any of the calibration methods. This development is extremely important and influences the scaling calculations immensely, as discussed in Chapter 5. This behaviour was found to be typical not only for hyperbranched poly(etheramide)s but also for 100% branched poly(etheramide) dendrimers (Figure 4.5).[4] However, due to molar masses lower than 5000 g mol^{-1}, the absolute molar mass of these dendritic molecules was successfully characterized by MALDI-TOF-MS.

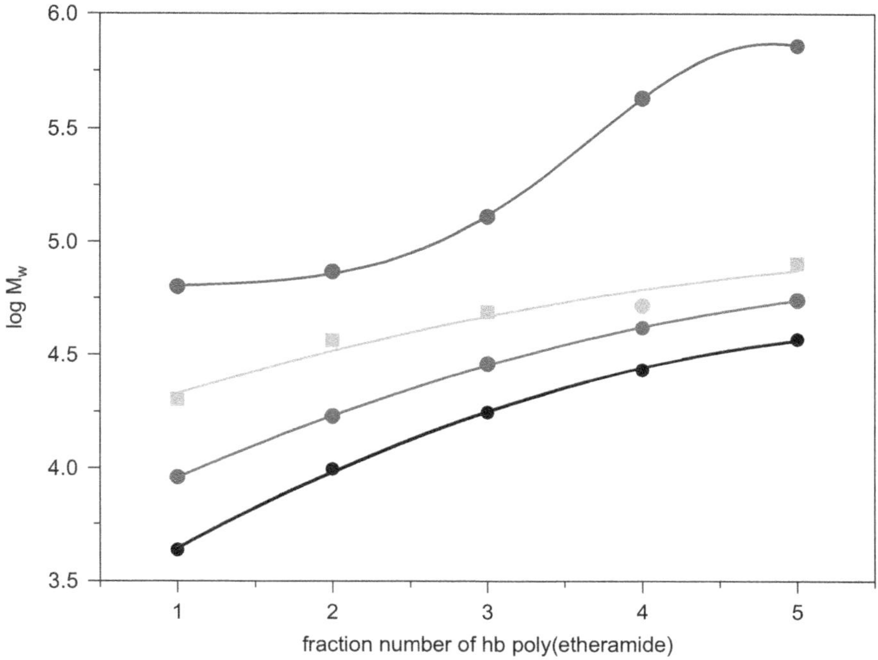

Figure 4.4 Deviation of the molar mass values for hyperbranched poly(etheramide) determined by different SEC interpretation methods: black symbols: SEC-RI detection and polyethylene oxide standard; dark grey symbols: SEC-RI detection and polystyrene standard; light grey symbols: offline SEC-viscosity detection and universal calibration; upper dark grey symbols: SEC-MALS (static light scattering detection).[3]
Adapted with permission from ref. 5. Copyright © 2009, American Chemical Society.

One might argue that the deviation of the universal calibration from the elution behaviour of dendritic polymers is a result of the strong interaction of the polymers with the column material, leading to separation governed by enthalpy, and exactly opposite to size exclusion.[6] However, using the absolute molar mass detection, separation of the polymer in size exclusion mode was confirmed. Therefore the observed deviation should be explained by the non-universal behaviour of the dendritic polymers: While for linear polymers the draining factor Φ of the Flory–Fox equation asymptotically becomes a constant with the molar mass, *i.e.* is not dependent on the size of the molecules,[7,8] in the case of branched polymers an increasing segment density leads to an increase of the draining parameter,[9–11] which contradicts the universality of this relationship (see also Chapter 5).

Hence, absolute detection is indeed the only reliable way to determine correctly the molar mass of dendritic structures. Yet, we have to critically consider the benefits and limitations of the two most commonly applied absolute methods for molar mass characterization in dendritic systems,

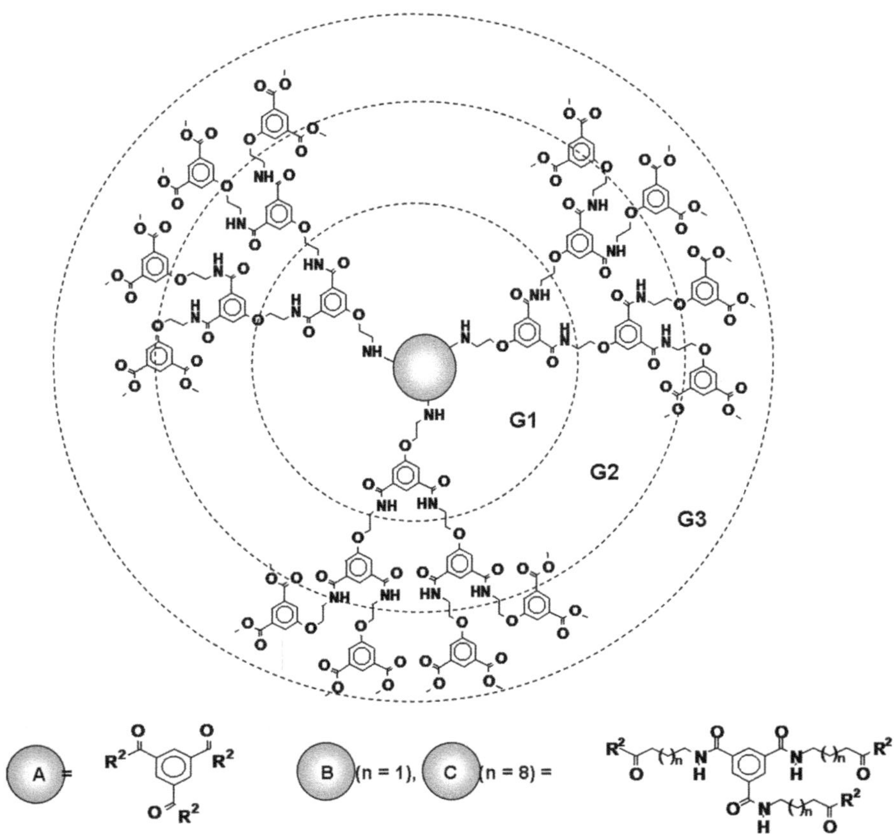

Figure 4.5 Poly(etheramide) dendrimers in different generations and with variation of the core molecule.
Reprinted with permission from ref. 4. Copyright © 2006, Springer.

static light scattering and matrix assisted laser desorption/ionization–time of flight–mass spectrometry (MALDI-TOF-MS). Static light scattering alone leads to reliable information about the weight average molar mass of polymers in the case that no aggregates are formed in solution,[12] but this information is not sufficient considering the broad polydispersities of hyperbranched polymers. In order to obtain complete evidence about the molar mass distribution of a hyperbranched polymer sample, the combination of SEC-RI-MALS is essential. According to eqn (4.1) the determination of the lower molar mass region by MALS requires high contrast of the polymer solution to the solvent, *i.e.* a high refractive index increment, d*n*/d*c*, or high concentration, though the maximum allowed concentration used in SEC is limited. Even if these conditions are fulfilled, reliable MALS molar mass determination is restricted to the high molar mass area. In order to overcome this problem, special interpretation methods must be applied for complete analysis of the molar masses.[13]

In contrast to static light scattering, MALDI-TOF-MS is a suitable technique for detection of lower molar masses and for the detection of absolute molar mass too. A significant drawback, however, is that it is limited to monodisperse samples. These limitations are because, in broadly distributed samples, lower molar mass species are preferably detected and, hence, higher molar masses are disregarded or their intensity is so low that they are ignored. Therefore information about the complete polydispersity of broadly distributed hyperbranched samples cannot be achieved. This technique is limited not only to samples or fractions with low polydispersities but also, in general, to low molar mass polymers which can be successfully detected quantitatively. Therefore, for well-defined dendrimers this is the technique of choice. An example are poly(etheramide) dendrimers (Figure 4.5) in the range of first to third generation and molar masses lower than 6000 g mol^{-1}. MALDI-TOF leads to complete information on the molar masses (Figure 4.6), which was confirmed even in mixtures of these structures.

In order to apply this method for broadly distributed dendritic polymers, it is reasonable first to separate it in nearly monodisperse fractions, *e.g.* by SEC, and to measure them subsequently by MALDI-TOF-MS.[14–17] MALDI-TOF-MS investigations on hyperbranched polyesters was dealt exhaustively by Montaudo.[17] By coupling to SEC Jaumann *et al.*[16] achieved complete characterization of hyperbranched polyalkoxysiloxanes. The MALDI-TOF results from these experiments show successful application of this technique up to molar mass values of more than 10^6 g mol^{-1} if the investigated fractions possess sufficiently narrow distributions (Figure 4.7).

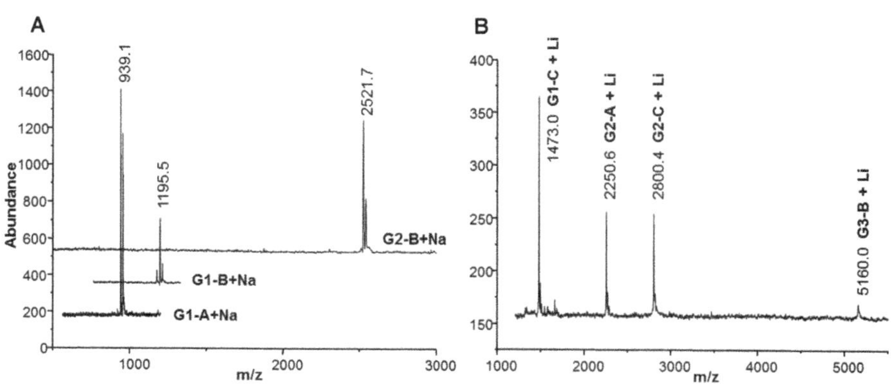

Figure 4.6 MALDI-TOF spectra: (A) molar masses of first and second generation dendrimer – the detected peaks correspond to the Na and Li adducts; (B) Mixture of three different generations dendrimers – with increasing molar mass the intensity is decreasing despite equal concentrations in the solution. The reason is the preferable detection of lower molar masses in polydisperse samples.
Reprinted with permission from ref. 4. Copyright © 2006, Springer.

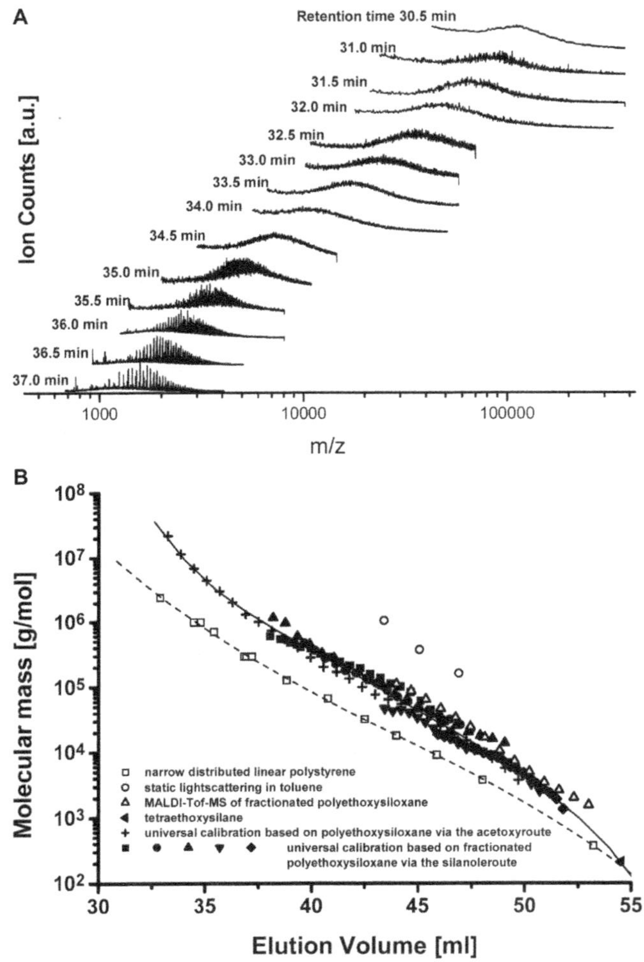

Figure 4.7 (A) MALDI-TOF mass spectra of polydiethoxysiloxane fractionated by SEC. For clarity the spectra were normalized and shifted vertically. (B) Calibration of the SEC curves of polyethoxysiloxane obtained through MALDI-TOF MS, static light scattering, and universal calibration. Reprinted with permission from ref. 16. Copyright © 2003 WILEY-VCH Verlag GmbH & Co. KGaA, Weinheim.

4.2 Suitability of the Separation Technique

Size exclusion chromatography, the usual technique for separation of polymers, is based on the well-balanced interactions between column material, solvent and polymer sample (Figure 4.8). In this separation mode the elution is governed by entropic interactions between the solvent, the column material and the macromolecules. If the interactions between these three components are governed by enthalpic effects, the macromolecules are not separated according to their size but according to adsorption to the column

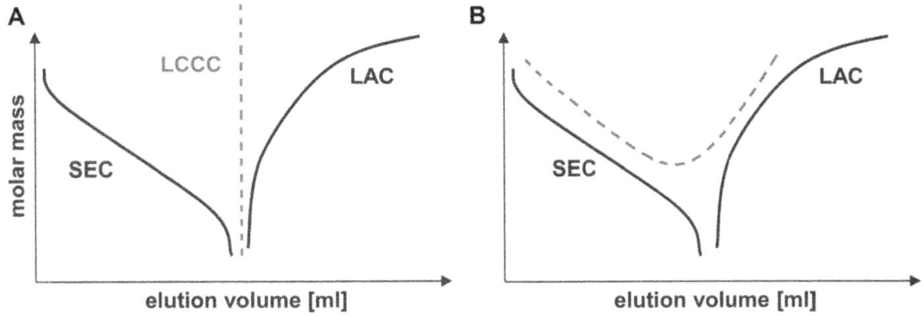

Figure 4.8 Different separation modes depending on the interactions governed by eluent, column material and polymer structure. (A) size exclusion chromatography (SEC), liquid chromatography under critical conditions (LCCC) and liquid adsorption chromatography (LAC). (B) the dashed curve shows combined separation according to both modes - SEC and LAC.

material. This separation mode is called liquid adsorption chromatography (LAC). There is a third possibility, which corresponds to critical separation condition (*e.g.* eluent composition) at which both effects are neutralized and neither separation according to molar mass nor to adsorption takes place. These conditions are specific for certain chemical structures and the separation mode is called liquid chromatography at the critical conditions (LCCC).[6]

In order to achieve complete separation according to size and to determine reliable polydispersity values, enthalpic interactions between sample and column material have to be excluded (Figure 4.8).

Only entropic interactions lead to SEC separation. This is not always possible in the case of dendritic polymers. Being multifunctional architectures, their interactions with the column material are predestined and simultaneous separation according to SEC and LAC could lead to molar mass dependency shown in Figure 4.8B. This effect is more severe for higher molar mass products in which the absolute number of end groups which are able to interact with the column material is higher as is the case for a statistically hyperbranched polymer, in which the number of end groups is nearly equal to the degree of polymerisation. Figure 4.9 shows one example of aromatic hyperbranched polyesters with different molar masses but equal degree of branching. The molar mass dependency on the elution volume shows clearly different separation regimes for the same polymer at different molar masses.

This phenomenon can be indicated only in the case of molar mass sensitive detection such as MALS. It cannot simply be explained by the fact that the absolute number of end groups per macromolecule increases with the molar mass, since the molar amount of end groups in the sample is constant. The most reasonable explanation is that there is an interplay between end-group interactions and the change in molecular shape with molar mass, as discussed in Chapter 5. Several methods are employed to reduce these

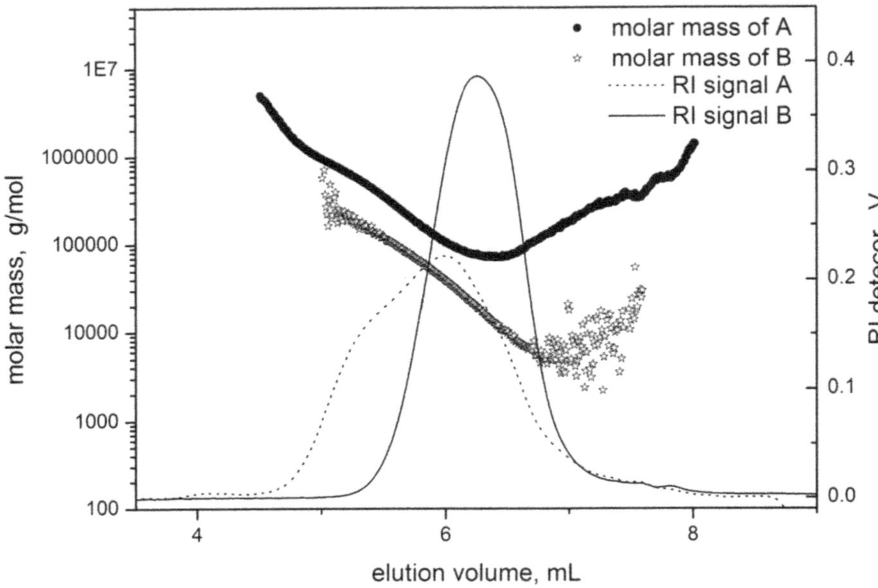

Figure 4.9 SEC-RI-MALS separation of the hyperbranched aromatic polyester with different molar mass: $M_w = 240\,000$ g mol^{-1} (A) and $M_w = 27\,800$ g mol^{-1} (B). Reprinted with permission from ref. 2. Copyright © 2009 Wiley Periodicals, Inc.

troublesome interactions, *e.g.* changing the solvent or the column material, or adding salts to the eluent.[3,12] However, these actions are often ineffective when it comes to multifunctional materials. A solution has been found for the separation of complex functional polymers using a column-free method: asymmetric flow field flow fractionation (AF4).[2,13,18] This recently commercialized technique separates molecules according to their size in a channel using the diffusion properties of the molecules and the external influence of an eluent cross-flow.[19,20] Figure 4.10 elucidates the principle of AF4 separation.

Interactions with the channel material do not affect the analysis and the only limitation is given by the membrane, which is permeable for molecules with molar mass lower than the cut-off of the membrane (approx. 1000–5000 g mol^{-1} the smallest). Therefore, this low molar mass region cannot be analysed directly. Figure 4.11 shows an AF4-MALS fractogram of aromatic polyesters with a high number of phenolic groups. The molar mass clearly has a linear dependence on the elution volume. In contrast to SEC, during AF4 separation the lower molar mass species elute first followed by the higher molar masses.[20–23]

4.3 Novel Separation Methods for Dendritic Polymers

Hyperbranched structures have a globular shape and pronounced multifunctionality in conjunction with broad and multidimensional

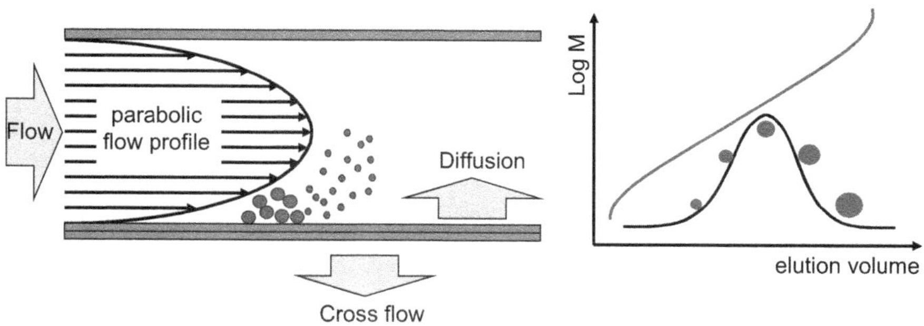

Figure 4.10 AF4 principle: The separation in AF4 is driven by the different diffusion of particles with different size. Large particles possess lower diffusion coefficients and remain closer to the bottom of the channel, while smaller macromolecules diffuse into the higher layers. A parabolic laminar flow drives the particles out of the channel in a size sequence from small to large molar masses/sizes.

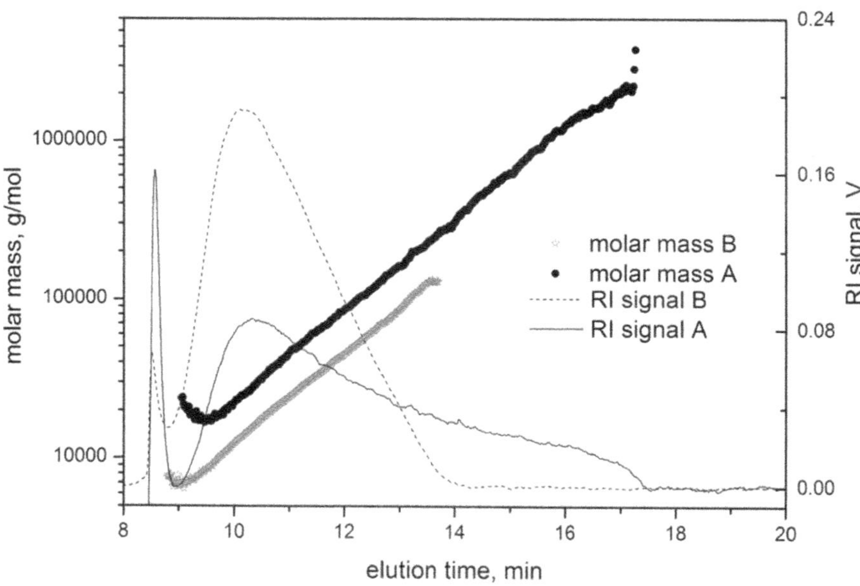

Figure 4.11 AF4-RI-MALS separation of the hyperbranched aromatic polyester with different molar mass: $M_w = 240\,000$ g mol^{-1} (A) and $M_w = 27\,800$ g mol^{-1} (B).
Reprinted with permission from ref. 2. Copyright © 2009 Wiley Periodicals, Inc.

distributions, *i.e.* in molar mass and degree of branching at the same time.[5,24] Therefore, improved methods or even new characterization techniques are required to study their molecular parameters and to correlate them to material properties. One possibility is to use SEC to separate polymers according to their hydrodynamic volume. Coupling SEC to a

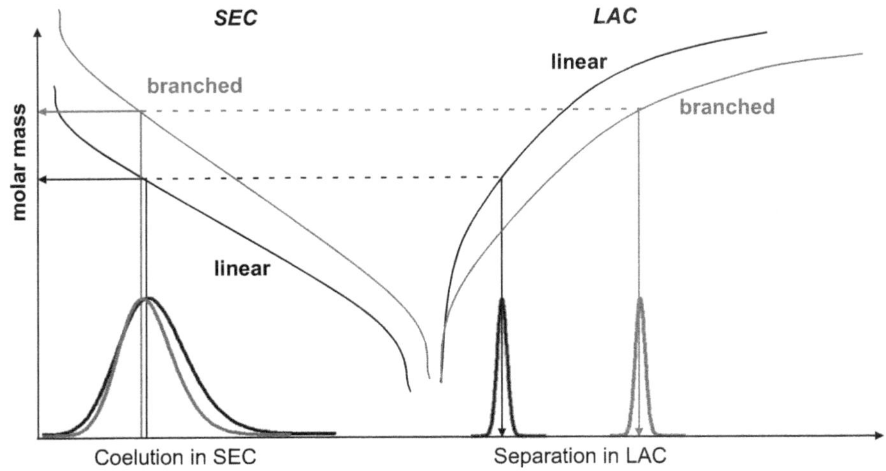

Figure 4.12 Elution behaviour of branched and linear polymer at fixed chemical character depending on the separation mode: SEC or LAC.

light-scattering detector provides reliable information about the molar mass distribution in the hyperbranched sample. However, co-elution of molecules with similar hydrodynamic volumes but different degrees of branching can never be excluded. Figure 4.12 shows the differences in elution of branched and linear polymers depending on the elution mode. The real-time detection of the elution behaviour of branched and linear polymers in both modes could give two-dimensional information about the branching and molar mass distribution. This is the basic idea behind the development of multidimensional separation techniques for characterization of branched polymers.

Two-dimensional liquid chromatography (2D-LC)[25–27] helps to obtain simultaneous information on molar mass and on structural characteristics, *e.g.* topology. Recently, several groups[28–31] reported on the separation of star and long-chain branched polymers from linear analogues, demonstrating the high potential of 2D-LC separations. The approaches were based on two different principles: the interaction of end groups with the column material[30] and topology-based separation in very narrow channels.[28] Reports on the application of 2D-LC are mainly focused on structures of lower branching density. Due to the rather dense branching in dendritic architectures, topology-based separation could not be applied to them. The 2D-LC proposed by Gerber and Radke[30] works successfully only on polymers with a limited number of end groups. In addition, the development of characterization techniques for branched polymers faces a lack of branched model compounds and linear polymers of identical chemical structure and defined number of functional groups. Tuned structural characteristics such as variation of the degree of branching, molar mass and functionality are prerequisites. The first branching separation by statistical branching topology was performed using a series of gradually branched polyesters with identical backbone and functionality. The synthetic background and structural characteristics of this polymer series are described in detail in Section 3.2. Another approach for separation could be based on the constant branching but variation in functionality. Since functionality in statistical polymers correlates with branching, this factor is also a possibility for estimation of branching. For the purpose of separation according to end-group type, hyperbranched polyesters with selectively modified linear or terminal functionalities are discussed in Section 4.3.2.

4.3.1 Separation According to Degree of Branching

The first separation of statistically branched polymers with a variation of the degree of branching was developed by Al Samman and colleagues[32] and was based on their previous studies on liquid adsorption chromatography (LAC).[30,33] LAC is based on the separation according to enthalpic interactions with the column material under the controlled elution with a solvent gradient as explained above. In the special case of LC, the liquid chromatography under critical conditions (LCCC), the separation is not sensitive to the molar mass, but only to the polymer structure. Hence, chemically identical structures would elute at the same time.[6] The idea behind this work is that separation according to molecular architecture should be possible under critical conditions, since the influence of the molar mass is excluded.

For this purpose and using a series of differently branched polyesters, the critical conditions for linear polymer (DB = 0%) were defined at a mixture of 6% THF in acetone. Under the same conditions, however, branched polymers elute depending on their molar mass (Figure 4.13). Moreover, a clearly increasing dependency of molar mass on the retention time is achieved at an increasing degree of branching, *i.e.* higher values of the degree of branching result in longer retention times. Since the type of monomer units and the number of end groups per monomer unit is identical for all polymers, independently of the degree of branching, this cannot be the reason for their different retention behaviour. The reason for that should be searched for in the special, branched architecture. With increasing degree of branching, a higher density of functional groups is concentrated on the surface of the polymer coil. Therefore, separation by liquid chromatography is obviously governed by the branching topology of the polyesters.

These first measurements were performed on the basis of samples with different molar masses. More interesting is the application to real, broadly distributed polymers. The indication of samples with different degrees of branching but same retention volume, however, would not be possible by this technique. Another technique, 2D-LC, *i.e.* simultaneous separation according to molar mass and to topology, is more sensible for broadly distributed hyperbranched samples. For this purpose, SEC was performed in the first dimension while gradient LAC was performed in the second dimension. The first successful 2D separation of hyperbranched polymers

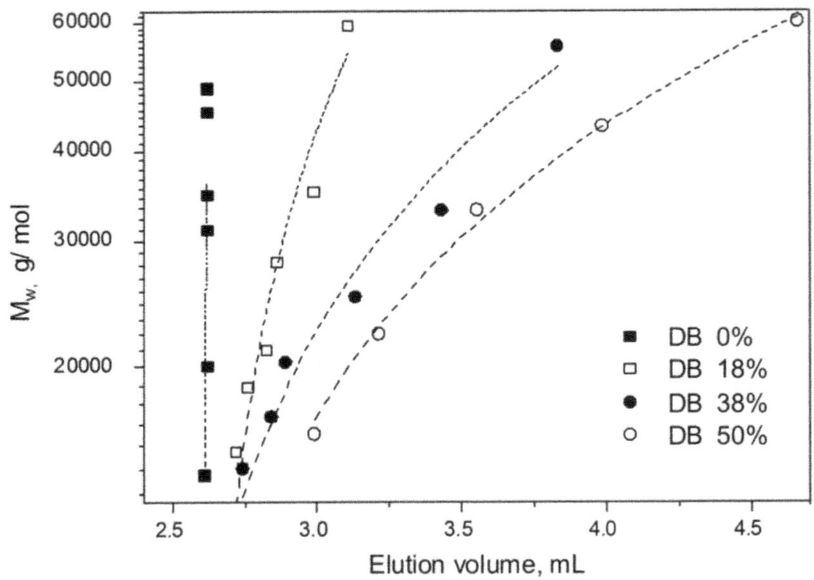

Figure 4.13 Molar mass versus elution volume for linear and branched polymers at critical conditions of the linear polyester (94% acetone and 6% THF). Reprinted with permission from ref. 32. Copyright © 2010 American Chemical Society.

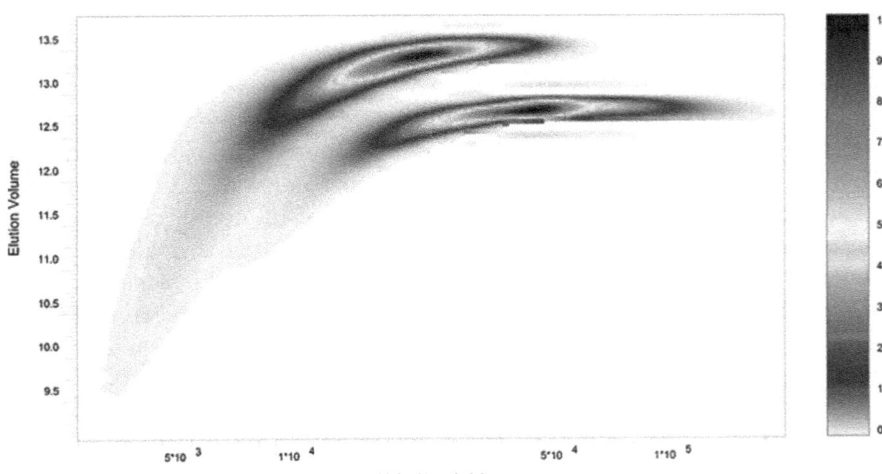

Figure 4.14 Separation of a blend of a linear and hyperbranched polyester by two-dimensional chromatography.
Reprinted with permission from ref. 34. Copyright © 2014 Elsevier.

was performed by Radke's group.[34] Hyperbranched aliphatic aromatic polyesters with 50% degree of branching and a linear polyester with identical chemical character and functional groups were separated from a mixture as shown in Figure 4.14. While in the region of lower molar mass no complete separation was observed, in the region of higher molar mass well-defined signals for both samples are visible.

Moreover, using the 2D contour plots reproducible quantification of the amounts of linear and branched structures in the mixtures is potentially possible. This approach is the basis for the development of fast and quantitative determination of branching structures and their distribution for future analysis of commercial, broadly distributed hyperbranched polymers. The development of such an approach would greatly improve our understanding of the structure–properties relationship in hyperbranched polymer materials.

4.3.2 Elution Properties Depending on Functionality

The separation of polyesters by liquid chromatography is obviously possible according to branching if all other parameters, *e.g.* number and type of functional groups, is kept constant. The interaction with the column material depends on the density of the end groups at the polymer surface. Theoretically, the degree of branching influences not only the molecular density but it also depends on the number of linear, terminal and dendritic units in the hyperbranched structure. Hence, suitable methods for the characterization of the degree of branching should be those which are sensitive to the number of linear, terminal and dendritic units. The relationship between the degree of branching and these numbers was proposed by Fréchet and Frey [eqn (2.9)–(2.11)]. The common way to determine the degree of branching according to these equations is the calculation based on the intensities of NMR signals which are characteristic for the different structural units. This pathway, however, provides only average values. The distribution of the degree of branching can be obtained only by appropriate separation. The question is, What is the influence of the end-group type on the elution behaviour? Is there a potential to separate hyperbranched polymers according to the number of linear and terminal units? In order to apply such separation, selective modification of the end groups of one given type should be carried out switching them from polar to non-polar. The purpose of this selective modification is to exclude the influence of the linear units and at the same time tune the solution and interaction properties. Hence, by systematically changing the solvent content it could be possible to separate hyperbranched polymers by their degree of branching. The potential of this idea was explored by investigating the elution properties of tailored, selectively modified polymers (Section 3.4).[35] The polymers investigated are presented schematically in Figure 3.27 and the changing polymer units are shown in Figure 4.15.

Gradient elution fractionation studies of the selectively modified polymers resulted in a series of narrowly distributed fractions with variation in the molar mass. The fractionation was carried out on a preparative scale[3,36] obtaining fractions in sufficient amounts (200 to 400 mg) for further investigations, *e.g.* viscosity and molar mass determination.

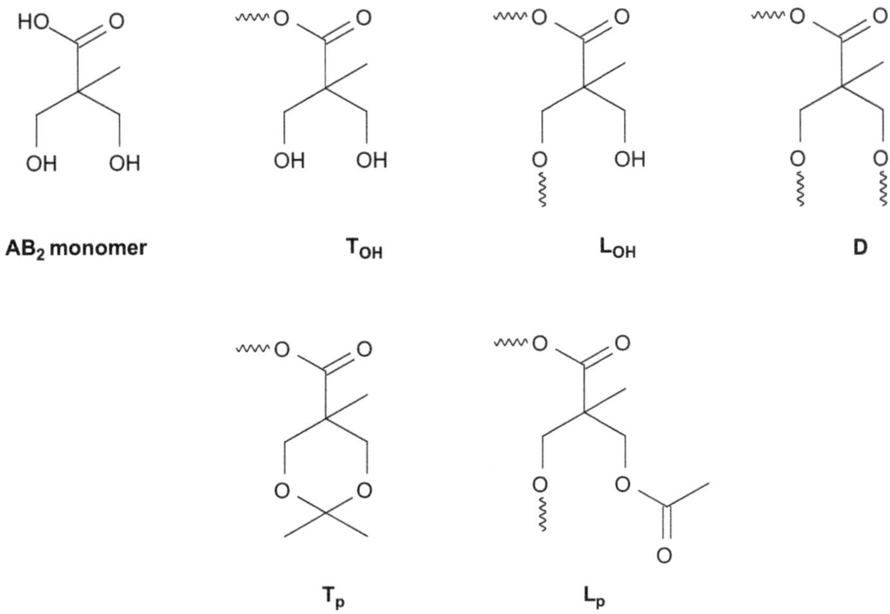

Figure 4.15 Schematic representation of the end group structures in selectively modified polyesters leading to polymers with variation in the polarity of the end groups depending on their type (see also Section 3.3). T, L, D stays for terminal, linear and dendritic units; OH – for OH-terminal groups; p – for protected, non-polar functions.

Table 4.1 Molar mass distributions and degrees of branching after fractionation of $T_{OH}L_{OH}$.

Sample	M_w (g mol^{-1})	M_w/M_n	$DB_{Fréchet}$	DB_{Frey}
Fractions	6000	1.43	0.42	0.34
	7400	1.48	0.42	0.34
	9500	1.64	0.41	0.34
	15 500	1.91	0.43	0.34
	21 100	1.69	0.44	0.39
	25 200	1.74	0.42	0.34
Starting material	27 300	7.38	0.43	0.39

In order to evaluate the elution behaviour of the polymers relative to each other, comparable solvent gradients of THF/*n*-hexane were used. Clear separation according to molar mass for all polymer types was found. Further information about the chemical identity and degree of branching was obtained by NMR investigations of all samples. Chemically identical structures have been found for the fractions of every polymer type. As an example, the molar masses and the degrees of branching of selected fractions of $T_{OH}L_{OH}$ are listed in Table 4.1. The molar mass distributions within the fractions are with $M_w/M_n < 2$, far below the initial polydispersity (see Table 4.1).

The NMR calculation shows clearly that within the experimental error (±2%), no fractionation according to the degree of branching took place. This fact does not exclude the possibility that different degrees of branching exist in one sample. It could be expected that elution with different eluents would lead to separation according to branching; however, comparable studies based on this particular hyperbranched system have not been performed so far.

Detailed comparison of $T_{OH}L_{OH}$ and $T_{OH}L_p$ during the fractionation demonstrates well-distinguishable differences in their elution behaviour. Figure 4.16 shows the dependency of the fraction yield on the solvent gradient during the elution fractionation.

Clear influence of the end-group polarity on the elution process can be identified. The OH end-groups show slower elution with THF/*n*-hexane gradient than the completely non-polar end groups of T_pL_p. As a result, at 100% THF content $T_{OH}L_{OH}$ is still not completely eluted, while T_pL_p achieves complete elution already at 64% THF. Interestingly, a well-distinguished behaviour of T_pL_{OH} and $T_{OH}L_p$ is observed, in which polar and non-polar end groups are combined in one molecule. This fact underlines the characteristic influence of non-polar polymer units, depending on their type: terminal (T_p) or linear (L_p). Consequently, the selective switching of the end-group polarity of hyperbranched aliphatic polyester leads to well-tunable elution properties

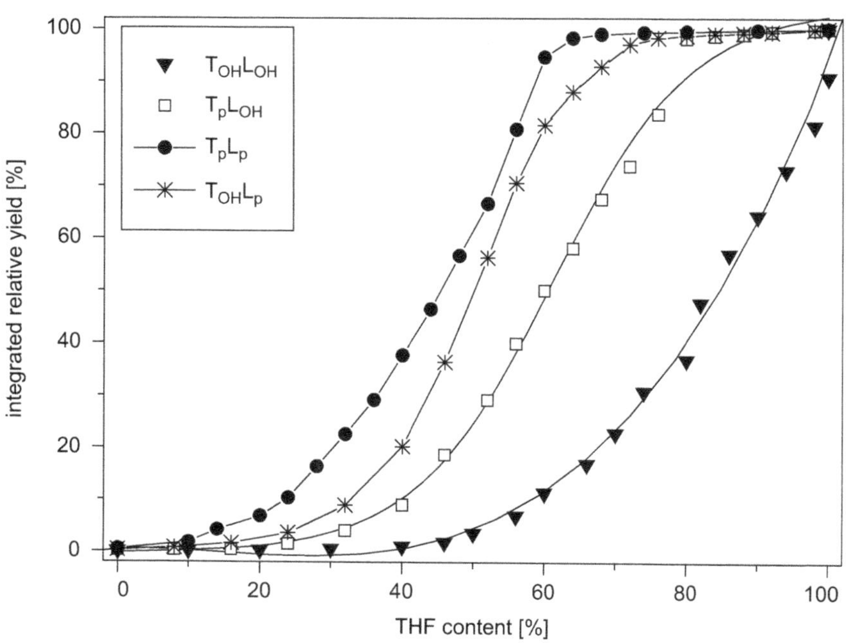

Figure 4.16 Elution behaviour of the different polymers during fractionation in a solvent gradient mixture of THF and *n*-hexane.
Adapted with permission from ref. 35. Copyright © 2010 Elsevier.

Table 4.2 Comparison of integrated, relative yields and molar masses of fractions from elution fractionation at fixed THF content of 60% (v/v) in the eluent.

Starting sample	Integrated relative yield (%)	M_w (g mol^{-1})
$T_{OH}L_{OH}$	11.3	1600
T_pL_{OH}	50.0	9600
T_pL_p	94.6	167 900
$T_{OH}L_p$	81.5	8900

of the macromolecule. This is supported by the values listed in Table 4.2, which correspond to a fixed THF content of the eluent at 60%.

Fully modified end functionalities (T_pL_p) lead to faster elution of higher molar masses under the given conditions, whereas with increasing numbers of OH end-groups, the eluted molar masses drastically decrease. Apparently, separation of discrete molecular sizes under the same conditions takes place, although separation depends on the strength of the eluent–polymer interactions correlating with the eluted polymer yield. These results are the first indication that separation by a solvent gradient is controlled by the type of end functionalities and builds the foundation for future determinations regarding the distribution of the degree of branching.

References

1. Z. Grubisic, P. Rempp and H. Benoit, *J. Polym. Sci., Part B: Polym. Lett.*, 1967, **5**, 753.
2. M. Erber, S. Boye, T. Hartmann, B. I. Voit and A. Lederer, *J. Polym. Sci., Part A: Polym. Chem.*, 2009, **47**, 5158.
3. A. Lederer, D. Voigt, C. Clausnitzer and B. Voit, *J. Chromatogr. A*, 2002, **976**, 171.
4. A. Lederer, D. Voigt, D. Appelhans and B. Voit, *Polym. Bull.*, 2006, **57**, 329.
5. B. I. Voit and A. Lederer, *Chem. Rev.*, 2009, **109**, 5924.
6. H. Pasch and B. Trathnigg, *HPLC of Polymers*, Springer-Verlag, Berlin, Heidelberg, 1999.
7. T. G. J. Fox and P. J. Flory, *J. Appl. Phys.*, 1950, **21**, 581.
8. T. G. J. Fox and P. J. Flory, *J. Polym. Sci.*, 1954, **14**, 315.
9. A. Lederer, W. Burchard, A. Khalyavina, P. Lindner and R. Schweins, *Angew. Chem., Int. Ed.*, 2013, **52**, 4659.
10. W. Burchard, *Macromolecules*, 2004, **37**, 3841.
11. W. Burchard, in *Branched Polymers II*, ed. J. Roovers, 1999, vol. 143, p. 113.
12. E. Žagar and M. Žigon, *Macromolecules*, 2002, **35**, 9913.
13. A. Lederer and S. Boye, *LCGC Ads*, 2008, (November/December), 24.
14. L. Chikh, M. Tessier and A. Fradet, *Polymer*, 2007, **48**, 1884.
15. L. Chikh, M. Tessier and A. Fradet, *Macromolecules*, 2008, **41**, 9044.

16. M. Jaumann, E. A. Rebrov, V. V. Kazakova, A. M. Muzafarov, W. Goedel and M. Möller, *Macromol. Chem. Phys.*, 2003, **204**, 1014.
17. M. S. Montaudo, *J. Am. Soc. Mass Spectrom*, 2004, **15**, 374.
18. Z. Fan, A. Lederer and B. Voit, *Polymer*, 2009, **50**, 3431.
19. S. Podzimek, T. Vlcek and C. Johann, *J. Appl. Polym. Sci.*, 2001, **81**, 1588.
20. J. C. Giddings, *Sep. Sci.*, 1966, **1**, 123.
21. J. C. Giddings, *Sep. Sci. Technol.*, 1984, **19**, 831.
22. J. C. Giddings, *J. Chem. Edu.*, 1973, **50**, 667.
23. K. G. Wahlund and J. C. Giddings, *Anal. Chem.*, 1987, **59**, 1332.
24. R. Hanselmann, D. Hölter and H. Frey, *Macromolecules*, 1998, **31**, 3790.
25. A. van der Horst and P. J. Schoenmakers, *J. Chromatogr. A*, 2003, **1000**, 693.
26. S. J. Kok, T. Hankemeier and P. J. Schoenmakers, *J. Chromatogr. A*, 2005, **1098**, 104.
27. G. I. Vivo Truyols and P. J. Schoenmakers, *J. Chromatogr. A*, 2006, **1120**, 273.
28. R. Edam, D. M. Meunier, E. P. Mes, F. A. Van Damme and P. J. Schoenmakers, *J. Chromatogr. A*, 2008, **1201**, 208.
29. K. Im, Y. Kim, T. Chang, K. Lee and N. Choi, *J. Chromatogr. A*, 2006, **1103**, 235.
30. J. Gerber and W. Radke, *Polymer*, 2005, **46**, 9224.
31. J. Gerber and W. Radke, *e-Polym.*, 2005, **045**.
32. M. Al Samman, W. Radke, A. Khalyavina and A. Lederer, *Macromolecules*, 2010, **43**, 3215.
33. W. Radke, K. Rode, A. Gorshkov and T. Biela, *Polymer*, 2005, **46**, 545.
34. W. Radke, *J. Chromatogr. A*, 2014, **1335**, 62.
35. S. Boye, H. Komber, P. Friedel and A. Lederer, *Polymer*, 2010, **51**, 4110.
36. A. Lederer, M. Abd Elrehim, F. Schallausky, D. Voigt and B. Voit, *e-Polym.*, 2006, **039**.

CHAPTER 5

Solution Viscosity

5.1 General Remarks

Dissolved polymer coils require a certain volume and contain solvent molecules surrounding the polymer segments to a different extent. To what extent the solvent penetrates into the macromolecule depends on the affinity between both components. As demonstrated by Flory this affinity does not mean pronounced attractive interactions between solvent and segments. In fact, predominantly repulsive interactions are measured and it is the entropy of mixing caused by the large difference in the size between the small solvent molecules and the large macromolecules which keeps the polymer in solution. The penetration of the solvent occurs, of course, if a porous structure is present. For polymers the penetration is more complex but certainly depends on the local density, which differs depending on spacial factors as topology (*e.g.* branching) or stiffness. This penetration of solvent is in direct correlation to the structural properties of the macromolecules. Hence, the solution properties are sensitive to the molar mass of the macromolecules and should remain in a relationship with their structure and shape in solution. The viscosity of dilute polymer solutions depends strongly on concentration, and at constant concentration it increases with the molar mass. Simple relationships have been suggested to give a direct indication of the state of the macromolecule in solution. First studies on the viscosity of polymer solution were made by Staudinger in the 1930s showing these dependencies empirically. With the development of the theory of polymer solution it was shown that they correlate with the scaling properties of polymers, which are usually studied by scattering techniques. In contrast to the rather complex physics of hydrodynamics, the measurement of viscosity is fairly easy, the instrumentation is inexpensive, which has led this method to being a widespread and versatile characterization technique. However,

RSC Polymer Chemistry Series No. 16
Hyperbranched Polymers: Macromolecules in between Deterministic Linear Chains and Dendrimer Structures
By Albena Lederer and Walther Burchard

Published by the Royal Society of Chemistry, www.rsc.org

some limitations should be taken into account for the correct understanding and interpretation of the solution properties especially in highly branched polymers.

Capillary viscometry is the classical technique for the determination of solution viscosity. According to Stokes, it is the friction between solvated molecular segments and free solvent molecules which influences the viscosity. Friction is the ratio between force and velocity of the moving particles. The force needed to overcome friction corresponds to the viscosity of the macromolecular solution. Hence, a higher viscosity will be found if the solvent flushes through the polymer coil and a lower viscosity will be measured if the interaction between segments and solvent is limited to the shallow outskirt area of the polymer coil.

The characterization of macromolecules in solution is not limited only to the determination of the overall solution viscosity. More interesting is its relative increase which is described by the specific viscosity, which divided by the concentration gives the reduced viscosity:

$$\eta_{\mathrm{red}} = \frac{\eta_{\mathrm{spec}}}{c} = \frac{\eta - \eta_0}{\eta_0 c} \tag{5.1}$$

Measurement of viscosity using capillary viscometry is based on the time which a liquid needs to pass a certain distance, as shown in Figure 5.1:

$$\eta_{\mathrm{red}} = \frac{t - t_0}{t_0 c} \tag{5.2}$$

Measuring the reduced viscosity for different concentrations and extrapolating the data to $c = 0$ gives the intrinsic viscosity which is specific for every polymer solution. The relationship between the specific and reduced viscosity gives a first indication about the quality of the solution, *e.g.* the interaction between polymer segments and solvent molecules expressed in the well-known Schulz–Blaschke and Huggins constants.

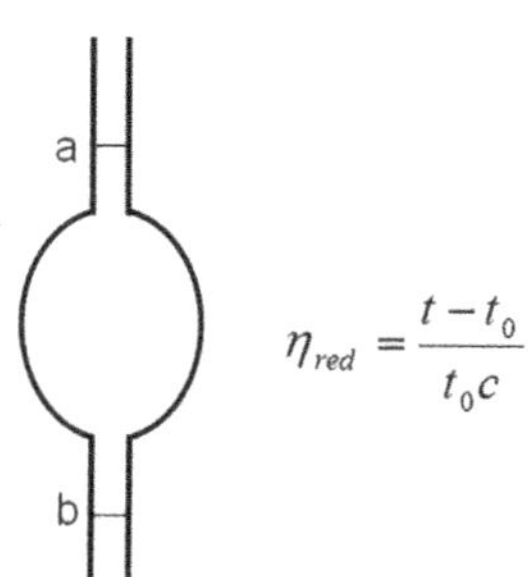

Figure 5.1 Determination of reduced solution viscosity takes place in a capillary viscometer by measuring the time t a solution needs to pass the volume from a to b taking into account the time the pure solvent needs (t_0) for this distance and the concentration of the solution.

Staudinger noticed that the intrinsic viscosity is strongly dependent on the molar mass of the macromolecules and at first he used it to determine the molar mass by considering that polymer chains are stiff rods. The fundamental treatment applied by Kirkwood and Riseman[1] corrected this view and a physical background was given. This dependency is used for the evaluation of global molecular structure in solution. The relation between molar mass and intrinsic viscosity is known as the Kuhn–Mark–Houwink–Sakurada (KMHS) equation:

$$[\eta] = K_{\eta} M^{\alpha_{\eta}} \tag{5.3}$$

Typically, the quantity of the KMHS exponent α_{η} is related to the shape and compactness of a polymer in a certain solvent. There are many investigations on this parameter and usually values of $0.3 < \alpha_{\eta} < 0.5$ for hyperbranched polymers were found.[2] For a linear statistical coil in a good solvent values of $0.5 < \alpha_{\eta} < 1$ are typical, and for hard spheres, values approximating zero are theoretically expected.

Actually, the KMHS equation is widely used to obtain information on scaling properties of macromolecules in solution in analogy to the more general scaling law with regard to the radius of gyration, which is described in detail in Chapter 6.

The correlation between intrinsic viscosity and size of the molecules has its origin in the Flory–Fox equation. This relationship was found between 1908 and 1910 when Albert Einstein[3] derived his famous equation on the increase of the solvent viscosity after adding spherical particles to the solvent. In this regard, the eqn (5.1) may be written in terms of volume and mass of the particles or it may also be expressed in terms of molar mass and the Avogadro number N_{A}:

$$[\eta] = \frac{\eta - \eta_0}{\eta_0 c} = 2.5\,\frac{V}{m} = \frac{10\pi}{3}\,N_{\mathrm{A}}\frac{R^3}{M} \tag{5.4}$$

where R is the radius of a hard sphere. Evidently, the left and the right sides in eqn (5.4) have the same dimensions, and therefore the parameter $\Phi_{\mathrm{Einstein}} = 10\pi/3$ is a dimensionless quantity. Twenty years later Staudinger showed that the intrinsic viscosity is an excellent quantity to describe solution properties of polymers and this was used by Fox and Flory in the viscosity equation for linear polymer chains [eqn (2.7)].[4,5] In this equation again $[\eta]$ and R_g^3/M have the same dimension, and consequently the Flory parameter Φ seems to fulfil the characteristics of a universal quantity, at least for linear chains (see also Chapter 8). Such universality became questionable by the observation that Φ slightly decreases if a good solvent was used instead of a poor solvent, and this decrease correlated with the increased swelling of the coiled chain.[6] Nevertheless, this observation is often neglected in experiments for molar mass determination. Hence, the Φ-parameter is widely used for characterization of macromolecules and is often denoted as a *universal constant*. The *universal constant* is the background of the development of the *universal calibration*[7] for calculation of the molar mass from size exclusion chromatography measurements as described in Chapter 4. The universal constant for this calculation is used with the value 2.5×10^{23} mol^{-1}. In fact the value of Φ_{Flory} corresponds to 2.86×10^{23} mol^{-1} as calculated from the end-to-end distance of the linear chain and as 4.20×10^{24} mol^{-1} as calculated from the radius of gyration,[8] while for hard spheres it is calculated as $\Phi_{\text{Einstein}} = 1.36\times10^{25}$ mol^{-1}. The Flory parameter for a non-disturbed linear chain is in fact not applicable for all polymer systems and in the case of highly branched polymers it leads to incorrect results. The discussion of this effect is the subject of the next section.

As a consequence of the Flory–Fox equation, the molecular size dependence of the intrinsic viscosity appears highly useful for calculation not only of molar mass but also of the dimensions of macromolecules. For this purpose, the Flory–Fox equation can be modified for calculation of the equivalent sphere radius out of the intrinsic viscosity and the molar mass (see Chapter 8).

The correlation between intrinsic viscosity and molecular volume has a special advantage in calculating the branching in macromolecules. In regard to the fact that branching leads to higher molecular density (at the same molar mass and chemical structure) in comparison to linear polymers, contraction of the size by introduction of branches leads to reduction in the intrinsic viscosity (Figure 5.2), which can be recalculated for quantification of branching. For this purpose the contraction factor g' is used [eqn (2.8)] together with eqn (2.1)–(2.5).

The interpretation of the molar mass dependency of intrinsic viscosity $[\eta]$, radius of gyration R_g, hydrodynamic radius R_h or second virial coefficient A_2 for a certain polymer sample can be performed by viscosity and scattering measurements (see Chapter 8) and provide key information about its molecular shape, density or self-similarity, which are strongly influenced by

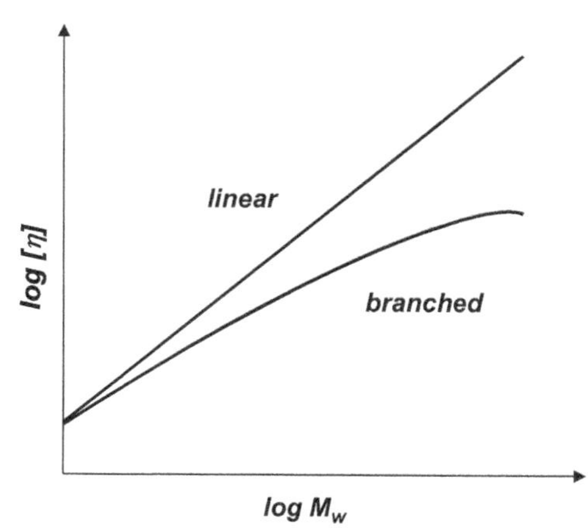

Figure 5.2 Schematic representation of the KMHS dependency for linear and branched polymer of similar chemical character.

branching. The solution behaviour of linear polymers is well described by general theories. The introduction of branches into a macromolecular structure, even if the branches are formed by linear segments, results in a different topology with a solution behaviour which is not governed by common universal laws. Moving from long- to short-chain and tree-like branches, as in the case of dendritic structures, the situation becomes more complex. Here, a combination of scattering and viscosity data is very useful for complete interpretation of branched structures.

There are couple of parameters which can be treated similarly to the KMHS equation and the scaling law for R_g. One of these parameters is A_2, the second virial coefficient extracted from scattering measurements together with R_g and M_w. Detailed discussion on the characteristics of the second virial coefficient is given in Section 6.3. One of the most sensitive parameters to branching topology is the one coupled to the coil–coil interpenetration function Ψ, which describes the ability of polymer molecules to interact with each other (Figure 5.3) and can be expressed in analogy to the Flory–Fox equation with the second virial coefficient as follows:[9]

$$A_2 = 4\pi^{3/2} N_A \left(\frac{R_g^3}{M^2}\right)\Psi \tag{5.5}$$

Combining eqn (5.5) with the Flory–Fox equation [eqn (2.7)], we obtain:

$$\frac{A_2 M_w}{[\eta]} = \frac{\pi^{1/2} 6\Psi}{5\Phi} \tag{5.6}$$

For linear polymers the value of $\frac{A_2 M_w}{[\eta]}$ approximates 1.07 and for hard spheres it should correspond to 1.6. With increasing branching density, this

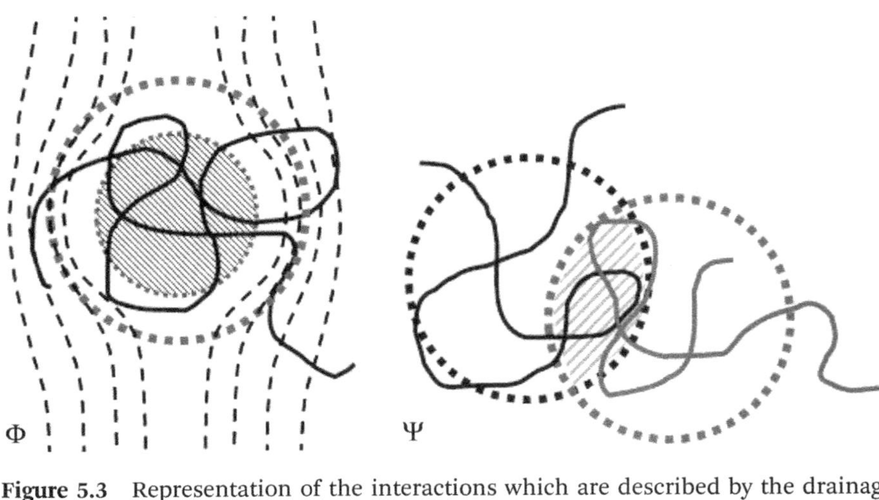

Figure 5.3 Representation of the interactions which are described by the drainage parameter (Flory parameter) Φ and the coil–coil interpenetration parameter Ψ.

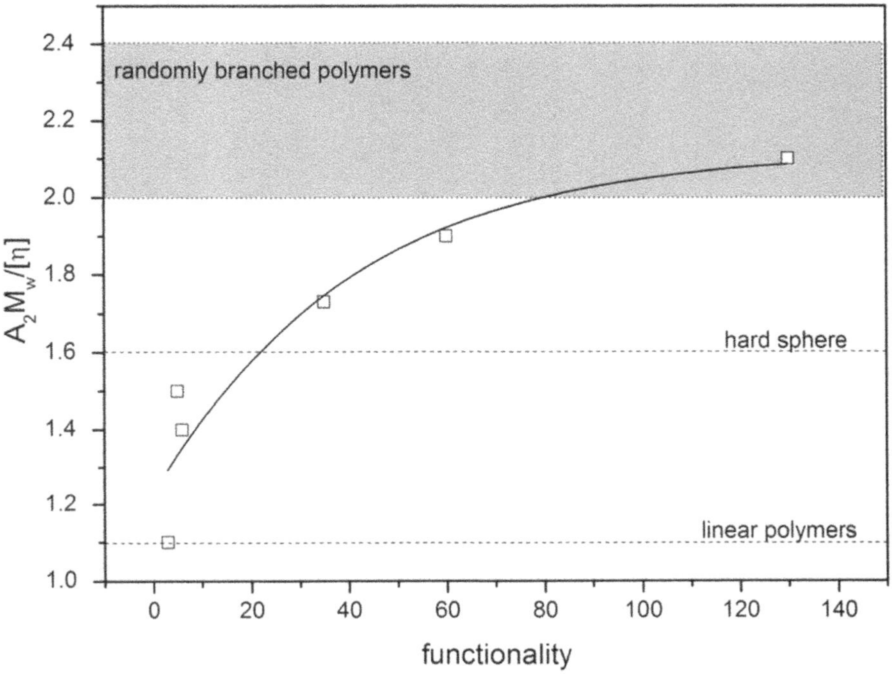

Figure 5.4 Dependence of $\frac{A_2M_w}{[\eta]}$ on the functionality for star and randomly branched (hyperbranched) polymers at high molar masses adapted from Burchard[9] and Galinsky.[10]
Reprinted with permission from ref. 11. Copyright © 2009 American Chemical Society.

value increases because the interpenetration ability of the molecular segments decreases rapidly with the higher branching density. At the same time the small solvent molecules still rinse the polymer segments. This effect is less influenced by the branched structure, which means that Ψ increasing stronger than Φ. This behaviour has been observed for stars[9,12] and was expanded to random starch polymers in the work of Galinsky and Burchard[10] (Figure 5.4), reaching values higher than 2.0. How this behaviour was found in synthetic hyperbranched systems will be discussed in the next section in detail. A detailed explanation of the $\frac{A_2M_w}{[\eta]}$ ratio is given in Chapter 8.

5.2 Determination of the Kuhn–Mark–Houwink–Sakurada Dependency

The KMHS equation [eqn (5.3)] is valid only for monodisperse polymers. Usually, polymers are polydisperse and for this reason, a molar mass average has to be used. In polydisperse samples, the intrinsic viscosity is a mass average and the KMHS relationship will be expressed as follows:

$$[\eta]_w = \frac{\sum_i w_i[\eta]_i}{\sum_i w_i} = K_\eta \frac{\sum_i w_i M_i^{\alpha_\eta}}{\sum_i w_i} = K_\eta M_\eta^{\alpha_\eta} \tag{5.7}$$

The viscosity average of the molar mass M_η depends on α_η:

$$M_\eta = \left(\frac{\sum_i w_i M_i^{\alpha_\eta}}{\sum_i w_i} \right)^{\frac{1}{\alpha_\eta}} \tag{5.8}$$

and it is equal to the weight average molar mass M_w only if $\alpha_\eta = 1$, which means that two parameters which have to be taken into account for a polydisperse sample: the polydispersity and α_η. This makes the calculation rather complicated and, therefore, for the correct determination of the KMHS exponent of macromolecules monodisperse fractions are needed. Low polydispersity fractions could be obtained using the elution fractionation described in Chapter 4. This method leads to fractions in sufficient amount for further characterization of the intrinsic viscosity by capillary viscometry and of the molar mass by SEC or static light scattering.

Another approach for measuring intrinsic viscosity of monodisperse fractions is the separation and detection of polymer samples using size exclusion chromatography with viscosity detection. SEC leads to separation according size (see Chapter 4) and by coupling the SEC column to a viscosity and molar mass sensitive detectors (*e.g.* multi-angle light scattering) online data analysis of intrinsic viscosity and of molar mass of the same fractions could be obtained. In this case the detection of viscosity is made online by measuring the pressure which a polymer solution exerts on a viscometer membrane compared to the pure solvent:

$$\eta_{\text{red}} = \frac{p - p_0}{p_0 c} \tag{5.9}$$

The different principle of measuring intrinsic viscosity does not make a significant difference to the result obtained. More significant is the type of separation technique for obtaining monodisperse fractions, based on different separation principles. While elution fractionation separates according to solubility, SEC separates according to elution volume of the macromolecules.

In order to investigate the influence of branching on the results, online measurements using SEC-MALS coupled to a viscometer and offline molar mass and viscosity determination of fractionated samples has to be compared. Additionally, comparison to linear samples of the same chemical origin has to be carried out. Figure 5.5 shows the offline and online

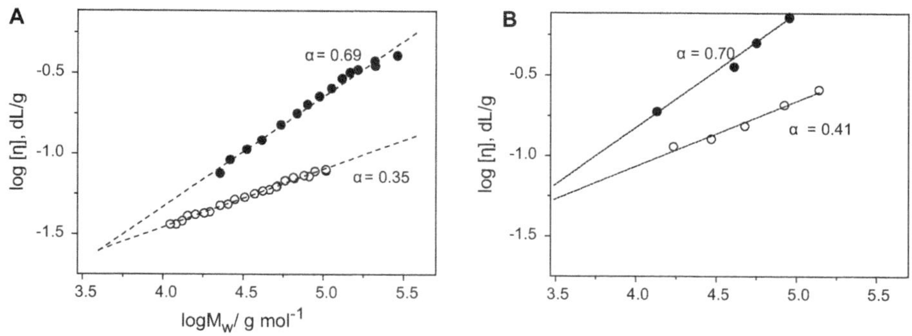

Figure 5.5 KMHS plots calculated from on-line (A)[13] and off-line (B) measurements[14] of linear and hyperbranched aromatic–aliphatic polyesters. Reprinted with permission from ref. 13. Copyright © 2013 Wiley-VCH Verlag GmbH & Co. KGaA, Weinheim.

measurements from a linear and a hyperbranched sample of the same chemical origin.

While for a linear polymer similar KMHS behaviour was found independently on applied analysis, for the hyperbranched sample a slightly different slope was found. This behaviour seems to reflect similarity for different pairs of linear and hyperbranched macromolecules of the same chemical structure. Table 5.1 shows values of the KMHS exponents obtained for two types of polyesters: aliphatic and aliphatic–aromatic. In order to enable a comparison based only on molar mass and degree of branching, the chemical structure of the polymers was kept constant.

The reproducibility of the α_η values for linear polymers is independent of the applied determination technique. In contrast to linear polymers, hyperbranched samples show significant differences between the values obtained from online and offline measurements. Additionally, linear KMHS dependencies were detected *via* the online method, supporting the data obtained in previous works on hyperbranched polymers but deviating from theoretical calculations. In contrast, in many cases, fractionated samples show a non-linear shape of the KMHS curve.[15–17]

Both effects in the hyperbranched polymers – the deviation of the α_η values from the online determination and the appearance of a maximum in the viscosity – could be ascribed to the differences in the separation mechanism in SEC and preparative fractionation. The reasons for the inconsistency of the online viscosity detection results and the results from the single fraction analysis are found in the influence of the multiple distributions (degree of branching, molar mass, *etc.*) present in an original hyperbranched sample. During SEC separation, simultaneous elution of molecules with similar hydrodynamic radii but different degrees of branching can occur, leading to overlap of distributions of different origin. Therefore, molecules with higher molar mass but low hydrodynamic volume and lower viscosity appear together with lower molar mass species. Applying preparative fractionation,

Table 5.1 Comparison of KMHS exponents of different types of polyesters obtained online, by SEC-MALS-Visco and offline, by viscosity measurements of single fractions.

Sample	M_w/M_n [a]	$\alpha_{\eta,\text{online}}$	$\alpha_{\eta,\text{offline}}$	
Linear aliphatic	2.0	0.71	0.73	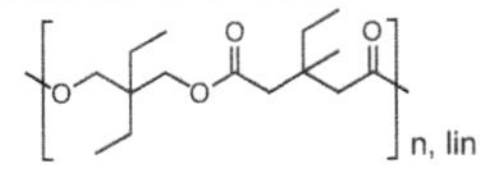
Linear aromatic–aliphatic	2.4	0.76	0.74	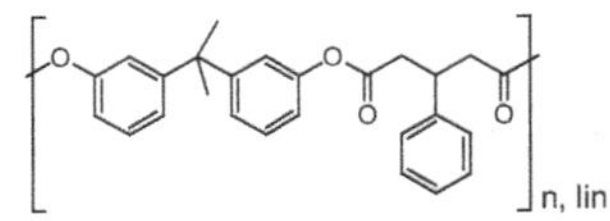
hb aliphatic	6.7	0.53	0.39[b]	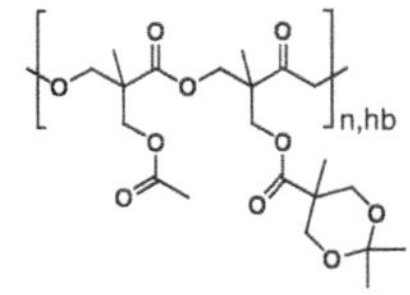
hb aromatic–aliphatic	3.2	0.35	0.47	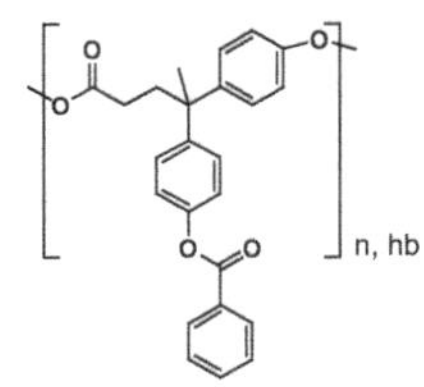
hb aromatic–aliphatic	1.9	0.36	0.38	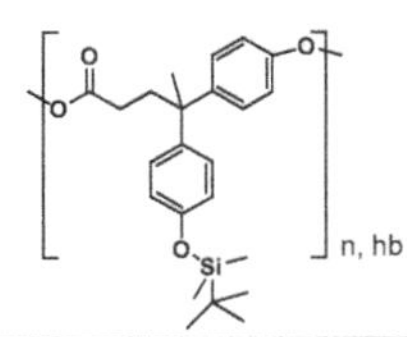

[a]Of non-fractionated sample.
[b]Slope of the linear part.[15]

separation according to one dimension (molar mass) can be confirmed if the averaged degree of branching between the fractions is comparable (see Section 2.2).

It should be noted that these deviations significantly increase at higher polydispersities of the hyperbranched samples, which are usually obtained in one-pot polymerization processes. At lower polydispersities, this deviation disappears due to the reduced co-elution in SEC, as observed in the case of a hyperbranched aromatic–aliphatic polyester with low polydispersity (Table 5.1).

In the discussion of the separation technique, the polydispersity of the fractions achieved by preparative fractionation should be mentioned too. While SEC delivers nearly monodisperse fractions (in hydrodynamic size!), elution fractionation could deliver fractions of different polydispersity, which is usually higher than unity. This fact should be also taken into account when interpretation of fractionated samples analysis is compared to online measurements. The polydispersity in molar mass after fractionation should be kept as low as possible in order to avoid calculation of average values as discussed above. In conclusion, the advantages and limits of both separation methods should be critically considered, taking into account also the specific structural and distribution properties in a single sample.

5.3 The Viscosity Behaviour of Dendritic Polymers

In order to understand the significance of the statistical branching for the solution properties of hyperbranched polymers, a comparison between the behaviour of samples with different degree of branching is essential. Depending on the type of dendritic branching, different and unique viscosity behaviour was observed.

5.3.1 Dendrimers

Linear polymer coils in a dilute solution, regardless of their interactions with the solvent (θ-solvent or a good solvent) exhibit a constant increase in viscosity with the molar mass in the KMHS double logarithmic plot (Figure 5.2). In contrast, dendrimers do not show a linear increase of the viscosity, but pass through a maximum at a certain molar mass as shown in a number of theoretical calculations[19,20] and computational simulations.[21] The intrinsic viscosity of the dendrimers increases in the first three generations and after the fourth generation it starts to decrease. Mourey *et al.*[18] supported this hypothesis by experiments on poly(ether) dendrimers (Figure 5.6) and showed a maximum in the viscosity depending on the molar mass. In this work as well as in later reports on poly(etheramide) dendrimers[22] a sharp increase of the viscosity was observed in the first three generations with a slope rather typical for a linear coil in a θ-solvent, although after the fourth generation the viscosity decreases again. In fact, the observed slopes in Figure 5.6A calculated for the first three generations (0.33) and the last generations after passing the maximum (−0.27) resemble very strongly the intrinsic viscosity behaviour of microgels in pre-gel (0.30) and post-gel states (−0.21).[23] This behaviour of dendrimers is a result of a significantly slower increase of molecular size than of molar mass with the generation number.[24] While the molar mass increases from 580 g mol^{-1} in G0 to 38 000 g mol^{-1} in the sixth generation, the difference in the hydrodynamic radius between the core and the highest generation is only 2 nm (Figure 5.6). This size collapse could be also a result of back-folding of higher generation dendrons, as shown by Ballauf and Likos.[25]

High-generation dendrimer synthesis is not an easy task as is the case of poly(etheramide) dendrimers. For these structures such a maximum was not observed, since only three generations were obtained and higher generation dendrimers were not available as a result of steric hindrances and H-bonding interactions not allowing their synthesis.[26] A variety of molar masses was achieved within the first three generations just by varying the

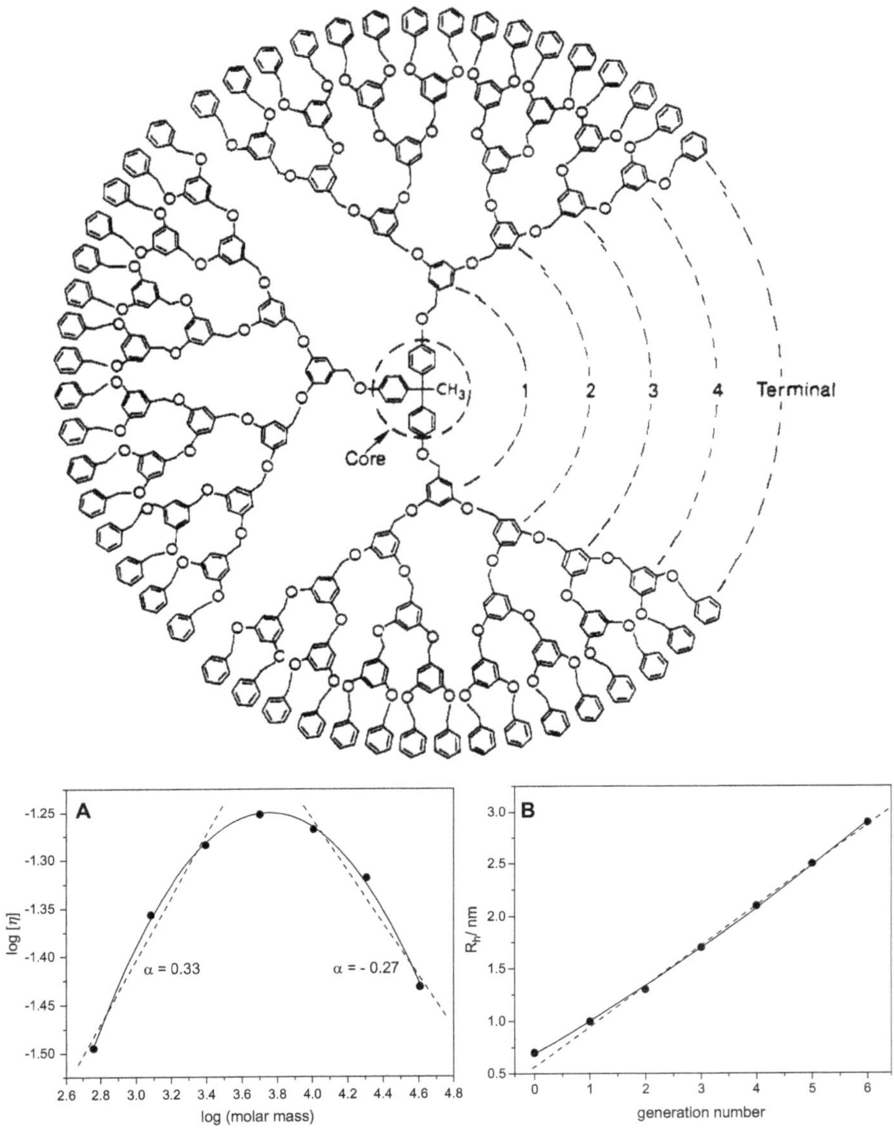

Figure 5.6 Dependence of the intrinsic viscosity on the molar mass for dendrimers. Chemical structure: polyether dendrimer in the fourth generation by Mourey *et al.*[18] KMHS plot of the core and six generations of the dendrimers (A) and development of the hydrodynamic radius with the generation number (B). The slope of linear fit in KMHS plot of G1–G3 corresponds to $\alpha_\eta = 0.33$ and G5–G7 corresponds to $\alpha_\eta = -0.27$.
Scheme reprinted with permission from ref. 18. Copyright © 1992 American Chemical Society.

size of the dendrimer core (Figure 4.5). Up to the third generation, the poly(etheramide) dendrimers show a linear behaviour in *N,N*-dimethylacetamide (Figure 5.7) as expected from previous investigations.[18,27]

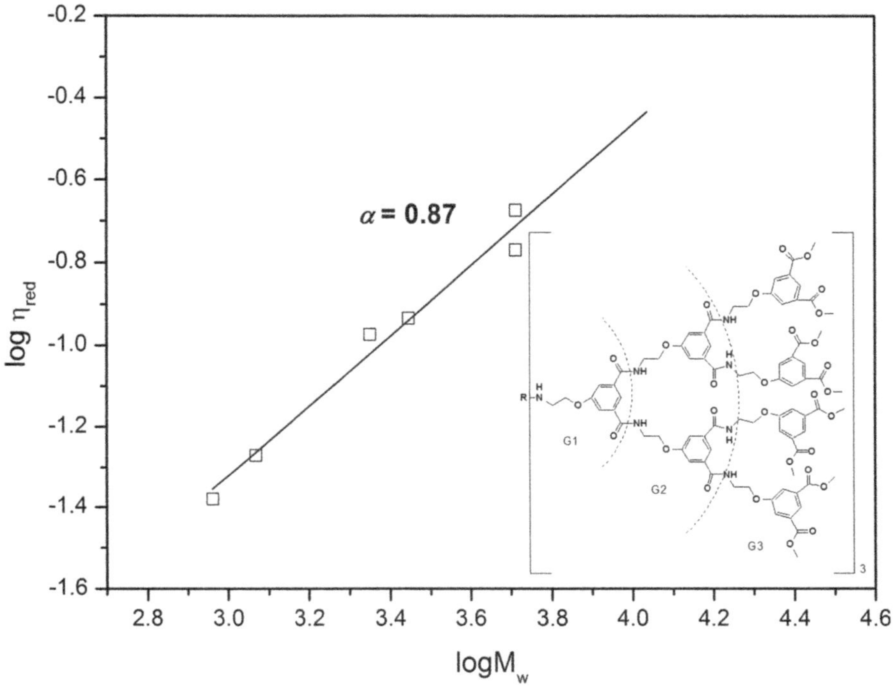

Figure 5.7 KMHS plot of poly(etheramide) dendrimers in three different generations and variation in the core size.[22]

The relatively high value of the KMHS-exponent, $\alpha_\eta = 0.87$, is rather unusual for dendritic polymers. Typically, dendrimers[18,27] and hyperbranched polymers[2] behave as globular structures with KMHS exponents below 0.5. These values correspond to the fractal dimensions between hard sphere ($\alpha_\eta = 0$) or randomly branched chain ($\alpha_\eta = 0.5$) in a θ-solvent. Indeed, former investigations on a series of the same poly(ether amide) dendrimers using their self-diffusion rates for calculating the relationship between hydrodynamic radius and molar mass[28] arrive at the conclusion that these species behave like hard spheres. However, the Euclidean dimensions of these dendrimers were found in dimethylsulfoxide (DMSO), a solvent most commonly used in NMR experiments. Obviously, dimethylsulfoxide cannot penetrate into the dendrimer molecules as thoroughly as can *N,N*-dimethylacetamide. The different behaviour of the poly(ether amide) dendrimers in both solvents leads to the assumption that the molecules are flexible, which allows the collapse of their structure in DMSO. This assumption is supported by investigations on dendritic molecules of chemically similar character – hyperbranched poly(ether amide)s[15] – discussed in the next section.

5.3.2 Hyperbranched Polymers

The common way to investigate polymer systems with regard to their KMHS exponent is by using online coupling of SEC with viscosity detector, a convenient technique, saving costly work time and sample amounts. Applying this technique to hyperbranched polymers, leads often to approximately linear dependencies, as shown in Figure 5.5.[2,29] However, simulations of this behaviour showed that the KMHS relationship should change exponentially for statistical dendritic branching[21,30] similar to dendrimers.[18,27]

The first experimental results, which supported these theoretical calculations, were obtained after elution fractionation of the hyperbranched poly(etheramide)s leading to narrow molar mass fractions with a very broad portfolio of molar masses of samples having the same chemical structure, degree of branching and branching topology (Figure 5.8). As theoretically expected, the maximum in viscosity was achieved for molar masses above 100 kg mol^{-1} and, subsequently, the open dendritic molecular structure ($\alpha_\eta = 1.5$) becomes denser ($\alpha_\eta = 0.1$).

Deviation of the viscosity/molar mass relationship from linearity was a subject of discussion in 1940s in an attempt to explain the behaviour of linear polymers at very high molar masses.[1,31,32] In these studies the non-linear trend is explained by the fact that in a linear polymer coil, the peripheral segments are better flushed by the solvent than the internal segments. The hypothetical reason for this behaviour was that the peripheral segments perturb the flow to the internal elements, leading to partial shielding of the internal part from the hydrodynamic interaction with

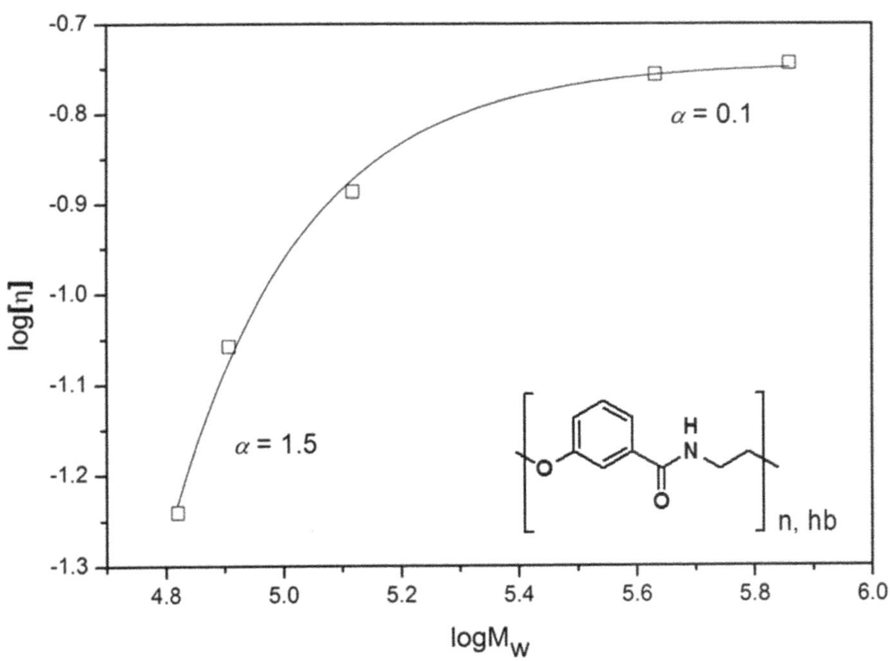

Figure 5.8 KMHS plot fractions of hyperbranched poly(etheramide) measured off-line by SEC-MALS and capillary viscometer for determination of molar mass and intrinsic viscosity.
Reprinted with permission from ref. 16. Copyright © 2002 Elsevier.

solvent molecules. This hydrodynamic shielding strongly increases with the molar mass, so that the internal part of the molecule cannot contribute to a constant change of the viscosity. In order to quantify this shielding effect, the so-called shielding ratio was introduced as the ratio between the hydrodynamic radius of the coil and the thickness of the external part of the coil, through which the solvent flows freely. The shielding ratio is well described in conjunction with a change of the KMHS exponent, α_η, from 1 to 0.5, values expected for linear coils in a good or θ-solvent. In a hyperbranched poly(etheramide) of high molar mass the values are far below 0.5 and strongly resemble the results of several simulation studies on intrinsic viscosity behaviour of dendritic molecules with a defined degree of branching and degree of polymerization. For these simulations different procedures were applied: the kinetic bead–rod model was used in the calculations by Aerts[30] and Lescanec and Muthukumar;[21] Brownian dynamics, including hydrodynamic interactions and calculations on the excluded volume, were carried out by Lyulin *et al.*;[33] and, finally, Widmann and Davies used RIS metropolis Monte Carlo simulations.[34] Regardless of the calculation procedure, the authors found decreasing, but clearly evident, dendrimer-like maximum with a decreasing degree of branching from 100% to 0% (Figure 5.9).

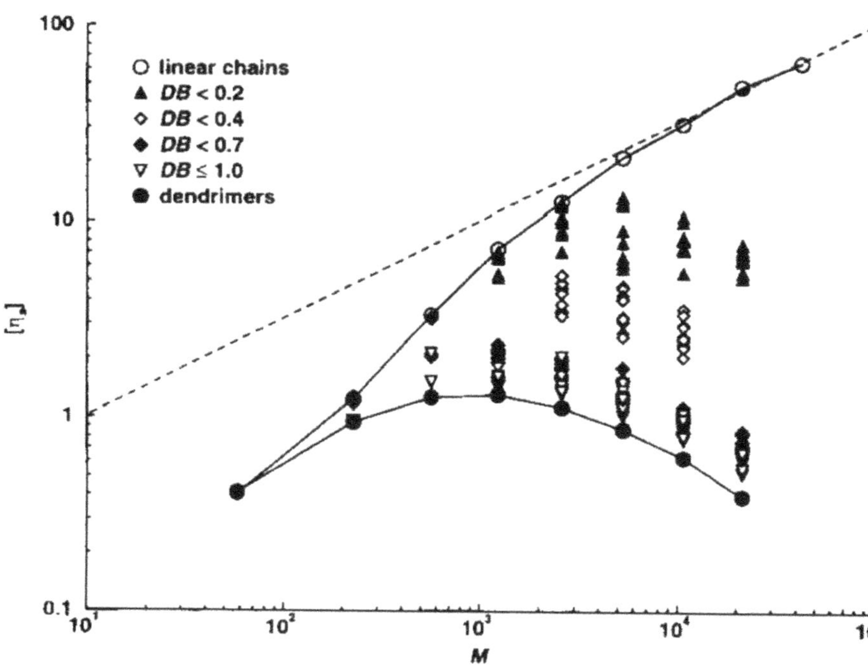

Figure 5.9 Intrinsic viscosity *versus* molar mass for statistically branched polymers with different degrees of branching, from linear chains (empty circles, DB = 0) to dendrimers (full circles, DB = 1). Dashed line corresponds to the theoretical behaviour of Gaussian chain in a θ-solvent. Each data point represents the simulation of $[\eta]$ of a single molecule.

However, in these studies, the influence of the branching distribution and polydispersity, as present in real samples, was not taken into account. With the introduction of the Wiener index[35,36] in the work of Sheridan *et al.*[37] first steps were made in an attempt to simulate real, non-monodisperse hyperbranched systems. Further studies on the complex influence of multidimensional polydispersity (degree of branching, molar mass and Wiener index) are needed to understand why polydispersity using different separation approaches sometimes does not reflect simulated KMHS behaviour. Obviously, this observation is a result of simultaneous elution/fractionation of molecules with distribution in degree of branching and/or molar mass.

5.3.3 Pseudo-dendrimers

Polymer hybrids between a dendrimer and a hyperbranched polymer are known as pseudo-dendrimers. Their synthesis is based on a hyperbranched polymer, which plays the role of a core, and is further modified using a divergent modification in different generations of the free functional groups with the AB_2 monomer, in analogy to the divergent dendrimer approach. The 100% degree of branching similar to those of the perfect dendrimers gave them the name pseudo-dendrimers,[38–40] but they still possess the irregular branching originating from the hyperbranched core. In fact, as pseudo-dendrimers were introduced, no direct conjunction was indicated with very similar viscosity properties to the dendrimers. First, with the synthesis of higher generation pseudo-dendrimers (see Section 3.2.3), the specific viscosity behaviour was recognized. It was found that with increasing generation the intrinsic viscosity drops evidently already in the first generation (SY-G1). Figure 5.10 shows the development of this parameter with the generation number. This behaviour was observed also independent on size and polarity of the end groups. For two types of functionalities, OH-terminated and protected, up to the fourth generation the viscosity decreases continuously.

While such a drop in $[\eta]$ is typical for perfect dendrimers after the fifth generation,[18,27] in pseudo-dendrimers it occurs just after the first modification step, *i.e.* in the first generation. This effect shows a significant potential advantage of the pseudo-dendrimers in comparison with dendrimers in regard to their synthesis and rheological properties. Taking into account protection and deprotection synthetic procedures, the decrease in viscosity needs at least seven synthetic steps for perfect dendrimers to achieve a compact molecular structure, which is responsible for an abrupt change in solution rheology.[24,42,43] Pseudo-dendrimers achieve this effect after only two synthetic steps: (1) self-condensation of AB_2 units to a hyperbranched polymer (50% degree of branching), and (2) modification to eliminate the linear units (100% degree of branching).

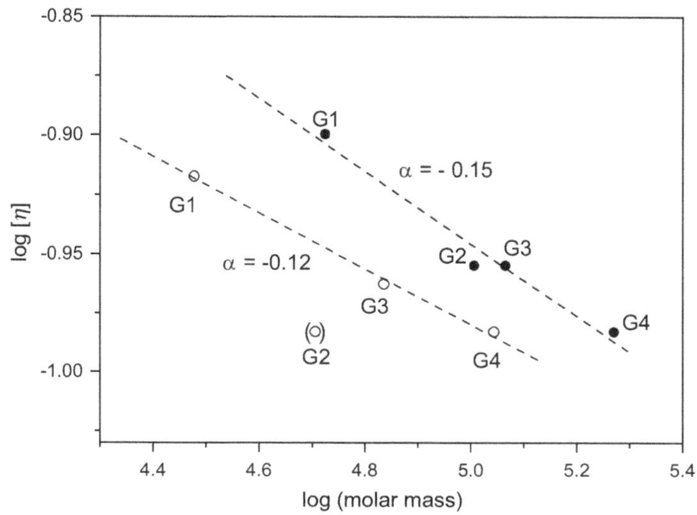

Figure 5.10 KMHS plots of pseudo-dendrimers depending on the generation number. The molar mass increase corresponds to the molar mass of the single generations.[41] Similar slopes of $\alpha_\eta = -0.12$ and -0.15 were observed for OH-terminated (unfilled circles) and for SY-terminated pseudo-dendrimers (filled circles).

The measured slope of the viscosity dependence *versus* molar mass from generation 1 to generation 4 is in the range of -0.12 to -0.15 and with this it is higher, *e.g.* less pronounced, than for dendrimers (see Figure 5.6), as has to be expected from the simultaneous influence of hyperbranched and dendritic components in the hybrid structure. In fact, the effect of pseudo-dendritic units on viscosity was not expected to such an extent, especially in the case of the aromatic–aliphatic polyester which features a rather stiff, open geometry, as discussed in the next sections.[11] This shows that the compactness of the molecules strongly improves with modification and slightly increases with the next generation numbers. This effect leads consequently to changes in the Kuhn–Mark–Houwink–Sakurada parameter α_η within the different generations. While the intrinsic viscosities in Figure 5.10 were determined in a batch, their dependence on the molar mass for every single generation was calculated after separation by size exclusion chromatography and subsequent static light scattering and viscosity detection (Figure 5.11).

The slope of the double-logarithmic dependence molar mass/intrinsic viscosity, α_η, clearly changes with the generation number decreasing from 0.37 to 0.18. This behaviour corresponds to the transformation of the molecular shape from a well-rinsed branched structure to a compact, nearly hard sphere.[9]

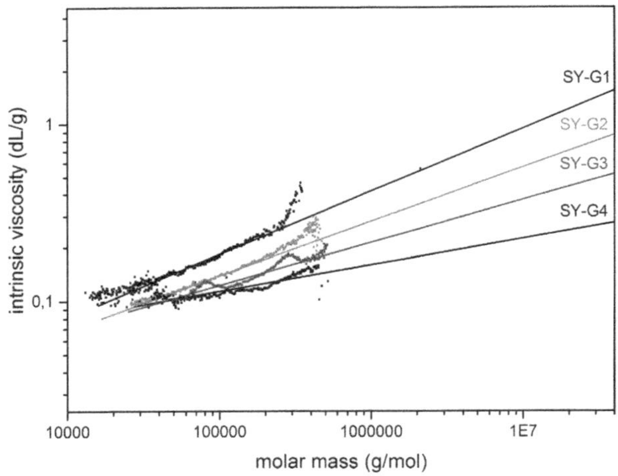

Figure 5.11 KMHS plot of the SY-terminated pseudo-dendrimers measured by SEC-MALS-Viscosity detection.
Reprinted with permission from ref. 41. Copyright © 2012 Wiley-VCH Verlag GmbH & Co. KGaA, Weinheim.

Figure 5.12 shows the dependence of α_η from the apparent density for the different generations, which can be calculated by modification of the Flory–Fox equation (eqn (2.7))

$$d_{\text{app}} = \frac{3}{4\pi}\frac{M}{R^3} = 2.5\,N_{\text{A}}\frac{1}{[\eta]} \tag{5.10}$$

Both tendencies, the KMHS exponent and the apparent density dependences from the generation number, refer to increasing compactness of the molecules. This effect is independent of the type and polarity of the end groups, which shows that the high branching density fully suppresses the influence of the end-groups type. This effect is additionally discussed in Chapter 9 on the basis of polyesters with different degree of branching, showing that the end-group type does not significantly influence the solution properties, in contrast to the behaviour in bulk.[44]

The hyperbranched core possesses a rather open structure and the functional groups are regularly distributed within the molecule. This fact helps to explain the higher density in the first generation, after filling the free space within the molecule by modification with the AB_2 monomer. The further increase of density in the next generations cannot be referred only to this effect. The behaviour clearly shows a lack of self-similarity. Likewise, for dendrimers, the decreasing viscosity with the generation corresponds to a minor increase of molecular size when increasing the molar mass from 23 000 g mol^{-1} in the hyperbranched core to 186 000 g mol^{-1}. Elucidation of the molecular structure on a local scale is possible by scattering techniques, which are the subject of Section 5.5 and Chapters 6 to 8.

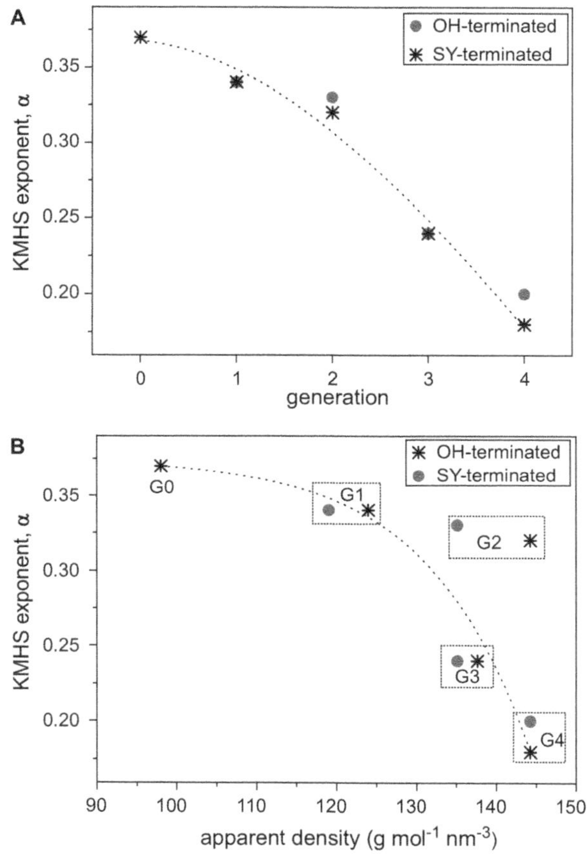

Figure 5.12 KMHS exponent in dependence on the generation number (A) and as a function of the apparent molecular density for different generation of pseudo-dendrimers (B).
Reprinted with permission from ref. 41. Copyright © 2012 Wiley-VCH Verlag GmbH & Co. KGaA, Weinheim.

5.3.4 Dendronized Polymers

Another type of hybrid structures are the dendronized polymers, a combination of linear chain and dendrons (Figure 1.6). In contrast to pseudo-dendrimers, dendrons are situated on every repeating unit of the linear chain and could feature different generation numbers. The density along the chain and its thickness increases with generation number while the contour chain length remains the same. The persistence length changes with generation number due to the simple fact that at higher dendron generations the main chain is more stretched due to steric reasons and the chains dynamics is limited.[45,46] Such behaviour was observed in several studies, *e.g.* the transformation of helical peptide-like structures into worm-like objects with increasing generation number (Figure 5.13).[47]

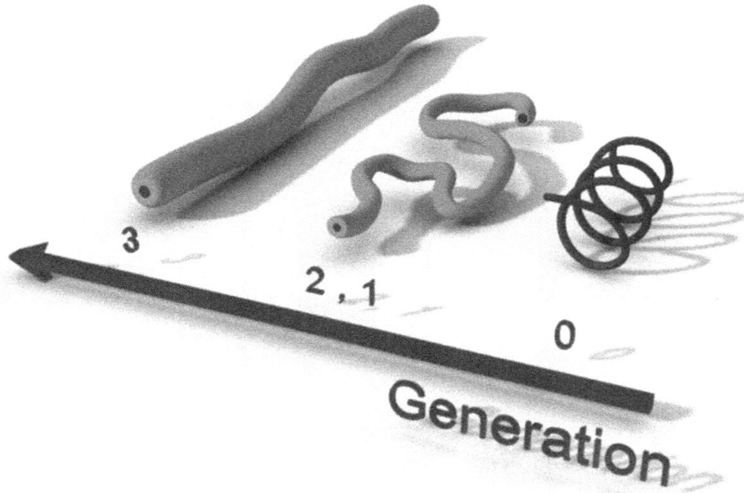

Figure 5.13 Molecular shaper transformation of a helical linear chain to a worm-like high-generation dendronized polymer.
Reprinted with permission from ref. 47. Copyright © 2012 American Chemical Society.

In dendronized polymers additional parameters compared to pseudo-dendrimers have to be taken into account, *i.e.* the increasing thickness of the chains and the increasing persistence length which lead to giant molecules in the higher generations with properties typical for colloidal objects.[48–50] For these dendronized structures special viscosity behaviour has to be expected if considering the interplay between a linear backbone and huge number of dendrons. Zhang *et al.* investigated the solution behaviour of dendronized polymethacrylate in which the degree of polymerization was kept constant.[51] The only structural parameter change was the generation number of the dendrons, which varied between one and four. In the fourth generation the extended polymer chain reaches approx. 1 μm length. The thickness of the polymer chain varies from 2.4 to 5.2 nm from G1 to G4. Consequently, the persistence length l_p changes from 3.4 to 12.2 nm. This fact shows the transformation of the dendronized polymers to stiff, dense objects with intrinsic viscosity which decreases significantly with the generation number and strongly depends on the persistence length (Figure 5.14).

The fact is that the slope of instrinsic viscosity/molar mass double logarithmic plot resembles very strongly those of denrimers, showing strong dependence on the generation number. In order to explain this behaviour, Zhang *et al.*[51] suggest taking into account the persistence length for evaluation of the intrinsic viscosity as follows:

$$[\eta] = K\left(\frac{M}{M_p}\right)^{\alpha_\eta} \tag{5.11}$$

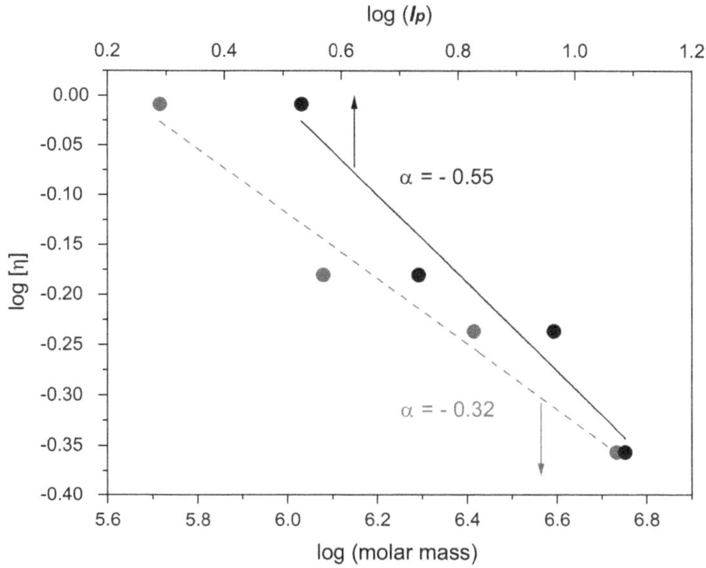

Figure 5.14 KMHS and persistence length dependence of dendronized polymers in four generations. Increasing molar mass corresponds to increasing generation number.[51]

with M the molar mass of the polymers and M_p the molar mass of the segment corresponding to the persistence length. This correction leads to modified viscosity law [eqn (5.11)] and α_η exponent of −0.55. The magnitude of the exponent after correction by l_p corresponds to the one of a polymer chain close to θ-conditions.

5.4 Influence of Chemical Structure on Viscosity Behaviour

5.4.1 Influence of the Backbone Character

Hyperbranched polymers in solution possess the typical intrinsic viscosity behaviour for dendritic polymers achieving a maximum with the molar mass, which corresponds to a very compact structure. This behaviour was theoretically predicted and experimentally observed for poly(etheramides) and aliphatic polyesters (Figure 5.8),[16,17] although for hyperbranched aliphatic–aromatic polyesters, a linear dependency of the viscosity on the molar mass in the double logarithmic plot (Figure 5.5) was observed. This linearity is also independent of the applied method, online or offline determination of $[\eta]$. It is evident that even by using elution fractionation in order to achieve low polydispersities at different molar masses and by extracting the KMHS behaviour of the hyperbranched sample, we often cannot obtain a relationship exactly matching the theoretical predictions,[21,30,33,34] as described in Section 5.2. In order to understand

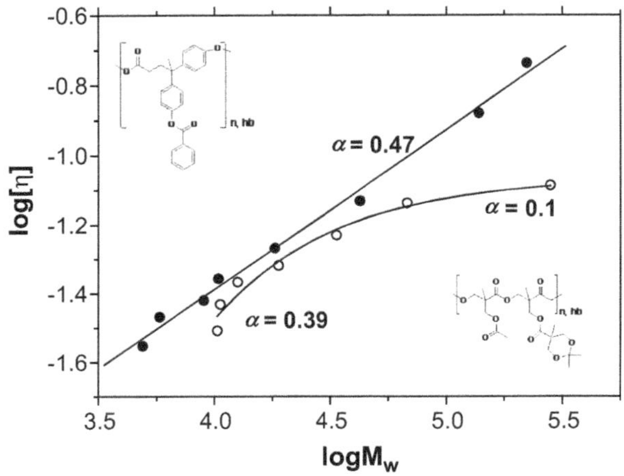

Figure 5.15 KMHS plot of aliphatic[15] (open symbols) and aliphatic–aromatic polyester[52] (full symbols) with non-polar end groups.
Reprinted with permission from ref. 11 (Slopes are corrected.). Copyright © 2009 American Chemical Society.

these deviations from theory, we have to compare hyperbranched polyesters with similar degree of branching but different backbone character: aromatic–aliphatic and aliphatic. In the case of hyperbranched aromatic–aliphatic polyester, a very open molecular structure in solution is obtained with $\alpha_\eta = 0.47$.[52] At the same time, an aliphatic hyperbranched polyester is changing its shape from open to dense (from $\alpha_\eta = 0.39$ to $\alpha_\eta = 0.1$), as theoretically predicted in the region of 50 000 to 100 000 g mol^{-1} (Figure 5.15).[17] In order to exclude influences of the polar functionalities on the solution behaviour, the polar end groups of both polymers were 100% capped by unpolar end groups.

Molecular dynamic (MD) simulations of both types of hyperbranched molecules shed light on the different behaviour of both structures. In Figure 5.16 the shape of both molecules is compared, assuming that there are no further interactions and supposing good solvent conditions. The different compactness of both polymer molecules neglecting solvent-interactions is clearly visible in Figure 5.16A and C. This difference is caused by the different sizes of the monomer units defining the distance between the branching points. However, even more pronounced is the different stiffness of the macromolecules. The rather stiff structure of the aromatic–aliphatic structure is brought about by the aromatic units and the tertiary carbon, not allowing for free rotation. This steric effect prevents the aliphatic–aromatic macromolecule from taking a more compact shape. Hence, interactions with a good solvent does not further influence the rather open structure, even at higher molar masses. This stiff structure was also derived from the SANS data of this polymer in a large q-regime, causing a scattering behaviour that corresponds to stiff structures at small distances.[53]

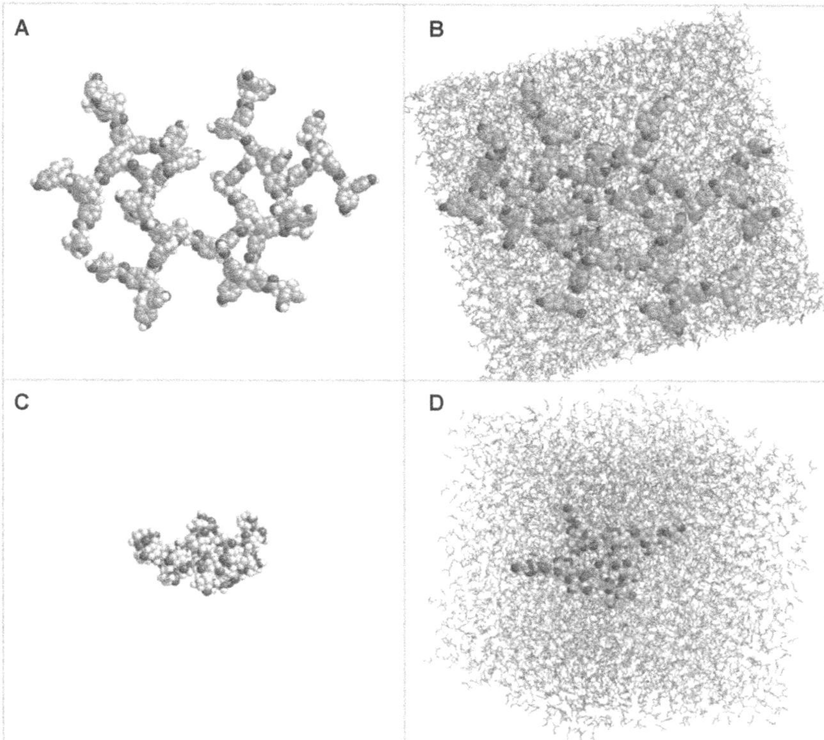

Figure 5.16 Molecular dynamic simulation of hyperbranched polyesters with 35 monomer units at 300 K. (A) Aliphatic–aromatic hyperbranched polyester without solvent; (B) aliphatic–aromatic hyperbranched polyester interacting with solvent *N*,*N*-dimethylacetamide; (C) aliphatic hyperbranched polyester without solvent; (D) aliphatic hyperbranched polyester interacting with solvent *N*,*N*-dimethylacetamide.
Reprinted with permission from ref. 11. Copyright © 2009 American Chemical Society.

The opposite case can be observed for aliphatic polyesters: interactions with a good solvent lead to a more open molecular structure compared to the model in the solvent-free state. This behaviour correlates with a better flexibility of the molecular segments, even though in total the branching point distance is much shorter and the macromolecules appear more compact. This is suspected to be the reason for the non-linear dependency of the shape in solution on the molar mass, leading to the expected maximum in the KMHS plot. In other words, the chemical structure of the monomer units plays an important role defining the solution properties of the hyperbranched polymer and cannot be excluded when searching for explanations for their unusual solution behaviour. However, we cannot exclude the possibility that at very high degrees of polymerization even the stiff aromatic–aliphatic structure could take a more compact shape as a result of the dense branching. In the 1950s such behaviour was observed for dextrans at very high molar masses,[54,55] which, however, cannot be experimentally reached for the hyperbranched aromatic–aliphatic polyesters.

5.4.2 Influence of Functionality

The chemical structure has a significant impact on the molecular shape of dendritic polymers in solution. This structure is identified not only by the backbone. In dendritic polymers the end-group functionalities affect most of their properties because of their extremely high number which rapidly increases with the degree of polymerization. These end groups can easily be modified for tailoring thermal, mechanical or solution properties of dendritic polymer materials. The influence of the functionality in hyperbranched aliphatic–aromatic polyesters can be shown on OH-terminated, silylated and benzoyl terminated samples, *e.g.* with hydrophilic or with hydrophobic end groups. The silylated polymers possess a slightly denser shape in THF as it could be concluded from the KMHS plots of OH-terminated and silylated samples presented together with the benzoyl-modified example in Figure 5.17. Both modified hyperbranched polymers show a lower KMHS exponent compared to the OH-terminated one. Since the aliphatic–aromatic backbone of this polymer is rather stiff, as discussed in the previous section on the basis of MD simulation results, this behaviour cannot be ascribed to a collapse of the molecule. Moreover, the somewhat non-polar solvent THF should rinse better the modified, less polar

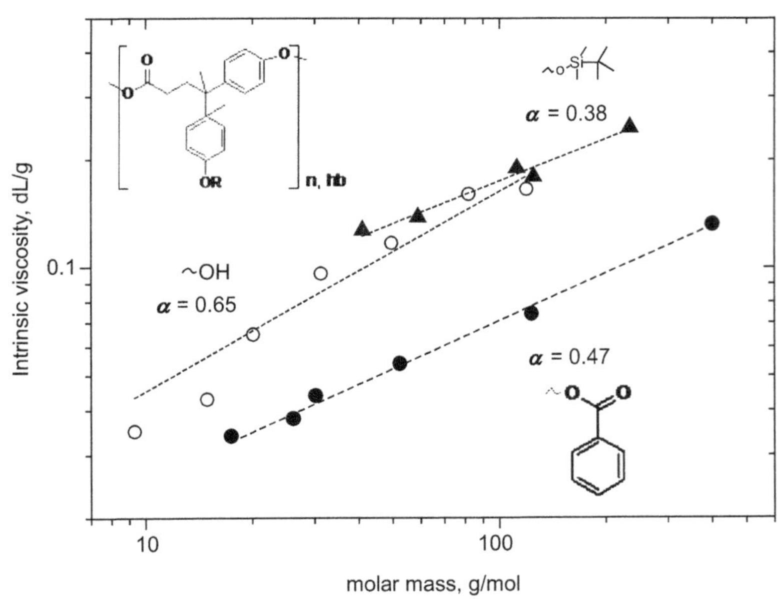

Figure 5.17 KMHS plot of fractions of differently modified hyperbranched aromatic–aliphatic polyesters: with benzoyl end groups (full circles), with silyl-end groups (full triangles), with OH-end groups (open circles).[14,52]

macromolecules. In fact, the opposite behaviour is observed, which should be ascribed to the size of the end groups used for modification. Both, silyl and benzoyl groups possess relatively large dimensions compared to the substituted H-atom in the OH-function. Furthermore, the modification in both cases is nearly complete, with a degree of modification higher than 99%. Since the functional groups (linear and terminal) in hyperbranched polymers are distributed not only at the surface but also inside the molecule, the observed effect can be explained by the occupation of free space in the rather stiff hyperbranched molecule by large functional groups, leading to an increase of the molecular density and compactness. This explanation is supported by the analysis of R_g in silylated structure compared to the OH-terminated polymers at similar molar mass, as calculated from scattering measurements and discussed in Section 5.5.

According to this explanation, the radius of gyration for hyperbranched OH-terminated and silylated polymers at similar DP should differ only slightly. Indeed, this is the case for a degree of polymerization of 110, for which a radius of gyration and hydrodanymic radius of both functionalities were in the range of 5 to 6 nm.[13]

In hyperbranched polymers there are two types of end groups: linear and dendritic. The interesting question arises as to what extent the different types of end group influence the solution behaviour of dendritic molecules (see also Section 4.3.2). For this purpose the dilute solution properties of selectively modified polyesters were investigated: OH-terminated ($T_{OH}L_{OH}$), completely non-polar modified (T_pL_p) as well as alternating, linear OH-functionalized units and non-polar terminal units (T_pL_{OH}) and *vice versa* ($T_{OH}L_p$) (Figure 3.27). The synthesis of these polymers is described in Section 3.4. Deeper understanding of the changes in the molecular parameters can be gained by dilute solution measurements of molar mass dependent viscosity and molecular size of fractionated samples. On the other hand, molecular dynamic (MD) simulations can provide valuable information and visualization of the shape of the polymers in solution.

The calotte models from the molecular dynamic simulations of $T_{OH}L_{OH}$ to T_pL_p in dilute solution of THF and identical degree of polymerization are shown in Figure 5.18.[17] Visually, only small differences can be noted between the dissolved molecules. The comparison of the monomer density distribution between the different macromolecular systems delivers more information about the internal scaling.

The monomer density in Figure 5.19A shows that $T_{OH}L_{OH}$ and $T_{OH}L_p$ is nearly identical with a maximum significantly higher than the distribution of T_pL_{OH} and T_pL_p. This indicates a reduced rinsing of the THF molecules into the macromolecule with polar terminal groups. Referring to Figure 3.27 it could be concluded that the acetal protection of the terminal groups has a stronger influence on the solvation of the hyperbranched structure in THF than the acetate group protecting the linear units.

Another evaluation of the simulation data leads to the radial distribution function of the monomers which describes the frequency of a given distance between the monomer units. In Figure 5.19B the corresponding radial

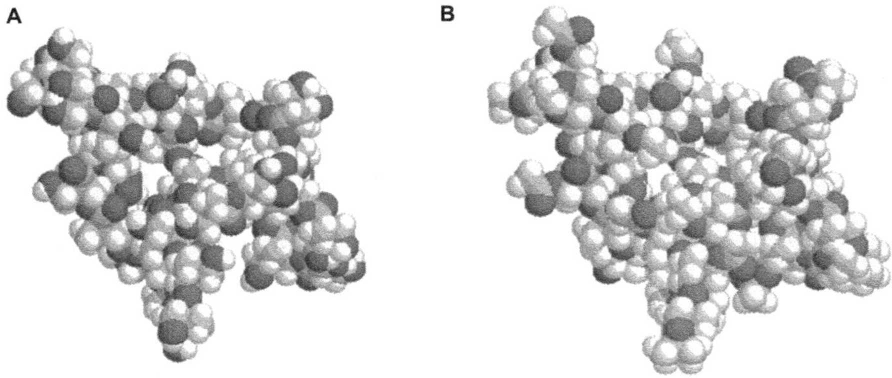

Figure 5.18 Calotte model of the relaxed state of $T_{OH}L_{OH}$ (A) and T_pL_p (B) after 185 ps simulation time. The solvent molecules of THF are omitted for better visualization of the polymer shape. Constant pressure $p = 101.3$ kPa and constant temperature $T = 298$ K as the thermodynamic standard state were used for statistical evaluation.
Reprinted with permission from ref. 17. Copyright © 2010, Elsevier.

distribution functions are presented which generally show four spheres at different distance. Slightly increasing degree of ordering with protection of the linear and terminal groups can be concluded. Summarizing the simulation results, polymers with similar, polar terminal end groups, $T_{OH}L_{OH}$ and $T_{OH}L_p$ possess similar density and internal ordering, both lower than in the case of polymers with polar terminal end groups, T_pL_{OH} and T_pL_p.

The MD simulation results are validated experimentally by studies of the intrinsic viscosity of the polymers. Fractionation of the four polymers enables molar mass dependent viscosity and molar mass measurements.

The dependencies of the intrinsic viscosities on the molar masses are presented in a double logarithmic plot in Figure 5.20. For all polymers, a linear KMHS dependency can be observed in the range up to approx. 50 000 g mol^{-1}. This behaviour changes at higher molar masses corresponding to a change in the molecular shape. A detailed analysis of the KMHS exponents in the lower molar mass range shows that depending on the polarity of the end groups the shape of the polymers is clearly changing. In Table 5.2, the α_η values are listed, showing that the OH-terminated T_{OH} L_{OH} possesses evidently a more compact shape in THF than the fully protected T_pL_p. This observation is in good agreement with the MD simulation, showing that with increasing number of non-polar end groups, the molecules become more expanded and are better rinsed by the solvent. However, the compactness determined by the KMHS calculation $T_{OH}L_p$ does not completely fit the trends observed by MD simulation results. The suspected reason for that is the impact of the size of the end groups on the molar mass leading to a shift of KMHS curves of the differently modified polymers according to their molar mass.

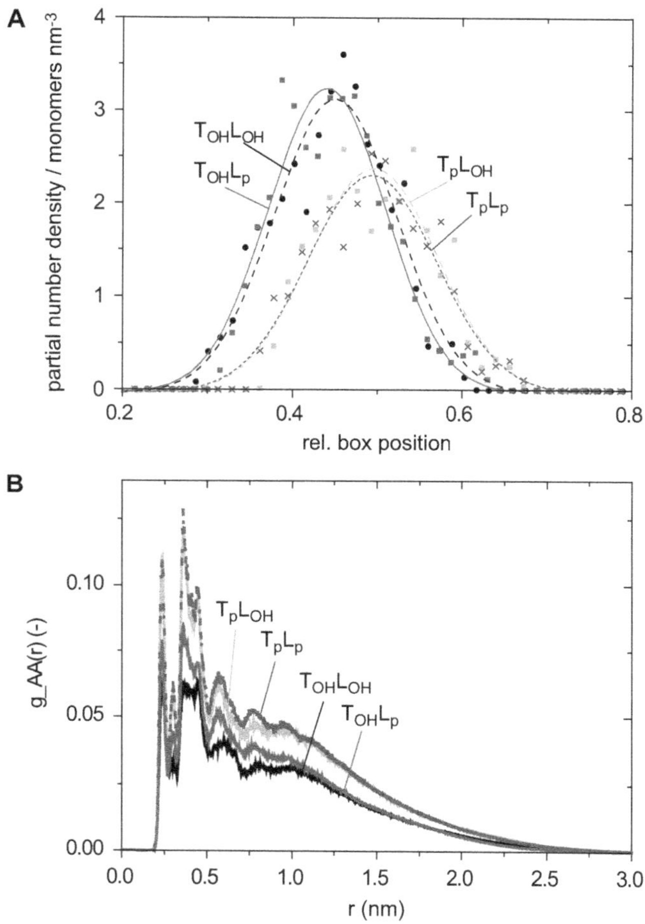

Figure 5.19 (A) Partial monomer number densities *versus* the relative position relaxed box length of 7.78 nm ($T_{OH}L_{OH}$), 8.17 nm (T_pL_{OH}), 8.18 nm (T_pL_p) and 7.18 nm ($T_{OH}L_p$); (B) Radial distribution function of the monomers *versus* the radial distance.
Adapted with permission from ref. 17. Copyright © 2010, Elsevier.

The different molar mass of the end groups makes the interpretation of the KMHS results rather complex. Therefore the change in the intrinsic viscosity should be analyzed as a function of the degree of polymerization (Figure 5.21). Slightly increased intrinsic viscosity after replacement of the terminal OH groups with acetals in T_pL_{OH} can be observed. However, a significantly stronger effect can be found after converting the linear OH groups to acetate groups. The decrease in the intrinsic viscosity can be explained by increased solubility of the non-polar macromolecules. It is remarkable that the intrinsic viscosity dependency is not influenced by the type of the terminal end groups as compared for T_pL_p and $T_{OH}L_p$. This fact shows the different roles of terminal and linear functional groups for the

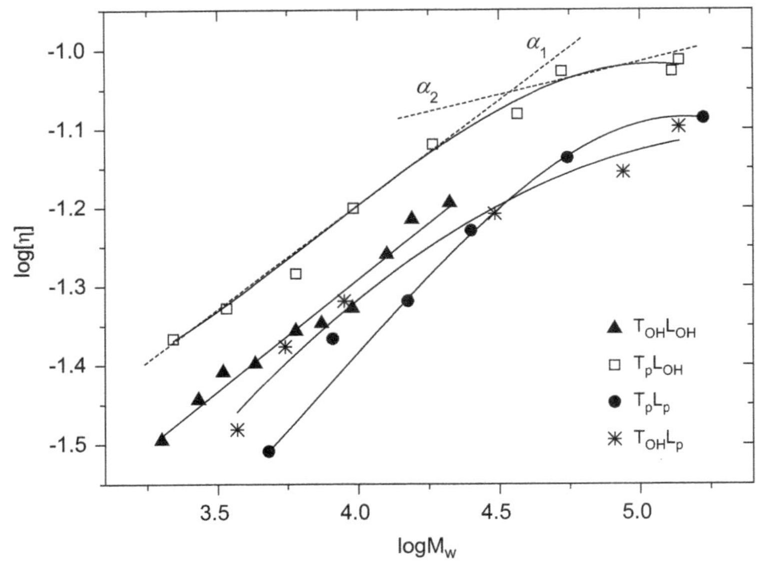

Figure 5.20 KMHS of the fractionated polymers *via* off-line determination of molar mass and intrinsic viscosity (non-linear fits should guide the eye); linear fits of both slopes in a non-constant KMHS behaviour lead to $\alpha_{\eta,1}$ and $\alpha_{\eta,2}$ in the different samples.

Table 5.2 KMHS exponents of differently modified aliphatic polyesters determined by offline fraction analysis. $\alpha_{\eta,1}$ is the first slope, in the low molar mass region, $\alpha_{\eta,2}$ is the second slope, in the higher molar mass region.

Polymer	$\alpha_{\eta,1}$	$\alpha_{\eta,2}$
$T_{OH}L_{OH}$	0.28	—
T_pL_{OH}	0.27	
T_pL_p	0.39	Approx. 0.1
$T_{OH}L_p$	0.36	

solution properties of hyperbranched polymers. In this case the affinity of the linear functional groups to the solvent evidently over-rides the influence of the terminal end groups independently of the shape of the molecules. An explanation of this effect could be the different content of linear (62%) and terminal groups (23%) as determined by ^{13}C NMR. However, the number of end B-functions in statistically branched AB_2 polymers is equal, which corresponds to comparable number of terminal and linear OH groups in $T_{OH}L_{OH}$.

The solution properties of dendritic polymers are affected by several parameters and interesting relationships have been identified. The

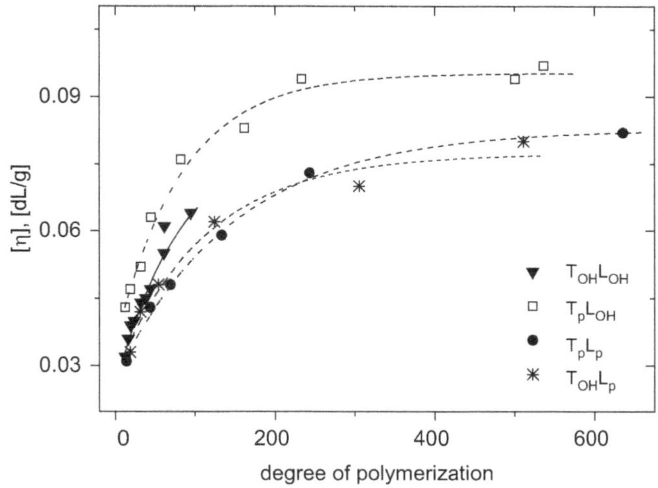

Figure 5.21 Intrinsic viscosity dependency on DP for the fractionated polymers. Reprinted with permission from ref. 17. Copyright © 2010, Elsevier.

development of a generally valid theory for dendritic polymer structures remains a big challenge, since not only the backbone type but also the influence of the very high number of end groups has to be taken into account.

5.5 Influence of Segmental Density on Viscosity and Scaling Characteristics

5.5.1 Gradually Increasing Statistical Branching

Comparison of the scaling parameters of polymers with different degrees of branching is the key to understanding the special solution behaviour of dendritic structures. However, there are usually at least three possible parameters which vary simultaneously with the branching: chemical structure, functional groups, and molar mass. Deliberate synthetic procedures have to be applied in order eventually to study the effect of one parameter, *e.g.* branching. Usually, dilution of the degree of branching by a bifunctional monomer leads to an increase of the KMHS exponent, as expected for increasing linearity of the structure. However, the chemical identity of the monomers is essential in order to obtain accurate dilution of the branches without further side effects which would change the overall properties of the system. Good examples are the gradually branched analogues of the hyperbranched aliphatic–aromatic polyester described in Section 3.2. Depending on the synthetic approach, we can obtain and compare two types of gradually branched polyesters (Figure 5.22) based on AB_2+AB and on AB_2+ABB^* polycondensation (Section 3.2, Figures 3.8 and 3.9).

The characterization of the AB_2+AB co-polymers in solution by online SEC-RI in combination with an online viscosity detector and a MALS detector

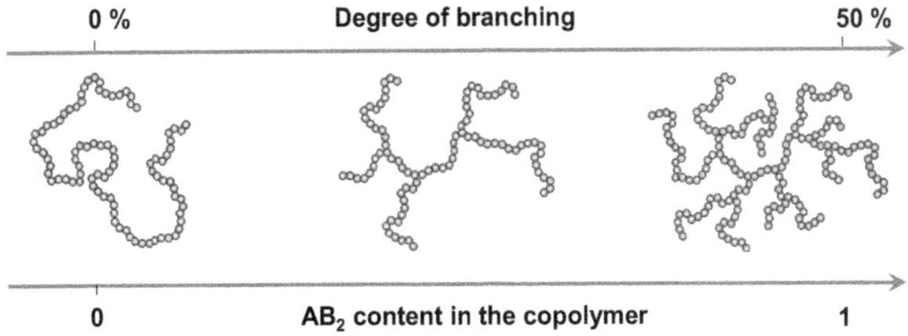

Figure 5.22 Variation of the branching degree from linear to hyperbranched structures for polyesters with different functional groups.

leads to the KMHS exponents for co-polymers with different degrees of branching. At 50% degree of branching a KMHS exponent of 0.38 was found, which is a typical value for a hyperbranched polymer in a thermodynamically good solvent.[2]

Introduction of a bifunctional AB monomer into the polymer structure brings about two different effects. The first is an increase of the molar mass of the statistical segment, which theoretically should not affect the conformation of the polymer. This is valid as long as the segments between the branching points are relatively short. When the spacer length becomes larger, a second effect of transformation into linear-like structure comes into play leading to higher KMHS values.

Surprisingly, the experiment shows an irregular scatter of the KMHS exponent within a range of 0.32–0.43 with increasing AB fraction (Figure 5.23). Obviously, with the introduction of the AB monomer the polymer structure significantly changes. One reason is the change in functionality: the number of OH end groups per monomer unit are reduced from unity in the hyperbranched sample to 0.39 for the 37% branched sample. This leads to different polarity of the macromolecules at different degree of branching and alters the tendency to form hydrogen bonds. This fact is also correlated to a decrease in glass transition temperature (T_g) with the degree of branching, which is discussed in Chapter 9.

Another reason could be the higher flexibility of the backbone with the increasing amount of AB units. Because of higher rotation freedom of the AB compared to the AB_2 units, steric hindrances are reduced and the polymer chain possesses reduced stiffness. This leads to two contrasting but overlapping effects: a reduction of the degree of branching and an increasing compactness.

Obviously, the dilute solution properties of the AB_2+AB polyesters depend on branching only to a minor extent, whereas other factors such as polarity and chain conformation play a more important role. Based on this, a comparison to the solution properties of homopolymers produced by the AB_2+ABB^* synthetic approach would be of interest.

The intrinsic viscosities of the gradually branched polyesters presented above with OH- and SY-terminal groups are measured by SEC-MALS coupled

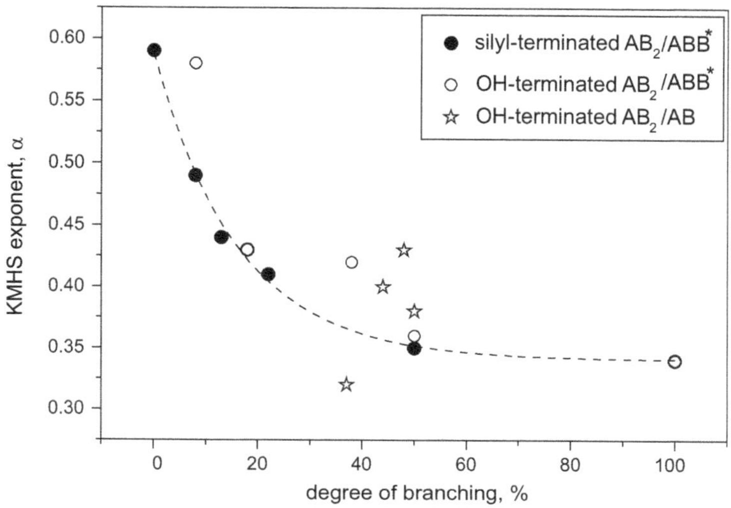

Figure 5.23 The KMHS exponent dependency on the degree of branching in series of gradually branched aromatic–aliphatic polyesters in THF at 24 °C.[41,44]

to viscosity detector. Figure 5.23 shows the dependency of with α_η on the degree of branching of differently branched homo- and co-polymers. In a series of samples prepared by the $AB_2 + ABB^*$ approach a clear trend is achieved for the dependency of KMHS exponents on degree of branching. The α_η values decrease with the degree of branching from 0.59 down to 0.3 for non-polar polymers, which indicates an increasingly compact shape of the macromolecules at higher degrees of branching. Interestingly, there is only a small difference between the OH- and SY-terminated samples with 50% degree of branching having KMHS exponents of 0.36 and 0.35, respectively. One would expect a denser macromolecular structure in the case of the SY-terminated sample due to its quite voluminous protective groups. At this point it should be mentioned that very small differences were detected by SANS measurements, indicating similar R_g after modification of the OH-terminated polymer with SY groups. This fact is additionally supported by the similar and positive A_2 exponents of 7.87×10^{-4} for the OH sample and 6.54×10^{-4} mol cm^3 g^{-2} for the SY sample.

The KMHS exponents calculated at 100% degree of branching correspond to fully modified end functionalities by AB_2 monomers which transform all end groups to terminal groups, *e.g.* a pseudo-dendritic structure of the first generation. The viscosity behaviour of this special type of dendritic polymers is discussed in Section 5.3.3.

Compared to the KMHS behaviour of gradually branched $AB_2 + AB$ polymers, for homopolymers with different degree of branching, all additional effects such as deviations in the amount of functional groups and different segment flexibility are eliminated. As a result, the behaviour of the KMHS exponents is in good agreement with the theoretical expectations and a comparison of the polymers solely by their branching architecture regardless of interfering side effects becomes possible.

Substantial information on the size of the molecules can be obtained by scattering techniques. Static light scattering gives the radius of gyration, R_g, and dynamic light scattering allows calculation of the hydrodynamic radius, R_h. At sufficiently large particles ($d > \lambda/20$) an angular dependence occurs and permits the determination of R_g. This means that the wavelength of visible light is too long to allow detection of small sizes, typical for hyperbranched macromolecules ($R_g < 10$ nm). This problem is solved by using scattering methods with short wavelengths, such as X-ray or neutron scattering. Small angle neutron scattering (SANS) with wavelengths between 0.5 and 2 nm and very small scattering angles[56] is used in the studies discussed. The interpretatioin of SANS and dynamic light scattering leads to valuable information on the scaling characteristics of the polymers. If different molar masses are available, the scaling parameter, ν can be calculated from the following equation:

$$R_g = KM^{\nu} \tag{5.12}$$

The same relationship is valid for the hydrodynamic radius, if the ratio between R_g and R_h

$$\rho \equiv \frac{R_g}{R_h} \tag{5.13}$$

remains constant. This ratio is another possibility to indicate branching in polymers without the influence of the molar mass, but just measuring the R_g and R_h of one sample, since ρ is related to the segmental density in the polymer.[57,58] Calculations predict that this value approximates 1.22 for hyperbranched polymers,[9] while $\rho = 0.78$ corresponds to hard sphere and $\rho = 1.78$ to linear coil in a good solvent.[9,59]

Measurements of the molecular size in dilute solution yield quantitative values for the degree of branching in long chain branched polymers, as was calculated by Zimm and Stockmayer and considered in detail in Chapter 2.[60] These considerations were based on the apparent shrinking of a linear chain when it transforms into branched one of the same molar mass and chemical character. This contraction can be determined either by the gyration radius or by the intrinsic viscosity of a branched and linear sample, g and g' [eqn (2.1) and (2.8)].

SANS and dynamic light scattering experiments lead to determination of the global parameters of the aliphatic–aromatic polyesters of $AB_2 + ABB^*$ type with degrees of branching from 0 to 50%. The radii of gyration and the hydrodynamic radii of both series are plotted in Figure 5.24. Independently of the functionality type, R_g and R_h differ significantly from each other at low degrees of branching, but at high degrees of branching the data approach similar values. This effect stands in direct correlation to the segmental density of the polymers and can be described by the apparent density of the

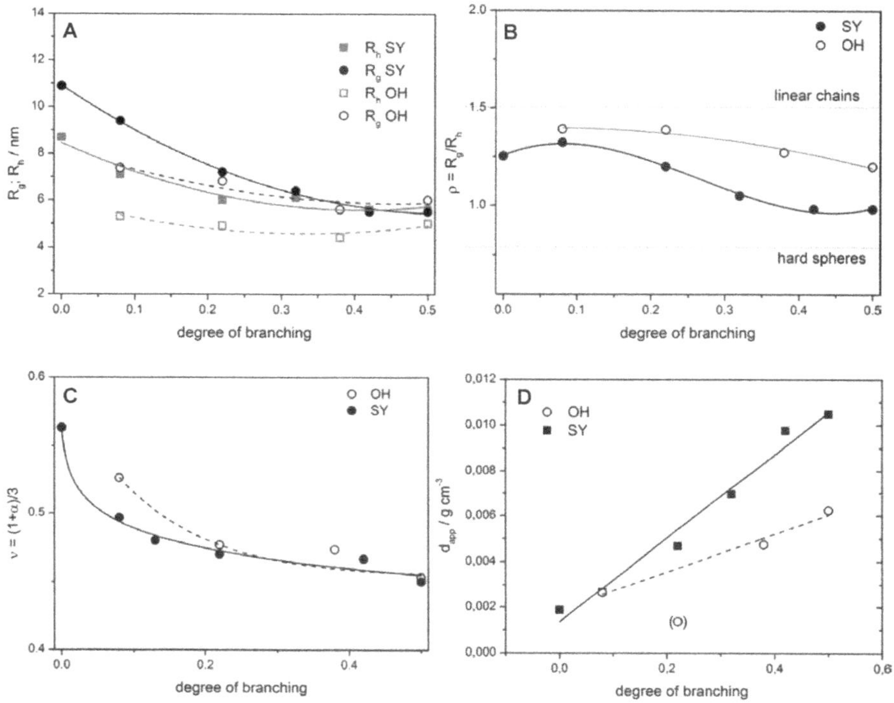

Figure 5.24 (A) R_g and R_h dependence on the degree of branching for SY- and OH-terminated samples. (B) $\rho = R_g/R_h$ dependence of the branching parameter on the degree of branching for SY- and OH-terminated samples. The lines correspond to tentative fits to the points of measurements. (C) Dependence of the scaling exponent and (D) the apparent density of different degrees of branching for SY- and OH-terminated samples. Reprinted with permission from ref. 13. Copyright © 2013 Wiley-VCH Verlag GmbH & Co. KGaA, Weinheim.

molecules or by the structure dependent ratio ρ. Values of ρ from 1.2 to 1.4 were obtained for the OH samples and 0.98 to 1.2 for the SY samples (Figure 5.24B). While the ρ-value for linear polymers lies in the expected range for linear coils, for hyperbranched OH-50 it approaches the theoretically predicted $\rho = 1.22$.[9] For SY-50, $\rho = 0.98$ is in agreement with the experimental values calculated for glycogen and amylopectin.[58,61]

Using the radius of gyration and the molar mass, the apparent density d_{app} can be calculated based on eqn (5.10). The apparent density increases with the degree of branching (Figure 5.24), but differences can be observed between samples with variation in the end functionality. In SY-terminated samples the density increases faster with the degree of branching than in OH-terminated. The interpretation of this observation brings us again to the volume effect of the end groups. The bulky SY group can fill free space within the molecule leading to dense packing, as concluded from the KMHS values (Figure 5.23). At low degrees of branching the same density was found independent of the functionality. This result is based on the different confinement of the internal segments, depending on branching. At higher degrees of branching the intramolecular mobility in branched structures is reduced, while linear chains could unfold and escape the confinement of bulky groups. At low degrees of branching the confinement caused by branching is released due to lower shielding of branching segments. Additionally, the functionality as well as the size of the end groups influences the conformation. This influence is increasing at higher degree of branching.

The scaling parameter from eqn (5.12) is a well-established function of the molar mass for interpretation of fractals, since $\nu = \frac{1}{d_f}$. The definition of fractals and the meaning of the fractal dimensions are discussed in detail in Chapter 6. Usually, this value is determined by measuring the molar mass dependence of the gyration radius as described above and discussed in detail in Chapter 8. If data from scattering measurements are not available, another opportunity to calculate this value is the KMHS exponent using the dependence $\nu = \frac{1+\alpha}{3}$. The scaling exponent of the linear polymers (Figure 5.24) calculated from both data corresponds to $\nu = 0.56$, showing evidently typical behaviour of a Gaussian coil in a good solvent (0.50 in a θ-solvent and 0.588 in a good solvent).[62] With increasing branching, the scaling exponent decreases as expected and the fractal dimension d_f increases. For statistically branched polymers d_f should approach theoretical values of 2 in a good solvent and 2.5 in θ-conditions.[63] Geladé *et al.*[64] found values for hyperbranched poly(esteramide)s in deuterated THF of 2.04 as measured by SANS; De Luca and Richards obtained 2.38 for hyperbranched polyester by static light scattering[65] and about 2.5 from SANS experiments.[66] Ioan *et al.*[58] studied naturally branched glycogens and found $d_f = 2.86$, which is closer to hard-sphere behaviour. For perfectly branched dendrimers $d_f = 3$ was obtained,[67] and for compact aliphatic hyperbranched polyglycerol structures, the same value was calculated on the basis of SANS measurements.[68] For the hyperbranched aliphatic–aromatic polyesters with an open, stiff structure values of $d_f = 2.2$ for both, SY- and OH-terminated polymers with DB = 0.5 are obtained.

The segmental density dependence on the degree of branching raises the question whether the universal parameter from the Flory–Fox equation exhibits a dependence on the branched topology. The Φ-parameter is calculated by eqn (2.7) using the intrinsic viscosity and radii of gyration. For aliphatic–aromatic polyesters the results are plotted in Figure 5.25 showing additionally the values for random coils[8] of linear chains, $\Phi_{Flory} = 4.20 \times 10^{24}$ mol^{-1} and for hard spheres, $\Phi_{Einstein} = 1.36 \times 10^{25}$ mol^{-1}.

The OH-terminated series show only slight variation with the degree of branching but for the SY samples the Φ-parameter clearly increases with the degree of branching. For both sample series a very close correlation with the behaviour of the apparent density can be observed (Figure 5.24). The data for the OH functions are difficult to interpret, yet most likely the polar groups

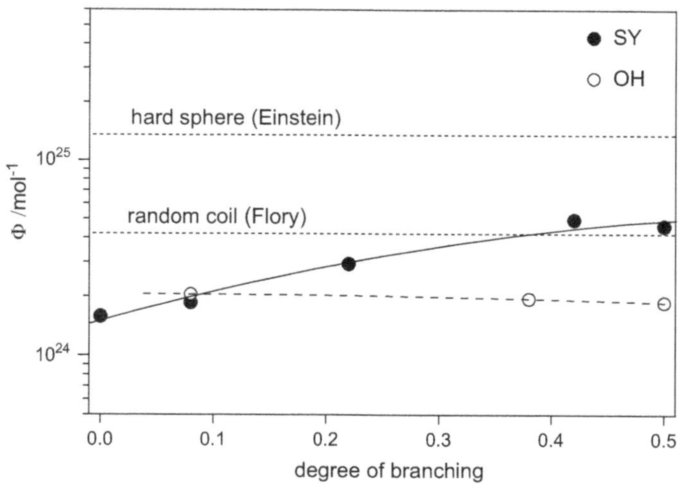

Figure 5.25 Dependence of the Flory–Fox parameter from the degree of branching for SY- and OH-terminated statistically branched aromatic–aliphatic polyesters.
Reprinted with permission from ref. 13. Copyright © 2013 Wiley-VCH Verlag GmbH & Co. KGaA, Weinheim.

can undergo strong intramolecular interactions in the rather non-polar THF solvent. Obviously, the Φ-parameter is not a universal constant and it depends on the topology of the macromolecules. Additionally, the observed tendency is far from the theoretically calculated values for random coil and hard sphere. This additional deviation is ascribed to the dependence of the Φ-parameter on the molar mass as is discussed in detail in Chapter 6. As a practical conclusion of this study, the application of the Flory parameter for calculation of molar masses using the universal calibration[7] could lead to misinterpretations especially for strongly branched macromolecules.

The non-universality of Φ has a consequence for another parameter, which was assumed to be constant for macromolecules: the correlation of the contraction factors to the degree of branching. A relationship which allows the re-calculation of g' to g would be very helpful for further interpretation of the branching number using eqn (2.2)–(2.5), because measurements of the intrinsic viscosity are easier to perform than scattering. The validity of this relationship was verified using calculation of g and g' from fractions of different molar mass and additional scattering and viscosity measurements or by online detection of these values (Figure 5.26). In these particular experiments not R_g but R_h was used in combination with the known ρ-parameter for the recalculation of R_h in R_g.

The relationship between both contraction factors was predicted by Stockmayer and Fixman,[69] who noticed that with the assumption of universality for Φ a power law correlation between g' and g should be observed. Such dependency would lead to $g' \propto g^{3/2}$, because the intrinsic viscosity $[\eta]$ is related to $R_g{}^3$ [eqn (2.7)] but g to $R_g{}^2$. This dependence was never

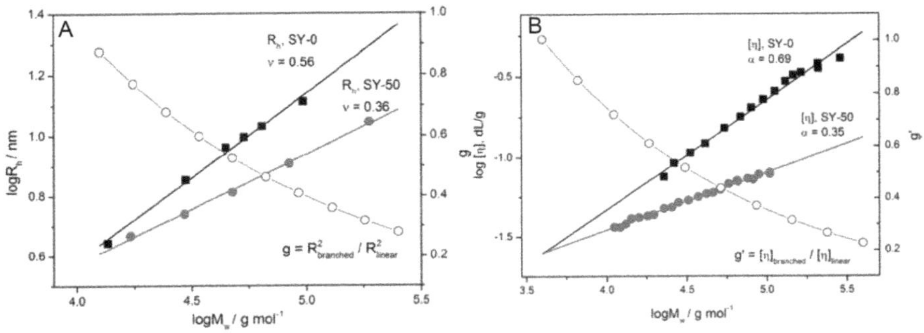

Figure 5.26 (A) Hydrodynamic radius and g dependency [eqn (2.1)] of the molar mass calculated for fractions of SY-0 and SY-50. (B) Intrinsic viscosity and g' dependency [eqn (2.8)] of the molar mass of SY-0 and SY-50. Reprinted with permission from ref. 13. Copyright © 2013 Wiley-VCH Verlag GmbH & Co. KGaA, Weinheim.

confirmed by experiments. Zimm and Kilb[70] tried to consider the hydrodynamic effects (e.g. draining) in branched polymers to solve the correlation between g' and g and found $g' \propto g^{1/2}$. These two values for the exponent $\varepsilon = 1.5$ and 0.5 reflect the two possible limits which can be found for branched samples, as it was assumed by Kurata and Fukatsu[71] introducing a power law:

$$g' \propto g^{\varepsilon} \tag{5.14}$$

The exponent ε was determined several times from experiment. The first experimental studies were performed on star molecules with up to 12 arms, giving ε of approximately 0.6.[72] Some years later Roovers *et al.*[73] performed accurate measurements on g' and g for star molecules and clearly demonstrated the increase of the exponent up to $\varepsilon \approx 1$ depending on the number of arms (from 3 to 128). The dependence of the draining exponent ε on the branching number led Stockmayer and Fixman to the assumption that the ratio $\Phi_{\text{branched}}/\Phi_{\text{linear}} \equiv \Phi^*$ in the equation

$$g' = \frac{\Phi_{\text{branched}} R^3_{\text{g,branched}}}{\Phi_{\text{linear}} R^3_{\text{g,linear}}} = \Phi^*(g) g^{3/2} \tag{5.15}$$

should be larger than 1. This presumes a dependency of Φ^* on the branching density and hence, ε depends on the branching even if it does not depend on the molar mass.

The calculation of the contraction factors g and g' shown in Figure 5.26 from the radii and viscosity of branched and linear chain of the *same* molar mass are performed for different degrees of branching. For the hyperbranched sample (DB = 50%) the ε exponent was calculated from the data in Figure 5.26 and the following dependence was obtained:

$$g' \propto g_n^{0.77} \tag{5.16}$$

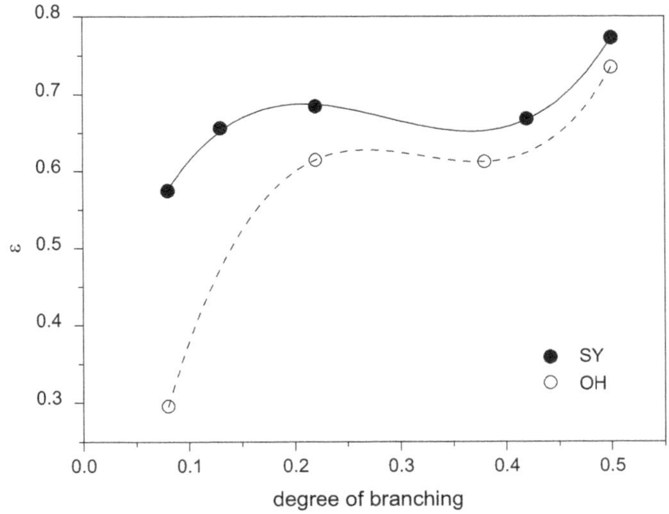

Figure 5.27 Dependence of ε as a function of the degree of branching for SY- and OH-terminated samples. The lines are not fitted curves but are chosen to guide the eye.
Reprinted with permission from ref. 13. Copyright © 2013 Wiley-VCH Verlag GmbH & Co. KGaA, Weinheim.

The extended evaluation of ε for the series of polymers with degrees of branching between 0 and 50% lead to different values of ε (Figure 5.27).

Between degrees of branching of 0.15 and 0.50 the exponent ε increases weakly with the degree of branching within an approximate value $\varepsilon = 0.72 \pm 0.05$. Below a degree of branching of 0.15 a strong decrease towards $\varepsilon = 0$ takes place. This observation shows that for higher degrees of branching a certain universality of ε could be stated, *i.e.* for hyperbranched structures. For weakly branched structures this value changes very rapidly. The explanation of this effect is obvious. A look at eqn (5.15) makes it clear that also the Φ*-parameter has a power law correlation of $\delta = \varepsilon - 1.5$ to the g factor. If $\varepsilon \rightarrow 0$ than $g' = g = 1$ which is valid for linear polymers and in this case also $[\eta] = \Phi_{\text{linear}}(R_g^{\,3}/M)$ is valid. In other words, for linear polymers (DB = 0) the ε exponent always becomes zero and the decrease of ε at lower degrees of branching has to be expected.

This dependency of ε on the branching type allows another comparison if a constant Φ_{linear} for linear chains is assumed. This leads to the following equation

$$\Phi^*(g) \propto \Phi_{\text{branched}} \propto 1/g^{(3/2-\varepsilon)} \qquad (5.17)$$

since $g \leq 1$. Increasing degree of branching causes the decrease of g and increase of Φ_{branched} which was already demonstrated in several experiments. De Luca and Richards[65] obtained $\delta = -1.22$, Kurata *et al.*[72] and

Weissmüller and Burchard[74] $\delta = -0.9$. In the present example of branched aliphatic–aromatic polyesters with 50% degree of branching $\delta = -0.76$ and -0.73 for the OH sample and the SY sample are found. These values lie within the frames discussed above (related to ε): Zimm and Kilb[70] predicted $\delta = -1$ and Stockmayer and Fixman[69] $\delta = 0$. These data seem to suggest that the approximate hydrodynamic calculations by Zimm and Kilb describe the limit of very high branching densities.[70]

In conclusion, the segmental density which perturbs the hydrodynamic flow is evidently controlled by the degree of branching. A number of investigations were focused on the validation and improvement of the universal calibration approach for interpretation of size determination by common analytical techniques, *e.g.* size exclusion chromatography.[75] These calculations show that the universal law cannot be directly applied for interpretation of high branching density polymers.

5.5.2 Pseudo-dendrimers with Increasing Generation

The combination of a hyperbranched core with different generations of dendrons is a relatively new approach leading to hybrid structures with an interesting viscosity behaviour (see Section 5.3.3). This behaviour resembles those of perfect dendrimers, which are known to be well-defined, unimolecular three-dimensional nanostructures. Based on this feature dendrimers have continuously aroused considerable attention in technological applications of biomedical and materials science.[76–78] Since the preparation effort for pseudo-dendrimers is advantageous, it would be an interesting issue to compare to which extent the segmental properties of pseudo-dendrimers resemble those of dendrimers.

In contrast to dendrimers, the functional groups in hyperbranched polymers are not positioned only at the terminal units of the dendrons, *e.g.* in a stretched dendrimer molecule this would be the outmost shell. In hyperbranched polymers they are distributed also inside the macromolecule. The position of the branching points and the length of the branches and the segments between two branching points are irregularly distributed, but defined by random statistics.[79] This structure causes a considerably more open shape in space and a lower segment density than observed for dendrimers. Similarly to dendrimers, pseudo-dendrimers contain only dendritic and terminal units (see Section 3.2.3). Both, dendrimers and pseudo-dendrimers are 100% branched structures (without linear units) but the dendrimers are regularly, while the pseudo-dendrimers are irregularly branched. The hyperbranched core is the starting structure with generation zero, G0 in a pseudo-dendrimer and the generation G1 to G4 are obtained after substitution of all terminal (and linear for G1) functional groups by AB_2 units using modification and deprotection procedures which are repeated similar to the divergent dendrimer approach (Figure 3.1).

Given that the degree of polymerization of the hyperbranched core is DP ~ 50, with this procedure the pseudo-dendrimers obtain already in the first generation a high molar mass and an end group number of 100, which increases to 800 in the fourth generation. At the same time, we observe the characteristic rheology behaviour for dendrimers right after the first modification (Section 5.2.3).[41] The question arises, which similarities and differences in the internal structure and the segmental density between pseudo-dendrimers and perfect dendrimers exist. For this purpose, the global properties should be analysed and compared: the molar mass M_w, radius of gyration R_g, hydrodynamic radius R_h, intrinsic viscosity $[\eta]$ and inter-particle interactions, *e.g.* the second osmotic virial coefficient A_2. First visualization by atomic force microscopy (AFM) shows that the generations G1, G2, and G3 possess similar size with estimated average diameters of 8.4 nm, 6.8 nm and 9.2 nm for G1–G3. This is also the result from the data of the accurately measured radii of gyration in solution (Figure 5.28).

The decrease of the radius of gyration of the pseudo-dendrimers with the generation number is strong especially in view of the significant growth of the molar mass from 53 kg mol^{-1} to 187 kg mol^{-1} for G1 to G4. In contrast, for dendrimers the radius of gyration increases with the generation number as has to be expected in accordance with the extreme molar mass increase

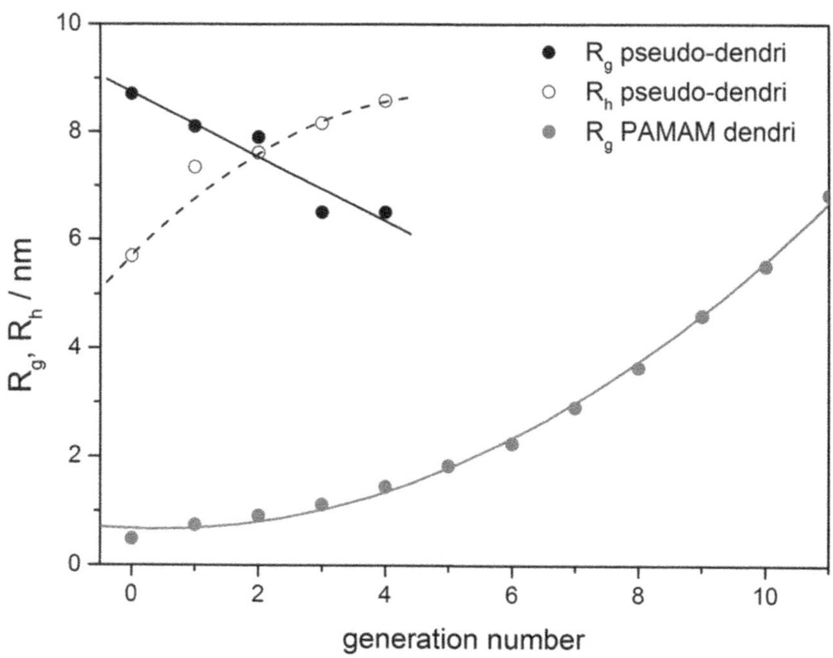

Figure 5.28 Radii of gyration calculated for 10 generations of a PAMAM dendrimer (filled circles)[80,81] compared to that of the pseudo-dendrimer of four generations (filled triangles).[82] The corresponding dependence of the hydrodynamic radius R_h is presented by the open symbols.
Reprinted with permission from ref. 82. Copyright © 2015 Wiley-VCH Verlag GmbH & Co. KGaA, Weinheim.

(Figure 5.28).[80,81] For the pseudo-dendrimers an increase only in the hydrodynamic radius is observed.

The apparent density calculation gives an insight into the apparently contradicting behaviour of the pseudo-dendrimer's radii. In fact, the molecular density distribution of dendrimers depends on the rigidity of the monomer structure.[83,84] It could be higher at the molecular core or at the molecular shell depending on the dendron backfolding. Comparing poly(amidoamine) (PAMAM) with aliphatic–aromatic polyester structures, the chemical difference exists, yet without over-riding the tendencies resulting from the increasing generation number. The overall segmental density of dendrimers can be compared to the pseudo-dendrimers by calculation of the apparent molecular density [eqn (5.10)]. Both dendritic structures follow a smooth increase of the segment density with the generation number, as expected, but with a much lower density of the pseudo-dendrimers. The lower density calculated for the pseudo-dendrimers is not necessarily a result of the branching topology but most probably it derives from the rather stiff segment conformation, as discussed in the previous section.

The ratio between the R_g and R_h, the ρ-value is interesting to interpret complementary to the molecular density (Figure 5.29), since it is a measure of the branching. The corresponding values of the dendrimers are very close to the values for a hard sphere independent on the generation number. Note that the value of a f-arm star-branched sample *e.g.* $\rho_{f\text{-arm star}} \rightarrow 1.33$ for functionality $f \gg 3$.[9] For the dendrimers the draining becomes negligible already in the first generations and approaches $\rho \cong 0.9$ in the higher generations. Nevertheless, the expected $\rho_{sphere} = 0.775$ for a hard sphere is not reached.[9,87] In contrast, the pseudo-dendrimers of fourth generation approach an even lower value, very close to that of a hard and solid sphere. This is an indication for a particle with an even smoother surface than it is apparently present in flexible dendrimers. The reason for this result can be searched in the origin of the evidently less ordered core structure of G0 with an open segmental arrangement. The attached dendrons fill the open space within the fringed surface and transform it into a smooth particle surface, which is in agreement with the lower R_g at higher generation numbers: The decrease of R_g does not necessarily indicate contraction, but it is caused by a condensed packing of the segments and it is *smaller* than the geometrical sphere radius. Opposite to the R_g the hydrodynamic radius R_h increases because the solvent no longer penetrates the particle to an equal extent but the penetration profile moves to the outer surface layers.

This interpretation is supported by the analysis of the physical factors which cause the difference in the dimensional values, which is a subject of statistical thermodynamics.[8,62,88,89] Linear chains, and even hyperbranched samples, are flexible and they could be deformed due to osmotic pressure or domains of two particles in contact can partially interpenetrate. The second virial coefficient A_2 can be used to describe the behaviour of open and flexible structures using [eqn (5.5)]. In this equation the interpenetration function Ψ increases with the molar mass reaching a constant Ψ^* in the

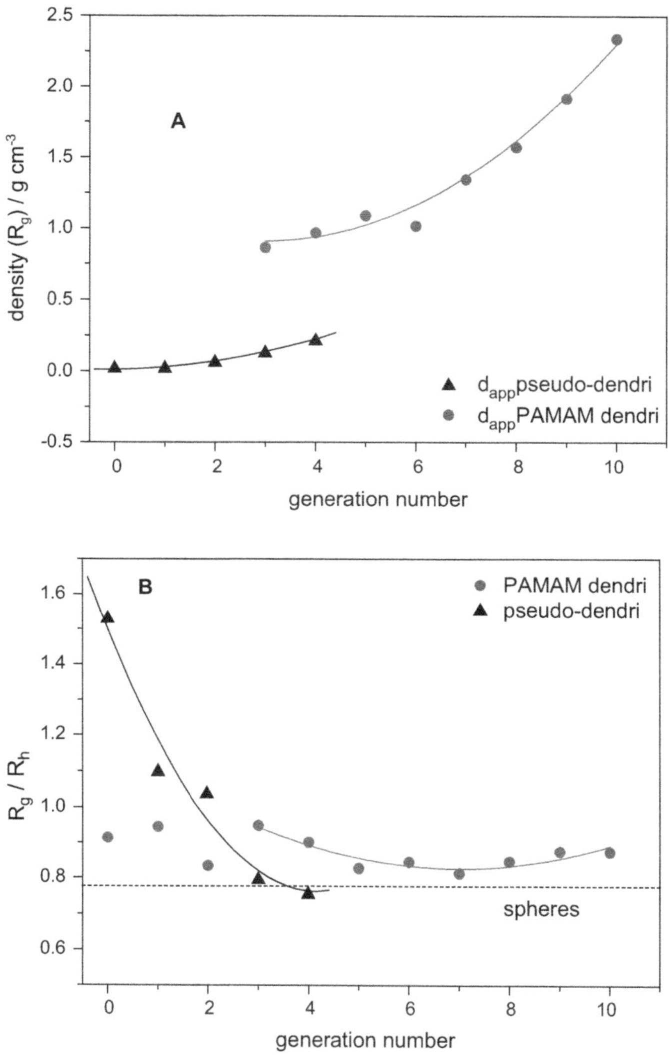

Figure 5.29 (A) Apparent density plotted for number of generations of pseudo-dendrimers from SANS in d_6THF (triangles)[82] and PAMAM dendrimers SAXS in methanol (circles).[85,86] (B) Behaviour of the ρ-parameter of pseudo-dendrimers compared to that of dendrimers of more than four generations. The data of R_g and R_h for PAMAM dendrimers are taken from Maiti *et al.*,[80] Prosa *et al.*[85] and Tomalia *et al.*[81]
Reprinted with permission from ref. 82. Copyright © 2015 Wiley-VCH Verlag GmbH & Co. KGaA, Weinheim.

limit of large molecules and for linear chains the renormalization theory predicts $\Psi^* = 0.269$.[62] For hard spheres $\Psi^*_{sphere} = 1.619$ which is the upper limit that can be expected. This value, Ψ^*, depends on the particle topology in a way that it increases with branching, showing shallow interpenetration.[62] For pseudo-dendrimers Ψ increases from 0.075 to 0.458 from G0 to G4 (Figure 5.30).

Taking into account the Flory parameter from the Flory–Fox equation [eqn (2.7)], a similar effect is observed but in contrast to A_2, the magnitude of Φ is not determined by interacting segment clouds but it is caused by the draining effect of the solvent through the particle due to hydrodynamic interaction. This fact depends on the segmental density, and therefore the draining is impeded more in the centre of the particle than in the outskirt of the particle (Figure 5.30).

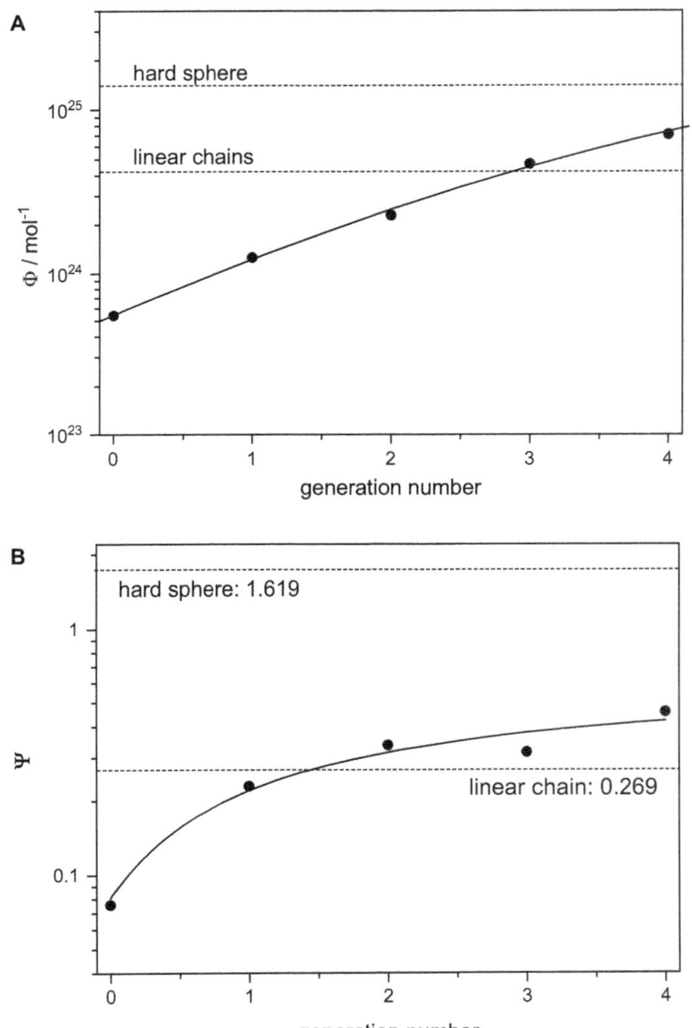

Figure 5.30 Change of the draining parameter Φ (A) and the interpenetration factor Ψ (B) with increasing pseudo-dendrimer generations.
Reprinted with permission from ref. 82. Copyright © 2015 Wiley-VCH Verlag GmbH & Co. KGaA, Weinheim.

As was suggested for the Ψ-parameter, the data for flexible linear chains and hard spheres can be taken as the two limits for the experimental data. Figure 5.30 shows increasing Flory parameter showing that beyond a certain generation number, the outskirt part of the molecule is kept swollen but not being drained any longer and the solvent moves around this obstacle as it is for hard spheres.

The data of both parameters could be interpreted in conjunction with each other as suggested by Oono and Kohmoto.[88,89] Eqn (5.6) includes the ratio $A_2M/[\eta]$ shown in Figure 5.31 for the pseudo-dendrimers, which solely depends on the ratio of Ψ/Φ. The generation dependence of this ratio is somewhat unexpected. A closer look at the plotted values in Figure 5.30 shows a smooth but pronounced increase of the draining parameter Φ with branching as expected in comparison to results from gradually branched polymers (Figure 5.25).[10,13,90–93] On the other hand the spatial interaction parameter Ψ is fairly low for G0 as was also found for other hyperbranched macromolecules,[10,90] but on stepping from G0 to G1 a sharp increase of Ψ occurs which is reduced again for the higher generations. The ratio now depends on the much stronger hydrodynamic hindrance than the effect of interpenetration such that the ratio Ψ/Φ decreases again.

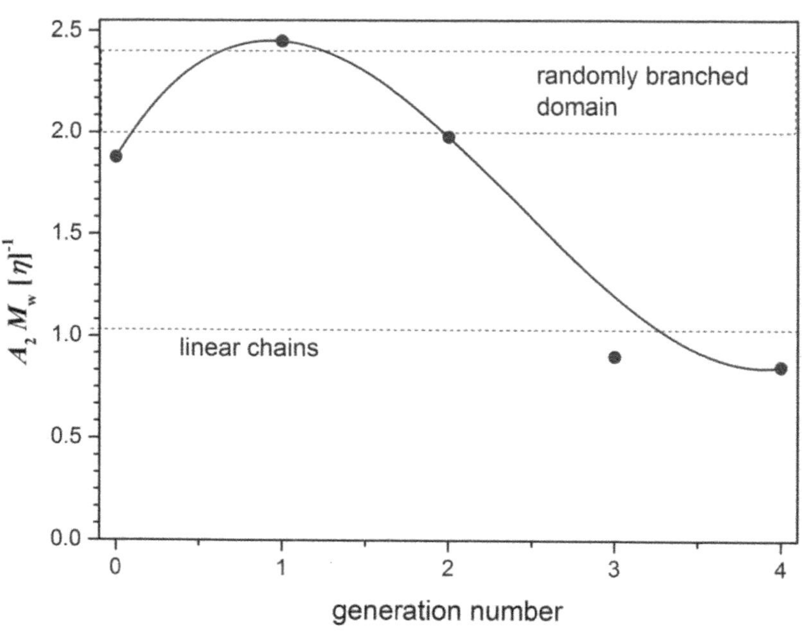

Figure 5.31 The graph demonstrates the ratio of interparticle interaction A_2M to the hydrodynamic-viscosity effect $[\eta]$ for different branching generations. The area above 2 is the regime of randomly branched samples observed so far;[10,90] the dotted line at 1 was predicted by the RG-theory[62] for linear chains in good agreement with experiments.[91–93]
Reprinted with permission from ref. 82. Copyright © 2015 Wiley-VCH Verlag GmbH & Co. KGaA, Weinheim.

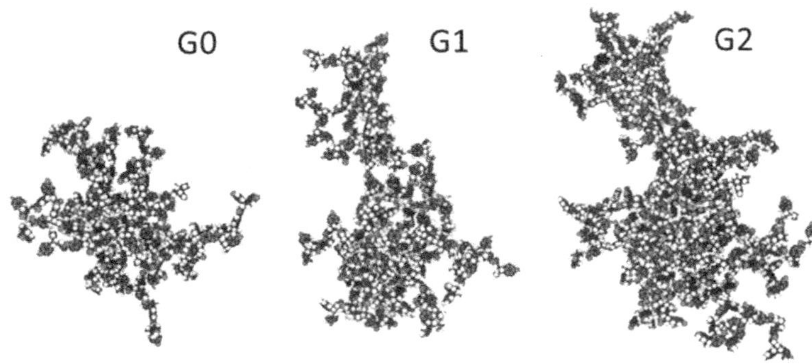

Figure 5.32 MD simulation in THF as a solvent (solvent molecules are hidden) of G0, G1 and G2 pseudo-dendrimers with deprotected OH end-groups[41] with DP 92 of the G0 hyperbranched core.
Reprinted with permission from ref. 82. Copyright © 2015 Wiley-VCH Verlag GmbH & Co. KGaA, Weinheim.

Obviously, the attachment of the dendritic elements causes a marked change in the draining effect. The interpenetration is strongly reduced only for the first AB_2-modification step in G1 compared to G0, which leads to a compact pseudo-dendritic structure. This situation should be compared with the decrease in viscosity of dendrimers with more than four generations to a nearly constant value which resembles that of hard spheres. Their behaviour is a result of significantly slower increase of molecular size than molar mass with the generation number, which is valid also for pseudo-dendrimers.[24] In contrast to pseudo-dendrimers, the compaction of flexible dendrimers occurs after reaching certain dendron size able to backfold (after G4), while the high compactness in pseudo-dendrimers starts in the first generation (G1) due to space filling modification. The observations of the ratio $A_2M/[\eta]$ are supported by molecular dynamic simulation studies of the pseudo-dendrimers in generations G0–G2 shown in Figure 5.32. The free space within the core molecule, G0 is essentially filled out right after the first modification to G1. In the next generation, dendrons with more open arms coupled to this core are formed which enable interpenetration between the segmental clouds, but lower flow of the solvent through the inner dense core. This observation is in agreement with the faster increase of Φ than Ψ with the generation number in G2–G4.

As discussed above the Flory parameter is in direct correlation with the drainage exponent ε, and it depends on the segmental density. The ε exponent is a function of topology and increases with the degree of branching (Figure 5.27). This feature of ε is confirmed by increasing the branching density from G0 to G1 (compare Figure 5.27 and 5.33B) which results in higher exponent of approximately 1.2 in the first generation compared to the core with 0.77.

In generating G2–G4, however, this value decreases again. The observed decrease of g and g' is expected (Figure 5.33A) but the change of the resulting exponent ε with the generation number shows an extraordinary trend.

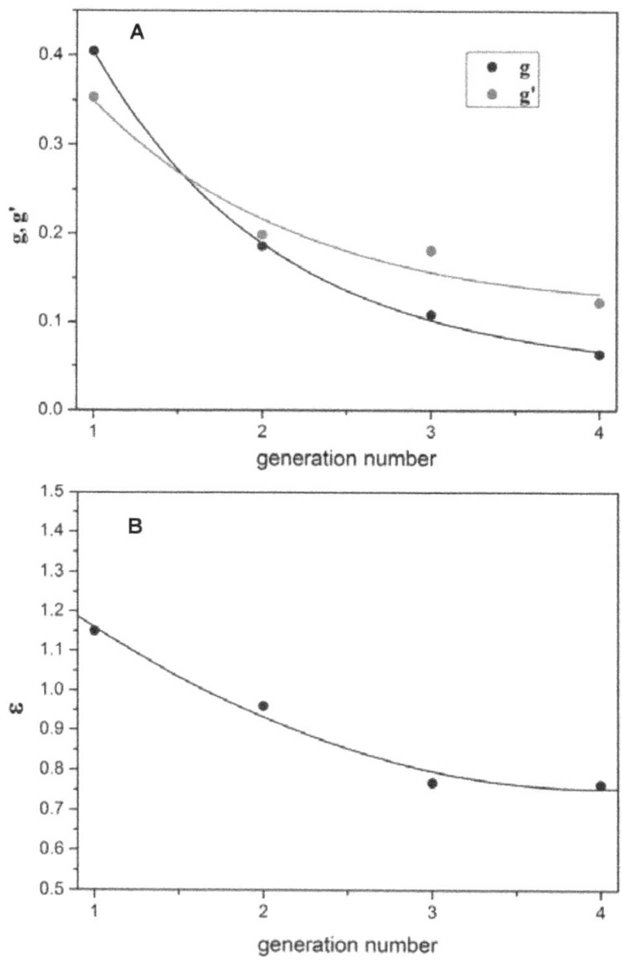

Figure 5.33 (A) Contraction factors $g = R^2_{g,br}/R^2_{g,lin}$ and $g' = [\eta]_{br}/[\eta]_{lin}$ in dependence of the branching generations. (B) The exponent ε in $g' = g^{\varepsilon}$ as a function of the generation number. All samples are measured in good solvent conditions ($A_2 > 0$).[82]
Reprinted with permission from ref. 82. Copyright © 2015 Wiley-VCH Verlag GmbH & Co. KGaA, Weinheim.

For generation G1 it would mean a behaviour close to the Stockmayer–Fixman supposition[69,70,73,94] of $\varepsilon = 1.5$, valid for high branching densities. For higher generations the exponent comes closer to the Kurata estimation of $\varepsilon = 0.78$ and, with this, close to the value found to the hyperbranched (DB = 50%) core. As explained above, g and g' are calculated from the ratios of the mean square radii and intrinsic viscosities of the branched and of the linear polymers of the same composition and molecular weight. It should be noted that in this case we are not comparing different degrees of branching as in Figure 5.27 shown. In fact, for pseudo-denrimers in each new generation a different topology is created. This leads to higher density in G1 but in G2, G3 and G4 the rather open dendrons are the reason for the new tendency of ε. This particluar pseudo-dendritic behaviour is most probably a result of the segmental density which significantly increases in the first generation, leading to a smooth, compact surface, which is transformed to a structure with a dense core and open dendritic shell in the next generations.

References

1. J. G. Kirkwood and J. Riseman, *J. Chem. Phys.*, 1948, **16**(6), 565.
2. H. Mori, A. H. E. Müller and P. F. W. Simon, in *Macromolecular Engineering: Precise Synthesis, Materials, Properties, Applications*, ed. K. Matyjaszewski, Y. Gnanou and L. Leibler, Wiley-VCH, Weinheim, 2007, vol. 2, p. 973ff.
3. A. Einstein, *Ann. Phys.*, 1910, **21**, 581.
4. T. G. J. Fox and P. J. Flory, *J. Appl. Phys.*, 1950, **21**, 581.
5. T. G. J. Fox and P. J. Flory, *J. Polym. Sci.*, 1954, **14**, 315.
6. W. R. Krigbaum and D. K. Carpenter, *J. Phys. Chem.*, 1955, **59**, 1166.
7. Z. Grubisic, P. Rempp and H. Benoit, *J. Polym. Sci., Part B*, 1967, **5**, 753.
8. H. Yamakawa, *Modern theory of polymer solutions*, Harper & Row, New York, 1971.
9. W. Burchard, in *Branched Polymers II*, ed. J. Roovers, 1999, vol. 143, p. 113.
10. G. Galinsky and W. Burchard, *Macromolecules*, 1996, **29**, 1498.
11. B. I. Voit and A. Lederer, *Chem. Rev.*, 2009, **109**, 5924.
12. J. Roovers, L. L. Zhou, P. M. Toporowski, M. van der Zwan, H. Iatrou and N. Hadjichristidis, *Macromolecules*, 1993, **26**, 4324.
13. A. Lederer, W. Burchard, A. Khalyavina, P. Lindner and R. Schweins, *Angew. Chem., Int. Ed.*, 2013, **52**, 4659.
14. A. Khalyavina, PhD thesis, Technische Universität Dresden, 2010.
15. A. Lederer, D. Voigt, C. Clausnitzer and B. Voit, *J. Chromatogr. A*, 2002, **976**, 171.
16. L. J. Hobson and W. J. Feast, *Chem. Commun.*, 1997, 2067.
17. S. Boye, H. Komber, P. Friedel and A. Lederer, *Polymer*, 2010, **51**, 4110.
18. T. H. Mourey, S. R. Turner, M. Rubinstein, J. M. J. Fréchet, C. J. Hawker and K. L. Wooley, *Macromolecules*, 1992, **25**, 2401.
19. C. Cai and Z. Y. Chen, *Macromolecules*, 1998, **31**, 6393.
20. M. L. Mansfield and L. I. Klushin, *Macromolecules*, 1993, **26**, 4262.
21. R. L. Lescanec and M. Muthukumar, *Macromolecules*, 1990, **23**, 2280.
22. A. Lederer, D. Voigt, D. Appelhans and B. Voit, *Polym. Bull.*, 2006, **57**, 329.
23. W. Burchard, *Polym. Bull.*, 2007, **58**, 3.
24. M. Ballauff, *Top. Curr. Chem.*, 2001, **212**, 177.
25. M. Ballauff and C. N. Likos, *Angew. Chem., Int. Ed.*, 2004, **43**, 2998.

26. D. Appelhans, H. Komber, D. Voigt, L. Häußler and B. I. Voit, *Macromolecules*, 2000, **33**, 9494.
27. D. A. Tomalia, D. M. Hedstrand and L. R. Wilson, in *Encyclopedia of Polymers Science*, Wiley, New York, 2nd edn, 1990.
28. S. Wong, D. Appelhans, B. Voit and U. Scheler, *Macromolecules*, 2001, **34**, 678.
29. S. R. Turner, B. I. Voit and T. H. Mourey, *Macromolecules*, 1993, **26**, 4617.
30. J. Aerts, *Comput. Theor. Polym. Sci.*, 1998, **8**, 49.
31. B. H. Zimm, *J. Chem. Phys.*, 1956, **24**(2), 269.
32. P. Debye and A. M. Bueche, *J. Chem. Phys.*, 1948, **16**(6), 573.
33. A. V. Lyulin, D. B. Adolf and G. R. Davies, *Macromolecules*, 2001, **34**, 3783–3789.
34. A. H. Widmann and G. R. Davies, *Comput. Theor. Polym. Sci.*, 1998, **8**, 191.
35. H. Wiener, *J. Am. Chem. Soc.*, 1947, **69**, 17.
36. D. Bonchev, E. J. Markel and A. H. Dekmezian, *Polymer*, 2002, **43**, 203.
37. P. F. Sheridan, D. B. Adolf, A. V. Lyulin, I. Neelov and G. R. Davies, *J. Chem. Phys.*, 2002, **117**, 7802.
38. R. Haag, J.-F. Stumbé, A. Sunder, H. Frey and A. Hebel, *Macromolecules*, 2000, **33**, 8158.
39. C. Lach and H. Frey, *Macromolecules*, 1998, **31**, 2381.
40. R. Haag, A. Sunder and J.-F. Stumbé, *J. Am. Chem. Soc.*, 2000, **122**, 2954.
41. A. Lederer, T. Hartmann and H. Komber, *Macromol. Rapid Commun.*, 2012, **33**, 1440.
42. E. M. M. de Brabander, J. Brackman, M. Mure-Mak, H. de Man, M. Hogeweg, J. Keulen, R. Scherrenberg, B. Coussens, Y. Mengerink and S. van der Wal, *Macromol. Symp.*, 1996, **102**, 9.
43. D. Poetschke, M. Ballauff, P. Lindner, M. Fischer and F. Voegtle, *Macromolecules*, 1999, **32**, 4079.
44. A. Khalyavina, F. Schallausky, H. Komber, M. Al Samman, W. Radke and A. Lederer, *Macromolecules*, 2010, **43**, 3268.
45. A. D. Schlüter, in *Topics in Current Chemistry*, ed. A. D. Schlüter, Springer, 2005, vol. 245, p. 151.
46. B. Zhang, R. Wepf, K. Fischer, M. Schmidt, S. Besse, P. Lindner, B. T. King, R. Sigel, P. Schurtenberger, Y. Talmon, Y. Ding, M. Kröger, A. Halperin and A. D. Schlüter, *Angew. Chem., Int. Ed.*, 2011, **50**, 737.
47. S. Boye, D. Appelhans, V. Boyko, S. Zschoche, H. Komber, P. Friedel, P. Formanek, A. Janke, B. Voit and A. Lederer, *Biomacromolecules*, 2012, **13**, 4222.
48. A. Kroeger, B. Zhang, C. Rosenauer, A. D. Schlüter and G. Wegner, *Colloid Polym. Sci.*, 2013, **291**, 2879.
49. E. Cordova-Mateo, O. Bertran, B. Zhang, D. Vlassopoulos, R. Pasquino, A. D. Schluter, M. Kroger and C. Aleman, *Soft Matter*, 2014, **10**, 1032.
50. J. Majoinen, J. S. Haataja, D. Appelhans, A. Lederer, A. Olszewska, J. Seitsonen, V. Aseyev, E. Kontturi, H. Rosilo, M. Österberg, N. Houbenov and O. Ikkala, *J. Am. Chem. Soc.*, 2013, **136**, 866.

51. B. Zhang, A. Kroeger, A. D. Schlüter and G. Wegner, *Macromol. Rapid Commun.*, 2013, **34**, 1537.
52. F. Schallausky, PhD thesis, Technische Universität Dresden, 2007.
53. W. Burchard, A. Khalyavina, P. Lindner, R. Schweins, P. Friedel, M. Wiemann and A. Lederer, *Macromolecules*, 2012, **45**, 3177.
54. C. E. Ioan, T. Aberle and W. Burchard, *Macromolecules*, 2001, **34**, 3765.
55. F. R. Senti, N. N. Hellman, N. H. Ludwig, G. E. Babcock, R. Tobin, C. A. Glass and B. L. Lamberts, *J. Polym. Sci.*, 1955, **17**, 527.
56. J. S. Higgins and H. C. Benoit, *Polymers and Neutron Scattering*, Oxford University Press, Oxford, 1994.
57. M. Daoud and J. E. Martin, in *The Fractal Aproach to Heterogeneous Chemsitry*, ed. D. Avnir, John Wiley & Sons, Chichester, 1989.
58. C. E. Ioan, T. Aberle and W. Burchard, *Macromolecules*, 1999, **32**, 7444.
59. W. Burchard, M. Schmidt and W. H. Stockmayer, *Macromolecules*, 1980, **13**, 1265.
60. B. H. Zimm and W. H. Stockmayer, *J. Chem. Phys.*, 1949, **17**, 1301.
61. G. Galinsky and W. Burchard, *Macromolecules*, 1995, **28**, 2363.
62. K. F. Freed, *Renormalization group theory of macromolecules*, Wiley, New York, 1987.
63. D. Stauffer, *Introduction to Percolation Theory*, Taylor and Francis, London, 1985.
64. E. T. F. Geladé, B. Goderis, C. G. de Koster, N. Meijerink, R. A. T. M. van Benthem, R. Fokkens, N. M. M. Nibbering and K. Mortensen, *Macromolecules*, 2001, **34**, 3552.
65. E. De Luca and R. W. Richards, *J. Polym. Sci., Part B: Polym. Phys.*, 2003, **41**, 1339.
66. E. De Luca, R. W. Richards, I. Grillo and S. M. King, *J. Polym. Sci., Part B: Polym. Phys.*, 2003, **41**, 1352.
67. R. Scherrenberg, B. Coussens, P. van Vliet, G. Edouard, J. Brockman and E. de Brabander, *Macromolecules*, 1998, **31**, 456.
68. V. M. Garamus, T. V. Maksimova, H. Kautz, E. Barriao, H. Frey, U. Schlotterbeck, S. Mecking and W. Richtering, *Macromolecules*, 2004, **37**, 8394.
69. W. H. Stockmayer and M. Fixman, *Ann. N. Y. Acad. Sci.*, 1953, **57**, 334.
70. B. H. Zimm and R. W. Kilb, *J. Polym. Sci.*, 1959, **37**, 19.
71. M. Kurata and M. Fukatsu, *J. Chem. Phys.*, 1964, **41**, 2934.
72. M. Kurata, M. Abe, M. Iwama and M. Matsushima, *Polym. J.*, 1972, **3**, 729.
73. J. Roovers, P. M. Toporowski and J. Douglas, *Macromolecules*, 1995, **28**, 7064.
74. M. Weissmüller and W. Burchard, *Polym. Int.*, 1997, **44**, 380.
75. L. K. Kostanski, D. M. Keller and A. E. Hamielec, *J. Biochem. Biophys. Methods*, 2004, **58**, 159.
76. J. M. J. Fréchet and D. A. Tomalia, *Dendrimers and other denritic polymers*, Wiley, New York, 2002.
77. G. R. Newkome, C. N. Moorefield and F. Vögtle, *Dendrimers and Dendrons. Concepts, Syntheses, Applications*, Wiley-VCH, Weinheim, 2001.

78. M. Calderón, M. A. Quadir, S. K. Sharma and R. Haag, *Adv. Mater.*, 2010, **22**, 190.
79. P. J. Flory, *Principles in Polymer Chemistry*, Cornell University Press, Ithaca, New York, 1953.
80. P. K. Maiti, T. Cagin, G. Wang and W. A. Goddard, III, *Macromolecules*, 2004, **37**, 6236.
81. D. A. Tomalia, A. M. Naylor and W. A. Goddard, *Angew. Chem., Int. Ed.*, 1990, **29**, 138.
82. A. Lederer, W. Burchard, T. Hartmann, J. S. Haataja, N. Houbenov, A. Janke, P. Friedel, R. Schweins and P. Lindner, *Angew. Chem. Int. Ed.*, 2015, DOI: 10.1002/anie.201504059.
83. S. Rosenfeldt, N. Dingenouts, M. Ballauff, N. Werner, F. Vogtle and P. Lindner, *Macromolecules*, 2002, **35**, 8098.
84. S. Rosenfeldt, N. Dingenouts, D. Pötschke, M. Ballauff, A. J. Berresheim, K. Müllen and P. Lindner, *Angew. Chem., Int. Ed.*, 2004, **43**, 109.
85. J. Prosa, B. J. Bauer and E. J. Amis, *Macromolecules*, 2001, **34**, 4897.
86. T. J. Prosa, B. J. Bauer, E. J. Amis, D. A. Tomalia and R. Scherrenberg, *J. Polym. Sci., Part B: Polym. Phys.*, 1997, **35**, 2913.
87. J. Roovers and B. Comanita, *Adv. Polym. Sci.*, 1999, **142**, 179.
88. Y. Oono and M. Kohmoto, *J. Chem. Phys.*, 1983, **78**, 520.
89. Y. Oono, in *Advances in Chemical Physics*, John Wiley & Sons, Inc., 2007, p. 301.
90. J. Bauer and W. Burchard, *Macromolecules*, 1993, **26**, 3103.
91. Y. Miyaki, Y. Einaga and H. Fujita, *Macromolecules*, 1978, **11**, 1180.
92. T. Matsumoto, N. Nishioka and H. Fujita, *J. Chem. Phys.*, 1972, **10**, 23.
93. G. Tanaka, S. Imai and H. Yamakawa, *J. Chem. Phys.*, 1970, **52**, 2639.
94. W. H. Stockmayer and M. Fixman, *Ann. N. Y. Acad. Sci.*, 1953, **57**, 334.

CHAPTER 6

Size Determination by Scattering Techniques

6.1 Introduction

This chapter is concerned with the various characterization methods, by light scattering and related techniques, whose principles are also described in all textbooks on polymers science more or less as routine techniques. Often application of these techniques remains restricted to linear chains of uniform or polydisperse size. In this chapter the discussion is extended to branched and polydisperse macromolecules. Clearly these polymers have a more complex structure and appear at first sight scarcely treatable by analytical equations. In Chapter 7 it will be shown that this is not true in such a generalized manner, and the unfractionated branched samples are fairly easy to handle, much easier than the uniformly branched materials. To make the understanding easier the quantities detected by the corresponding instruments are at first explained with the conformation of *unperturbed* chains. The question whether the solvent used is a good or poor one is mainly expressed by the second virial coefficient and from this one learns that this parameter has some characteristic influence on the intrinsic viscosity. At that point a preparative chemist will often be satisfied with this information, and the question of what really is going on under the influence of excluded volume interaction may remain for him or her a special topic for theoreticians.

This rather light-hearted treatment of characterization methods is no more applicable if more complex structures have to be investigated. No doubt, the branching phenomenon appears at first sight to be a really complex topic, which upon a closer look is not justified. Chemists like it to get in a first step control of the polyreaction, such that uniform structures are obtained, and

RSC Polymer Chemistry Series No. 16
Hyperbranched Polymers: Macromolecules in between Deterministic Linear Chains and Dendrimer Structures
By Albena Lederer and Walther Burchard

Published by the Royal Society of Chemistry, www.rsc.org

they have been successful, for instance with the preparation of the perfectly constructed dendrimers. However, in order to reach these types of perfect structure much effort has to be invested, as well as common sense, caused by the counter-action against entropy. Even in biological processes nature prefers the statistical route whenever a specific reaction is not required. The ample diversity of structures with polysaccharides is an impressive demonstration.

The investigation of such complex structures is a real challenge to scientists, and it requires a deeper and more detailed understanding of the characterization techniques. Therefore, the following report goes much more into these details than described in textbooks. It is hoped that this outline will become helpful to gain insight into the main structural features of branched materials.

6.2 The Search for Non-invasive Structure Characterization

The preparation of new polymer structures is an exciting challenge for chemists and chemical engineers. With the first successful synthesis of hyperbranched samples[1,2] this topic became especially interesting because of the many possibilities for chemical modification and the reaction of a polymer with special properties. The topic is still growing. The chemistry seems to be fairly well developed but the realization of well-defined special structures remains difficult and a delicate task.[3] The straightforward synthesis as designed by a sketch on paper often proved not to be feasible, and other routes have to be elaborated. In the end a synthesis may be successful, though the prepared sample did not fully agree with the original ideal of the envisaged model but still, it is considered being worth publishing. At present this activity has become a real flood of recent papers illustrating by wonderful pictures the structures obtained. This is impressive but the phenomena are far too complex to illustrate the system with only one selected structure, and no realistic impression on structures in an ensemble is obtained by such cartoons. Chemists have a good feeling for a sensible interpretation but this is often incomplete and sometimes even misleading. A proof of the structure and the resulting properties by physicochemical means remains an essential issue.

A large number of analytical techniques are available which should make the required characterization easy. The reality is different for simple reasons.

One reason arises from the fact that all chemical reactions are statistical processes, and it is virtually impossible to prepare a sample in which all macromolecules in the ensemble have the same ideal structure. There has been much progress in getting control over the randomness of the reactions but it will not be possible to achieve really uniform samples in a one pot or batch reaction. Only with the preparation of fairly small sizes, like dendrimers, will the perfect structure be achieved by much work in well-controlled step-by-step reactions.

Another reason is that polymer samples are not rigid or semi-rigid like glass; the majority of the chains are highly flexible. For characterization the samples have to be dissolved in a good solvent to obtain the sample molecularly dispersed. Here the molecules are exposed to the thermal action of Brownian motion and the effect of the finite volume causes an exclusion of conformations originating from repulsion.

These two facts make clear that only *averages* can be measured not solely over the ensemble of differently sized molecules, but also the time dependence and the corresponding *time averages* have to be taken in consideration. It is common use to show an instructive cartoon of the sample, but it is not emphasized that such a graph is actually considered only as a typical representative from the ensemble. In science we believe what we see, and therefore microscopy pictures are taken as undoubtedly being true. In general we dislike the statistical feature of nature (often Einstein's comment is quoted: 'God does not play dice'). Also theoreticians prefer the derivation of equations for a description of uniform structures. Indeed, such nearly uniform samples can be obtained by the well-defined separation methods, *e.g. via* size exclusion chromatography (SEC). Nowadays, this separation technique can be combined with several in-line detectors for relevant physical quantities.

These data from in-line fractionation are of great value, but we will follow here a probability approach, since this is directly correlated to the kinetic process of chemical structure formation. The corresponding theory which we will apply is not familiar to chemists and not even to theoreticians although the mathematics of this theory were basically well developed about 150 years ago.[4,5] A detailed outline of this technique and the applicability to polymers will be given in a separate section of Chapter 8. Before discussing the method of static light scattering for characterization a short consideration of other techniques will be presented. The advantages and drawbacks of these methods will be made apparent. The first information we wish to obtain is the molar mass of the molecules and their size in solution. Concerning the size of the particles microscopy may be the best method.

6.2.1 Microscopy

Various types of microscopy can be used, *i.e.* visible light microscopy, the various techniques of electron microscopy, TEM, SEM, cryo-TEM, and atomic force microscopy (AFM) depending on the size of the objects. Light microscopy can be applied to structures larger than 1 μm, which are rare with macromolecules but adequate if supramolecular structures are present, or if natural particles like bacteria and large viruses or active organelles are studied. Next, higher resolution is obtained by electron microscopy which covers sizes between 10 nm to several μm. The technique is appropriate for the study of colloids and their ordering to clusters, but single

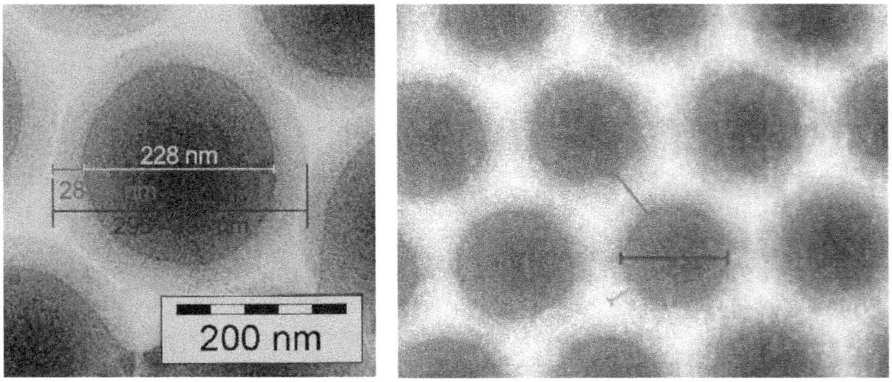

Figure 6.1 Cryo-TEM photographs from a core–shell colloid particle, obtained by emulsion co-polymerization of a monomer poorly soluble in water with a water soluble macromonomer (surfmer) prepared without added surfactant. The transmission of the shell is higher than that of the core and this makes the shell visible (because of less contrast) and appears in the two pictures as blurred grey part. The shell thickness of 28 nm, as indicated, and the core diameter was 228 nm. The micrograph to the right shows the very uniform size of these particles.[6]
We gratefully acknowledge the kind permission of Dr R. Schneider and Professor E. Bartsch to use this photograph, which has not yet been published.

macromolecules in solution cannot be sensibly studied because of segmental mobility and since during the drying process under vacuum the chains collapse or form aggregates of larger sizes. Certainly electron microscopy gives us a useful impression on the spatial structure of mesoscopic particles. The shrinking of dimensions and the collapse can be avoided if the solution is abruptly frozen in a time interval shorter than the fastest relaxation time. This cryo-TEM technique gives us realistic pictures of the structure but it requires much skill and is, so far, not much applied. Still, mostly only the shape of the particle can be made visible with the exception of cryo-TEM, by which technique also some insight into the internal structure can be obtained. The following picture from a core–shell colloidal particle gives an example[6] (Figure 6.1).

The core–shell structure was obtained by co-polymerization of two monomers in water. One is strongly hydrophobic and other is a macromonomer (surfmer), which has a hydrophobic polymerizable double bond and a water soluble polyether tail that acts like a common surfactant.

6.2.2 Rheology

Characterization can also be made by rheological measurements[7] and is most frequently applied to polymer melts. A simple method of sample characterization in solution by rheology is made by capillary viscometry, for instance in

Oswald viscometers. With the data of the solvent and the solution viscosity the intrinsic viscosity can be determined which is given by eqn (6.1)[8,9]

$$[\eta]_{c\to 0} = \frac{\eta - \eta_0}{c\eta} = \frac{t - t_0}{ct_0}. \tag{6.1}$$

Since the viscosity is proportional to the time the solvent or solution needs to run through a capillary the time can be used instead of the viscosities η and η_0 of the solution at concentration c and the solvent visosity (see also Chapter 5). The intrinsic viscosity is a valuable quantity that depends on the molar mass and a hydrodynamically defined molecular radius. The disadvantage of this parameter is the complex hydrodynamics, which does not permit a clear definition of the average molecular radius. This problem is considered at greater detail in a separate section. The main disadvantage in rheology is that a strong external shearing force is applied which can cause a deformation of the particle. However, it allows for a study of relaxation processes.

6.2.3 Sedimentation by Ultracentrifugation

This technique was first developed by Svedberg and Pedersen[10] to determine the molar mass of colloids and macromolecules. Again, a strong gravitational force is applied which can cause some deformation of the particles due to the high pressure created. The effect is certainly less pronounced than in rheology, but the mean disadvantage of this technique is the expensive instrument and the required large energy needed to build up the high sedimentation force which is about 1000 to 10 000 times the earth's gravitational force.

6.2.4 Application of Light

Although impressive progress was achieved by optimizing the sensitivity of the above-mentioned methods there remains the question of whether a non-invasive characterization method would be available, *i.e.* a method which has essentially no impact on the structure at equilibrium in solution. The search leads to the physics of light.[11–16] It is a well-known observation that a dust particle looks like a tiny glittering star when illuminated by a small beam of light that enters a dark room through holes in a shutter. The glitter is brighter with a large dust particle. Christiaan Huygens (1624–1695), a Dutch scientist and mathematician, thought of this observed fact and came to a conclusion that is known as Huygens' principle, which states: 'Every point hit by light will become the source of a secondary wave of light which is emitted as a spherical wave in all directions of space.' In contrast to Isaac Newton (1643–1727), who retained the idea of light as particles, Huygens considered light as waves, and developed a first theory of travelling waves which resulted in the setting up essential principles of geometric optics. It took a further 200 years before Maxwell disclosed the electro-magnetic nature of light, and Lord Rayleigh[17] recognized that these electro-magnetic waves of light are responsible for the excitation of scattered light. Many valuable observations of scattered light were made previously by Faraday and Tyndall (1868) which Lord Rayleigh used as the background for a derivation of the theory of scattered light (1872, 1899).[17] His scattering theory was later further developed by R. Gans and L. Hapel (1909, 1920)[18,19] to enclose anisotropy in scattering. With the development of the quantum theory it became clear that the wave concept of light can be used for neutrons, which are generally considered as particles, conversely the waves of light can be taken as particles denoted as photons. The photons from visible light are of low energy and induce oscillations of electrons in the outmost shells of atoms which resemble oscillating dipoles, and these vibrating dipoles emit scattered light of the same wavelength as the primary beam. The low energy of visible light does not cause changes in the properties of the dissolved particles. Thus, the light is a non-invasive tool for the study of molecules and colloids in a liquid medium.

The non-invasive property is not the only reason why light scattering has become an eminently important method of analysis. Swollen molecules and even small colloid particles (<500 nm in diameter) are mostly fully transparent. Thus the scattered light not only arises from the surface of the particles but also from the interior. No other characterization method provides such a possibility.

The shape and details of the internal structure can be probed if the size of the particle is larger than 1/20 of the wavelength of the light, *i.e.* >10 nm in the radius of gyration, if a He–Ne laser is used as primary light source. X-rays and cold neutrons have about 1000 times smaller wavelength such that also distances down to 0.2 nm are detectable. Small angle neutron scattering (SANS) is a suitable method for small particles such as proteins or soap micelles as well as for the study of the local structure in a particle. There still exists a gap of non-overlapping angular dependence. Such studies cannot be made in common laboratories and have to be made at places where the appropriate reactors for neutron sources are available. These sources are used by many applicants, and access is possible after submission of an application and the approval by a committee. The discussion of the scattering techniques will be in connection with the static light scattering (SLS).

6.3 Static Light Scattering[20–23]

Light scattering is often considered by chemists as a crypt characterization method, although this is based on light which we see with our eyes. The reason for this feeling of mystery is based on the fact that the scattered light from a particle causes a transformation of the structure in space into a mathematically defined inverse space which corresponds to a Fourier transformation. These pictures of the transformed structure into the inverse space cannot be directly understood but the mathematical correlations in this inverse space are much simpler and allow for solution of complex structures which could not be treated in Euclid space. This concept is indispensable for the understanding of the basics of this transformation, which was elucidated by the Maxwell and Herz, Fresnel, Kirkwood, Debye and Fourier, to mention only some of the scientists among many others. With regard to scattering, the work by Fourier on the use of special integral transformation, is of essential importance. The mathematics behind this transformation allow us to interpret the scattering phenomena we observe in terms of structure parameters in space. In the following the equation required for the structure analysis will be discussed in three steps concerning three main questions:

- How great is the scattered intensity if the particles are small compared to the wavelength of the light if a primary light beam hits N point-like scattering units per volume?
- Polarization has been observed in early experiments by Tyndall and Faraday. It will be shortly reviewed in Section 6.3.2.
- Have polarization effects to be discussed?
- What is the behaviour of particles with dimensions of about $d > \lambda/20$?

6.3.1 Scattered Light from Point-like Scattering Units: Correlation to Density and Concentration Fluctuations

Let $n = N/V$ be the particles found in the scattering volume. Each of these particles may be excited to oscillation of dipoles by the incident primary beam. Then the Maxwell theory of the electro-magnetic nature of light leads to a scattering amplitude ($A(\lambda)$) that is proportional to the number of dipoles per volume and defines the polarizibility α. Since the intensity is the product of amplitudes $A \times A^* = A^2$ (with $\mathbf{A}^*$ the conjugated complex amplitude to A) one has $i \propto \langle \alpha^2 \rangle$, where the $\langle \ldots \rangle$ brackets indicate the average overall possible changes of the dipoles. The pre-factor determines the magnitude of the scattered intensity and is determined by the behaviour of oscillating dipoles which result from electro-magnetic theory.[17] Omitting details of derivation some intuitive conjectures can still be made. No scattering can be expected if in a solution the refractive index of the sample is the same as that of the solvent, and therefore we expect $\alpha \propto \mathrm{d}n/\mathrm{d}c$, which is the refractive index increment, and might be expected to be proportional to the squared number of dipoles per volume N/V. In his first derivation Lord Rayleigh considered only the dipole oscillation which led to a result that the scattering intensity would be proportional to the square of molar mass. With Einstein's work on Brownian motion it became clear that also *concentration* and *density fluctuations* have to be taken into account which led to a mean square polarizibility $\langle \alpha^2 \rangle$ and the mean square average of the particle number fluctuations $\langle \Delta N \rangle^2$ has to be considered. This average $\langle \Delta N \rangle^2$ had to be calculated with thermodynamic potentials. The simplest form of such a potential is to use only the change in entropy, which can be represented by a Gaussian distribution. With this distribution the mean value of the squared fluctuation $\langle N^2 \rangle = N$ is obtained. Actually, the interaction potential contains besides the change in entropy also that of the enthalpy, *e.g.* the Gibbs free energy ΔG or, more correctly, the fluctuation of the chemical potential $\Delta\mu = \partial(\Delta G)/\partial c$. This, in turn, can be expressed by the osmotic pressure Π. Taking all facts of the fluctuating dipoles together one finally arrives at the normalized scattering intensity of[20,22,23]

$$\frac{i(\theta)}{I_0} r^2 \equiv R_\theta = \frac{4\pi^2}{\lambda_0^4 N_A}\left[\left(\rho n_0 \frac{\partial n}{\partial p}\right)^2 RT\kappa + \left(n_0 \frac{\partial n}{\partial c}\right)^2 RT \frac{Mc}{\partial\Pi/\partial c}\right] F(\theta). \qquad (6.2)$$

The *first term* contains the isothermal compressibility $RT\kappa$ and the change of the refractive index due to *density fluctuation*; the *second term* contains the contribution of concentration fluctuations expressed in term of the *osmotic compressibility*, the square of the differential refractive and the weight concentration $c = (N/V)(M/N_A)$ used in common experiments with M the molar mass of the particles and N_A Avogadro's number (the function $F(\theta)$ will be defined below). Eqn (6.2) requires some additional comments:

- Intensities cannot be measured in absolute values; the scattering intensity has to be expressed by the ratio of the scattering intensity to that of the primary beam, $i(\theta)/I_0$ where θ is the scattering angle (see below).
- We have to take into account that the scattering intensity of a spherical wave decreases quadratically with the distance r of the detector from the scattering volume. To compensate this fact the ratio of scattering to primary beam intensities has to be multiplied by r^2. The quantity $r^2 i(\theta)/I_0$ on the left side is denoted as the *Rayleigh ratio* $R(\theta)$. The use of this parameter now permits comparison with results from other laboratories and the use of differently constructed scattering goniometers.

The index '0' reminds us of a wavelength defined in vacuum; in solution the wavelength is shorter and is $\lambda = \lambda_0/n_0$ with n_0 the refractive index of the solvent, ρ is the mass density of the liquid and $\kappa = -(\partial V/\partial p)_T$ is the isothermal compressibility of the solvent. Similarly, the ratio $(\partial c/\partial\Pi)_{T,p}$ can be denoted as osmotic compressibility. This measurable function is of considerable importance as it gives information on the interaction between the particles or molecules in the solution.

Eqn (6.2) can now be expressed in a much simpler manner. Liquids are almost incompressible. The isothermal compressibility is low, but not negligible, and consequently, also, the scattering intensity of the density fluctuations is very low compared to the concentration fluctuations but not negligible. Furthermore, the density and concentration fluctuations are not coupled to each other. It is sensible to measure the solvent scattering intensity separately and subtract this value from the measured scattering intensity. This leads to

$$\Delta R_\theta = R_{\theta,\text{solution}} - R_{\theta,\text{solvent}} = RT\frac{Kc}{(\partial\Pi/\partial c)} \qquad (6.3)$$

with

$$K = \frac{4\pi^2}{\lambda_0^4 N_A}\left(n_0 \frac{\partial n}{\partial c}\right)^2 \tag{6.4}$$

The refractive index increment has to be measured separately by special refractometers. For convenience eqn (6.3) may be expressed in its reciprocal version[22,23] $Kc/R_\theta = (1/RT)(\partial\Pi/c)$ because then the osmotic pressure appears in the denominator and can be expressed in a virial series:

$$\frac{Kc}{\Delta R_\theta} = \frac{1}{RT}\left(\frac{\partial\Pi}{\partial c}\right)_{p,T} = \frac{1}{M_w} + 2A_2c + 3A_3c^3 + \cdots nA_nc^{n-1} + \cdots \tag{6.5}$$

Throughout this book the notation ΔR_θ will be written as R_θ. In all examples considered the Rayleigh ratio means the value after subtraction of the solvent. Eqn (6.5) presents the basis to application. Chemists are essentially interested in the molar mass which could be determined from measurement of Kc/R_θ for a series of concentrations and after extrapolation towards $c = 0$. In this way the effects of the virial coefficients are omitted and M_w is obtained. The light scattering method has the advantage that the scattering intensity *increases* with M_w whereas the osmotic pressure *decreases* with M_n. The difference between the number average molar mass M_n and the weight average M_w occurs if a certain length distribution of chains is present, which is the case for all statistically progressing polymerization processes. The ratio $M_w/M_n = PDI > 1$ is the **P**oly-**D**ispersity **I**ndex (according to IUPAC Đ).

The effect of the virial coefficients was not given much attention for a long time and was mainly taken as a qualitative indication of whether a solvent is good or poor. In time elapsed the investigation was extended to moderate concentration up to the overlap concentration c^*, and even beyond this point. It turned out that the value of A_2 is a key parameter for understanding inter-particle interactions. In addition, A_2 contains valuable information on the structure in space and permits the estimation of an effective particle sphere radius. This correlation will be discussed in Chapter 8.

6.3.2 Effect of Polarization

In eqn (6.2) a scattering angle θ is attached as an index to the Rayleigh ratio although in eqn (6.3) and (6.5) no angular dependence was assumed. In general, a function $F(\theta)$ has to be taken into account, which is related to the polarization of the primary beam light. The polarization can have two main positions described by $F_\perp(\theta)$ or $F_\parallel(\theta)$ depending on a polarization vertical to the plane of the scattering angles or within this plane. Laser light is mostly nearly fully polarized, and this polarization direction is conveniently arranged vertically to the plane. In such an arrangement no angular dependence is observed, the scattering intensity remained the same for all scattering angles, *i.e.* $F_\perp(\theta) = 1$. A very different behaviour is found if the direction of the polarization lies in the plane of detecting scattering angles.

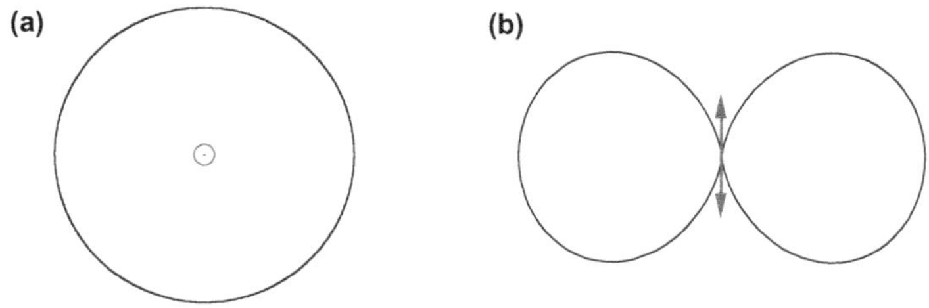

Figure 6.2 (a) Angular dependence of scattered light intensity excited by a vertically polarized primary beam. The detector is arranged in the paper plane and the dipoles vibrate vertically to this plane indicated by the point in the small circle. The scattering intensity is independent of the scattering angle. (b) Angular dependence of scattered light intensity excited by a horizontally polarized primary beam coming from the left. The dipoles (arrows) vibrate in the paper plane which also is the plane of the detector position. The scattering intensity has its maximum in forward and back direction and decreases to zero at 90° and 270°.

In this arrangement one has maximum scattering intensity at $\theta = 0°$ and 180° but this cannot be observed since these two angles lie in the direction of the primary beam with an intensity that is about 10^5 times higher than that from the scattered light. Another special case is the choice of $\theta = 90°$ or 270° where no scattering can be detected (see also Figure 6.2b). The full angular dependence is given by $F_{\|}(\theta) = \cos^2(\theta)$. Figure 6.2 elucidates schematically the effect of the two discussed arrangements of the polarization direction.

In summary, the following angular dependencies are observed at the three special choices of the orientation of the polarization direction:

$$
\begin{aligned}
F_{\text{vertical}} &= 1 \\
F_{\text{horizontal}} &= \cos^2(\theta) \\
F_{\text{unpolarized}} &= (1 + \cos^2(\theta))/2
\end{aligned}
\tag{6.6}
$$

The use of vertically polarized light is the preferred choice because then no angular dependence has to be considered. Additional effects appear if the particle is partially anisotropic. This special feature will not be discussed since branched samples are mostly isotropic, unless no special anisotropic stiff segments were incorporated in the structure. Note, the behaviour discussed so far remains valid only for *small particles*.

6.3.3 Scattering from Particles of Dimensions $> \lambda/20$

If the particles have dimensions larger than $\lambda/20$ the scattering intensity comes from different points of the particle, not only from the surface of the particle but also from the interior. These rays have different pathlenghts before reaching the detector, and their difference in phase causes interference. The magnitude of this effect depends on the scattering angle θ

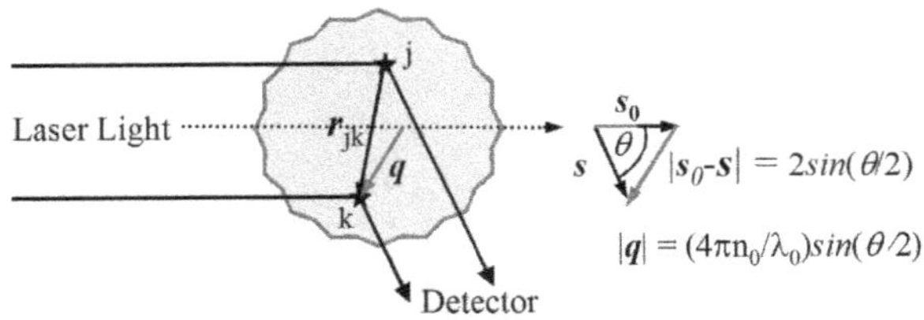

Figure 6.3 Geometric condition for the angular dependent scatter. $\boldsymbol{s_0}$ and $\boldsymbol{s}$ are unit vectors in directions of the incident and scattered rays, and $\boldsymbol{q}=(4\pi/\lambda)(\boldsymbol{s_0}-\boldsymbol{s})$ is the scattering vector of magnitude $q=(4/\lambda)\sin(\theta/2)$. The two scattered rays have a phase difference of $\Delta\varphi_{jk}=qr_{jk}=qr_{jk}\cos(\angle q, r_{jk})$ where $\angle \boldsymbol{qr}_{jk}$ indicates the angle between the scattering vector $\boldsymbol{q}$ and the distance vector $\boldsymbol{r}_{jk}$.

which is the angle between the forward direction of the primary light and the position of the detector. From Figure 6.3 it becomes clear that in the forward direction all scattered rays remain in phase. At higher scattering angles and at larger distances r_{jk} between two scattering centres the phase difference increases with increasing scattering angles.

The effect of interference can be described mathematically. This may be demonstrated with the example of the two scattering units as shown in Figure 6.4a and b. In this example four cases have to be considered: (1) the interference of the light i_{12} coming from unit 1 and unit 2; (2) conversely the i_{21} of the wave from unit 2 interfering with that from unit 1; (3) in addition the two units 1 and 2 emit light but without interference since the two units are taken as dimensionless scattering points, $S_1(0)=1$ and $S_2(0)=1$, but (4) for the two other situations where $qr>0$ the interfering waves are $S_{12}(qr_{12})=S_{21}(\boldsymbol{qr}_{21})=\exp(iqr_{12}\cos(\psi))$ if we assume fixed positions where ψ is the angle between the vectors $\boldsymbol{q}$ and $\boldsymbol{r_{jk}}$. In solution the thermal motion causes a random orientation of the individual molecules in the ensemble, thus only the average over all orientations can be measured. The integration of $S_{jk}(\boldsymbol{qr}_{jk})=\exp(iqr_{12}\cos(\psi))$ over all angles ψ is

$$\langle S_{jk}(\mathrm{q}r_{jk})\rangle = \frac{\sin(qr_{jk})}{qr_{jk}} \tag{6.7}$$

Therefore, the scattering intensity arising from the two selected scattering centres is

$$\frac{i_{jk}(q)}{I_0} = 2\left(1+\frac{\sin(qr_{jk})}{qr_{jk}}\right) \tag{6.8}$$

Note, the product of $qr=x$ has no dimensions since $q\propto 1/r$. The dependence of the value in the brackets of eqn (6.8) is shown as the solid line in Figure 6.4a. The curve fluctuates around a value of zero and finally approaches zero. Such periodic and damped fluctuations are typical for uniform structures which in the simple present example correspond to a

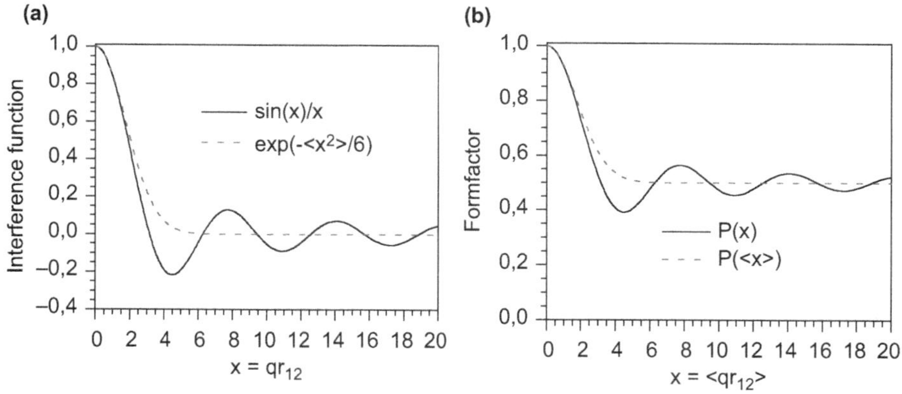

Figure 6.4 (a) The effect of interference of the rays coming from two scattering units in the distance $r_{12} = |r_1 - r_2|$. The solid curve refers to a fixed distance, while the dashed curve displays the behaviour of fluctuating distances with an average length $\sqrt{l^2}$. (b) Angular dependence of the corresponding *formfactor* $P(q)$. The solid curve corresponds to a fixed distance $l = r_{ij}$ where the dashed curve fluctuates randomly around an average value > 0. Note the difference to Figure 6.4a where the fluctuations are around the zero value.

fixed length l. However, due to Brownian motions the two selected points in Figure 6.3 fluctuate in their distance such that an average length is expected which corresponds to the mean square fluctuations. This average can be calculated with a Gaussian fluctuation distribution and is

$$\left\langle \frac{\sin(qr_{jk})}{qr_{jk}} \right\rangle = \int_0^\infty W(r_{jk}) \frac{\sin(qr_{ij})}{qr_{ij}} \mathrm{d}r_{jk} = \exp\left(- \left\langle l_{jk}^2 \right\rangle q^2/6 \right) \tag{6.9}$$

where $\sqrt{\langle l_{jk}^2 \rangle}$ is the root mean square distance between the two scattering units j and k.

This function in eqn (6.9) decreases *smoothly* and at large $q\sqrt{\langle l_{jk}^2 \rangle}$ approaches zero (dashed curve in Figure 6.4b). If the sample consists only of these two scattering units, the total intensity in eqn (6.7) would reach a value of 4, because $\sin(x)/x = 1$ at $x = 0$. It is convenient to normalize this intensity to unity at $q = 0$ and obtains

$$P(qr_{jk}) = \frac{1}{4} \frac{i_{jk}(q)}{I_0} \tag{6.10a}$$

The angular, or better, the q-dependence of the intensities from the two interfering particles has four components, two from the point-like units which show no angular dependence and two other ones which arises from the distance between the two points, with the angular dependence as shown in Figure 6.4a. Thus at $q = 0$ (*i.e.* forward scattering at $\theta = 0$) the sum is 4.

It is convenient to normalize the scattering intensity to unity at $q=0$ which leads to the function

$$P_2(q)=\frac{1}{4}\frac{\sum_j i_{jk}(q)}{I_0}=\frac{1}{4}\left(2+2\left\langle\frac{\sin(qr_{12})}{qr_{12}}\right\rangle\right)=\frac{1}{2}\left(1+\exp(-q^2l^2\ |j-k|)\right) \tag{6.10b}$$

Figure 6.4b shows the corresponding q-dependence. At a first sight one might have expected a decrease of curves in Figure 6.4b towards zero. Instead the asymptotes approach a constant value. This constant plateau results from the two point-like scattering units, which show no angular dependence. The point-like scattering units are mostly assumed to have no detectable structure compared to the distance between the two scattering units j and k. In the following the monomer repeat units will always be considered as the individual scattering units in chain-like structures.

Actually point-like structures do not exist; the monomer, of course, has a finite dimension but compared to the comparably long distances in large macromolecules the dimensions of the repeat units becomes negligibly small, still the effect of the monomer size has to be taken into account if distances are probed comparable to the size of the monomers. This situation appears sometimes in small angle neutron scattering (SANS) measurements and requires a special consideration.

If the particle contains n alike repeat units one has to consider N^2 pairs, which enclose N elements $j=k$, considered as point scatterers, and $N(N-1)$ pairs with $|j-k|=n>0$. If we assume that the different monomer units are linked randomly to each other the Gaussian statistics can be applied which leads to the equation

$$P(q)=\frac{1}{N^2}\sum_{j=1}^{N}\sum_{k=1}^{N}\left\langle\frac{\sin(qr_{jk})}{qr_{jk}}\right\rangle=\frac{1}{N^2}\sum_{j=1}^{N}\sum_{k=1}^{N}\exp(-q^2l^2n) \tag{6.11}$$

The function $P(q)$ is denominated as the particle scattering factor, sometimes also called the form factor. $P(q)$ displays structurally characteristic features. To gain an impression one has to solve the double sum in eqn (6.11) for a few instructive examples. These are (1) spheres of uniform density,[17] (2) random coils of linear chains,[20,22,23] and (3) rigid rods of very thin diameter compared to the length of the rigid thread. Clearly the sphere is a three-dimensional object with $R^3 \propto M$, and the rigid rod can be considered as a one-dimensional object with $R \propto N$. Between these two examples the random coil has an apparent dimension of 2. This was shown in 1936 by Werner Kuhn[24,25] who, for the first time, studied the structure of flexible chains of N independently linked units and found that the mean square end-to-end distance $\langle R_N^2\rangle$ is proportional to the molar mass of the chain.

Of course, the structure of a thin disc[26] is a real two-dimensional object. The choice of a random coil is of particular interest, also for hyperbranched samples because this dimension is defined by a statistical model of unperturbed chains with a 'fractal' dimension of 2. Actually the four mentioned particle scattering factors are:[17,24,25]

- For a sphere[17]

$$P(q) = \left[\frac{3}{X^3}(\sin X - X\cos X)\right]^2 \qquad X = qR \tag{6.12}$$

  where R is the radius of sphere
- For a random coil[20,22,23]

$$P(q) = \frac{2}{X^2}[X - 1 + \exp(-X)] \qquad X = q(\langle L^2\rangle 6) \tag{6.13}$$

  where L is the chain length
- For a disc[26]

$$P(q) = \frac{2}{X^2}\left[1 - \frac{J_1(2X)}{X}\right] \qquad X = qR, \tag{6.14}$$

  where R is the radius of disc, and $J_1(X)$ is a Bessel function of the first kind
- For a thin rod[26]

$$P(q) = \frac{2}{X^2}\int_0^X \frac{\sin t}{t}\mathrm{d}t - \left(\frac{\sin X/2}{X/2}\right)^2 \qquad X = qL, \tag{6.15}$$

  where L is the rod length.

The corresponding curves are shown in Figure 6.5 in terms of a parameter $q^2\langle R_g^2\rangle$ where $\langle R_g^2\rangle$ is the mean square radius which is defined below.

The graphs in Figure 6.5 give a first impression of what type of structure was obtained in an experiment. It is advisable to confirm this impression by other plot types. Most important is proved to be the Kratky plot[26] which originally was suggested for flexible linear chains. Here the particle scattering factor is multiplied by the square $(qR_g)^2$ and $(qR_g)^2P(q)$ is plotted against qR_g.

Figure 6.6 shows the same structures in the Kratky plot version. This plot represents a marked amplification by $(qR_g)^2$ to the asymptotic part at large q-values, which is the structure characterizing part of the particles scattering factor. This fact is clearly demonstrated with the special feature of the flat disc compared to that of the random coil of flexible chains. Another choice, often used in SANS, is a log–log plot representation as shown in Figure 6.7.

The plots of Figures 6.5 to 6.7 present a framework according to which the experimentalist can make a first conjecture to which type of structure the

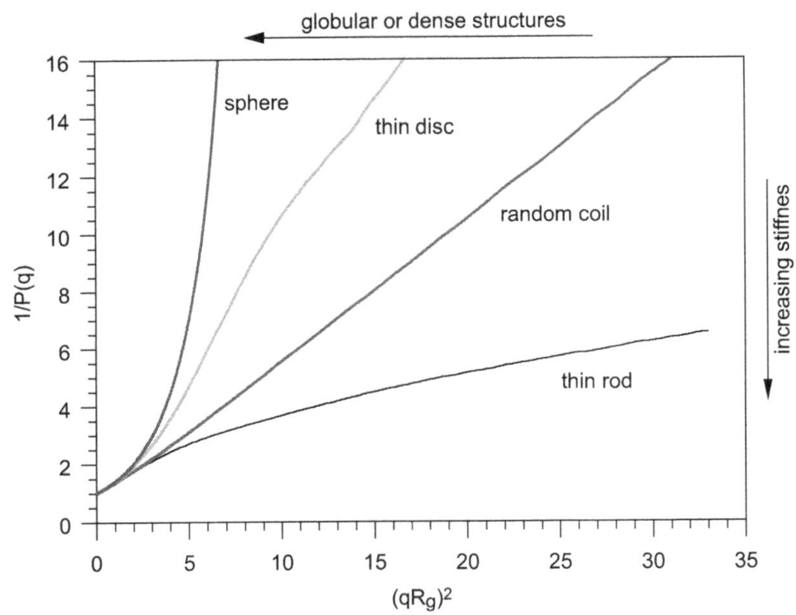

Figure 6.5 The dependence of $(qR_g)^2$ of the inverse particle scattering factors for four characteristic topologies.

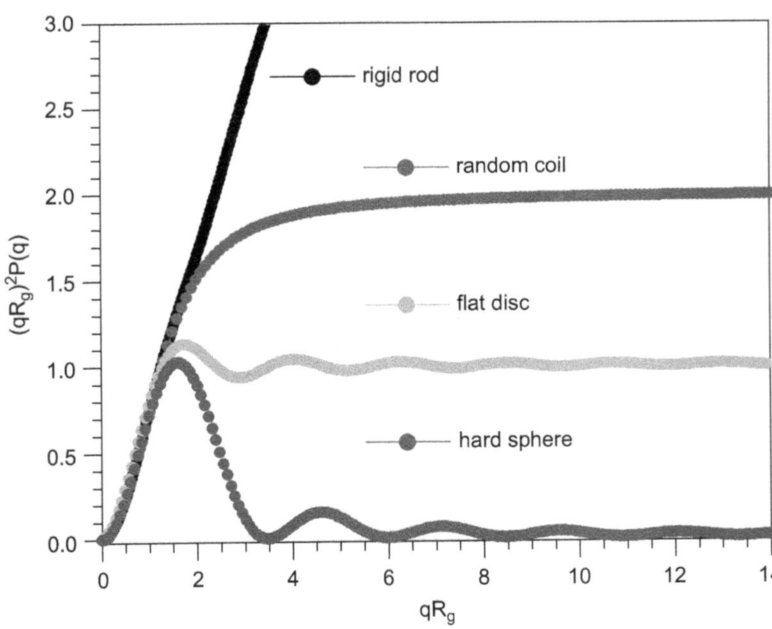

Figure 6.6 Kratky presentation of the same structures as shown in Figure 6.5.

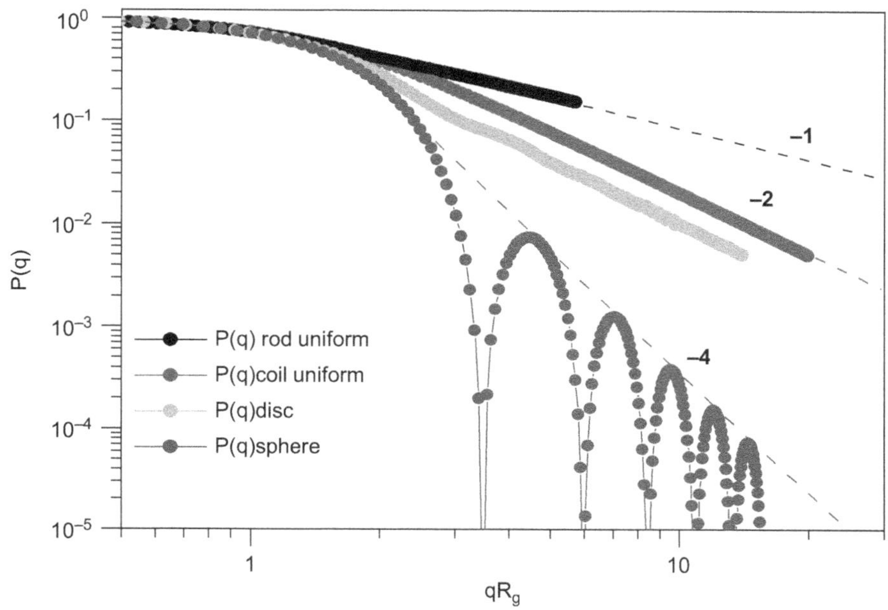

Figure 6.7 Log–log plot of the same $P(q)$ data as shown in Figures 6.5 and 6.6.

measured light scattering from an unknown material can be assigned. After this estimation a special model calculation can be made with additional parameters obtained by other techniques, *e.g.* dynamic light scattering, intrinsic viscosity, the molar mass M_w, radius of gyration R_g and the second virial coefficient A_2. Also, as many data as possible, known from the chemical synthesis, should be incorporated before carrying out a specified calculation.

6.3.4 Molar Mass and Radius of Gyration

As already mentioned, for convenience, the light scattering measurements are made with vertically polarized laser light since then the vibrating dipoles exhibit no angular dependence due to polarization [see eqn (6.6)] and the observed angular dependence arises solely from the dissolved particles. If their dimensions are larger than $\lambda/20$ ($\lambda = \lambda_0/n_0$ with n_0 the solvent refractive index) an angular or q-dependence is observed. The total angular dependent scattering intensity, expressed in terms of Rayleigh ratios, is the product of eqn (6.3) by eqn (6.11):

$$R_\theta = K\frac{cMP(q)}{1 + 2A_2Mc + 3A_3Mc^2 + \cdots} \tag{6.16a}$$

or in the limit of $c \to 0$:

$$\frac{R_\theta}{Kc}\bigg|_{c\to 0} \to MP(q) \tag{6.16b}$$

The Rayleigh ratio depends on the concentration and for large particles also on the scattering angle, *i.e.* on the value of the scattering vector q. To find the molar mass a twofold extrapolation of the Rayleigh ratio has to be made, one for $c \to 0$ and the other for $q \to 0$ where $P(q) \to 1$. For sufficiently low concentrations the influence of the third virial coefficient can be neglected but at higher concentrations a linear extrapolation towards $c = 0$ would be desirable, but this cannot be done with eqn (6.16). Debye suggested using the reciprocal version of eqn (6.16) which, after minor rearrangement, is

$$\frac{Kc}{R_\theta} = \frac{1}{MP(q)} + 2A_2c + 3A_3c^2 + \cdots \tag{6.16c}$$

with

$$K = \frac{4\pi^2}{\lambda_0^4 N_A}\left(n_0 \frac{\partial n}{\partial c}\right)^2$$

[which is eqn (6.4)], and a linear extrapolation to $c = 0$ can be done for solutions where the contribution of a third virial coefficient can be neglected. Another point is the extrapolation of the angular dependence towards $q = 0$. Again it would be good to apply a linear extrapolation. Such behaviour is found if the double sum in eqn (6.11) is expanded in terms of q^2. The first two terms in the expansion of $\sin(x)$ is $\dfrac{\sin(qr_{jk})}{qr_{jk}} = 1 - (qr_{jk})^2/6 + \cdots$, and with eqn (6.11) one obtains

$$\frac{1}{P(q)} = \frac{1}{1 - \frac{1}{6}\frac{1}{N^2}\sum_{j=1}^{N}\sum_{k=1}^{N}\left\langle r_{jk}^2\right\rangle q^2 + \cdots} = \frac{1}{1 - \frac{1}{3}\langle R_g^2\rangle q^2 + \cdots} \tag{6.16d}$$

where the double sum is the mean square radius of gyration $\langle R_g^2\rangle = \dfrac{1}{2N^2}\sum_j^N\sum_k^N \langle r_{jk}^2\rangle$, which still appears on the denominator, but for small values of q^2 the ratio in eqn (6.16d) can again be expanded which leads to the final result of

$$\frac{Kc}{R_\theta} = \frac{1 + \frac{1}{3}\langle R_g^2\rangle q^2}{M_w} + 2A_2c + 3A_3c^2 + \cdots$$

$$q = (4\pi/\lambda)\sin(\theta/2) \tag{6.17}$$

This is Zimm's equation.[20] The more general equation will be discussed somewhat later in this chapter. Zimm suggested a special plot for the double extrapolation of $c \to 0$ and $q^2 \to 0$. Such a typical Zimm plot is shown in Figure 6.8.

There Kc/R_θ is plotted as a function of $q^2 + kc_1$, $q^2 + kc_2$, $q^2 + kc_3$ *etc.* where c_1, c_2, c_3 *etc.* denote the concentration at which the Rayleigh ratios were

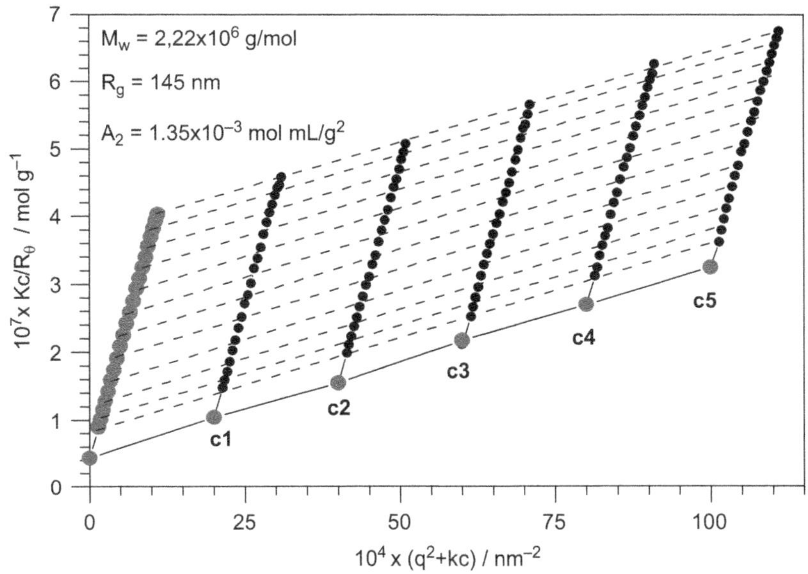

Figure 6.8 Zimm plot from Linters (cellulose from cotton) in aqueous metal complexing solvent.[27,28] The extrapolated q^2 curve at $c = 0$ is the curve closest to the ordinate, and the extrapolated data towards $q^2 = 0$ for the various concentrations are marked by the points closest to the labelling c1 to c5.

measured as a function of q^2. The constant k can be chosen *at libitum*, but this should be done such that a collapse of the various curve onto each other is avoided. The graph in Figure 6.8 demonstrates the required extrapolation to $c = 0$ and $q^2 = 0$. The Zimm plot has the advantage that it simultaneously shows the effect of the second virial coefficient and the effect of the interference expressed in terms of $(qR_g)^2$. Before computers became available the construction of a Zimm plot was a demanding task and had to be done manually. Even with help of modern computers the set-up of a Zimm plot requires some experience and includes some somewhat tedious work. On a computer the extrapolation towards $q^2 = 0$ can be done for each concentration separately with high precision. The slope may be denoted as $1/3R^2_{g,app}(c)$ and the ordinate as $1/M_{app}(c)$. These data then can be extrapolated separately against $q^2 = 0$, and the extrapolation of these data of $1/M_{app}(c)$ and $1/3R^2_{g,app}(c)$ towards $c = 0$ then gives the desired values for the molar mass M_w and the mean square radius of gyration $\langle R^2_g \rangle$. With these data the construction of the Zimm plot causes no problem.

The radius of gyration is the most important global parameter in the field of polymers. It can be determined from the initial slope of the particle scattering factor in a plot against q^2, and this initial slope can be determined without knowing anything about the structure. Furthermore, theory on the particle scattering factors demonstrated that the whole q-dependence can uniquely be described in terms of this radius of gyration.

6.3.5 Summary: Static Light Scattering

The scattering of light is a well-known phenomenon in daily life. Examples are: the tiny water droplets in mist, smog in big towns due to air pollution, the twinkling of dust particles in a ray of sunshine in a darkened room, or the trail of a visible light beam passing through an almost clear colloidal dispersion. The physics behind these phenomena is understood as an electro-magnetic phenomenon of light. Seven important features may be emphasized:

1. The scattering intensity arises from concentration and density fluctuations.
2. The amplitudes of these fluctuations are controlled by the osmotic and isothermal compressibility.
3. The scattering intensity displays an angular dependence if the particles have dimensions larger than 1/20 of the used light wavelength. This angular dependence originates from interference effects among scattered rays from different parts of the particles. The scattering intensity has its maximum at zero angle, *i.e.* in direction of the primary beam and decreases at larger angles.
4. The strength and form of the angular distribution depends on the size and shape of the dispersed particles, and also on the internal structure, *i.e.* the segmental arrangements.
5. Three essential parameters can be extracted from measurements without knowing details of the structures. These are (1) the molar mass M_w, (2) the radius of gyration R_g and (3) the second osmotic virial coefficient A_2, which gives a first impression on the strength of interactions among the dissolved particles.
6. The radius of gyration is an essential parameter for a quantitative interpretation of the q-dependent scattering curves and is a valuable scaling parameter for the whole angular dependence.
7. The second viral coefficient plays a similar role as a scaling parameter to describe the concentration dependence and of the interaction among the particles, even at moderate concentrations.

6.4 Global Parameters

6.4.1 Mean Square Radius of Gyration

The mean square radius of gyration is defined as the mean square distance of the various scattering elements from centre of mass or gravity. For compact particles, hard spheres, rigid cylinders, discs and other compact particles the calculation is easily performed. For flexible structures and randomly arranged repeat units the derivation is not a straight forward issue because in general the centre of mass is not positioned on a repeat unit. However, the mean square radius of gyration can be derived from the

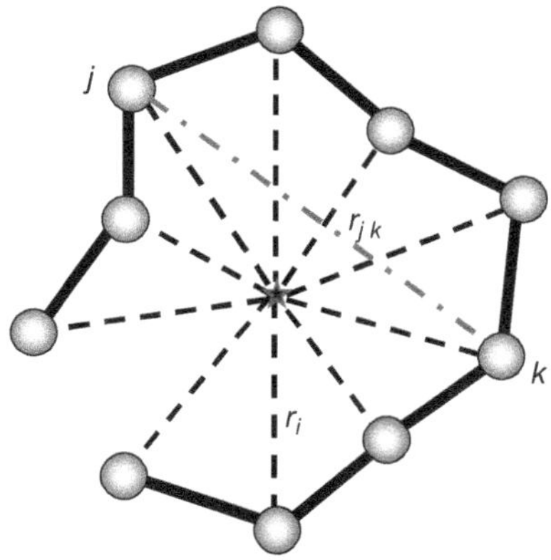

Figure 6.9 Scheme for the derivation of the mean square radius of gyration. The centre of mass is marked with a 'star' at the centre of the figure, and is not positioned on any of the repeat units. The dashed lines indicate the distances from the centre of mass to the various units of the chain, and the dot-dashed line denotes the distances between two repeat units. The mean square radius then is defined by the eqn (6.18) and (6.19).

positions of the repeat units without knowing the position of the centre of mass. Figure 6.9 elucidates the situation.

$$\langle R_g^2 \rangle = \frac{1}{N} \sum_{j=1}^{N} \langle r_j^2 \rangle \tag{6.18}$$

If the position of the centre of mass is known, this equation is equivalent to

$$\langle R_g^2 \rangle = \frac{1}{2N^2} \sum_{j}^{N} \sum_{k}^{N} \langle r_{jk}^2 \rangle \tag{6.19}$$

as was shown by Zimm and Stockmayer[29] and results from the fact that the position of the centre of mass can be eliminated by using the condition $\sum_{j}^{n} \langle \boldsymbol{r}_j \rangle = 0$ where the sum runs over all orientations of the vectors $\boldsymbol{r}_j$. Eqn (6.18) is of special value since the mean square radius of gyration can be derived from the positions of the beads and it does not need the position of the centre of mass. This definition is the mean square radius of gyration and is universally valid. Problems arise from the distance distribution between two endpoints of a segment, needed for the derivation of the mean squared distance of segmental end points. The required distribution is known only for a few examples. One is the Gaussian distribution for linear chains or segmental chain sections which is valid if a random arrangement of the units is assumed. It is given by the following equation:

$$W(r_{jk}) = \left(\frac{3}{2\pi\, r_{jk}^2} \right)^{1/2} \exp\left(\frac{3\, r_{jk}^2}{2 \langle r_{jk}^2 \rangle} \right) \tag{6.20a}$$

The average squared distance of the segment endpoints is obtained from the integral

$$\langle r_{jk}^2 \rangle = \int_0^\infty W(r_{jk}^2) 4\pi r_{jk}^2 \mathrm{d}r_{jk} = \lfloor j - k \rfloor b^2 \tag{6.20b}$$

Inserting the equation into eqn (6.18) gives for a linear chain a mean square radius of gyration of

$$\langle R_g^2 \rangle = Nb^2/6 \tag{6.21}$$

Eqn (6.20a) and (6.21) define the *unperturbed* dimensions of a chain. Unperturbed means that the scattering points have no dimension, and there will be no perturbation in the distribution if two units come together, because they were assumed to have no volume. Such assumption cannot be valid for a real system. If we admit that the individual repeat units have a finite impenetrable volume then there must exist a space around such bead which will not be accessible for another bead in contact. Thus, there should be an effect of the excluded volume on the chain dimensions. Figure 6.10 elucidates what is meant with this notation. The effect of the finite volume is a swelling of the coil and an expansion of the two chain ends. This leads to an increase of the radius of gyration. The actual effect is not as dramatic as might be expected because even in a good solvent some attractive interaction remains effective which reduces the effect of hard sphere volume exclusion. On the other hand the effect should become more pronounced if *via* branching the segment density is enlarged.

The theory of the excluded volume effect is complex and cannot accurately be expressed by an analytical function.[30–32] For linear chains Flory and Fox[33–35] developed on the basis of two assumptions a simple analytical

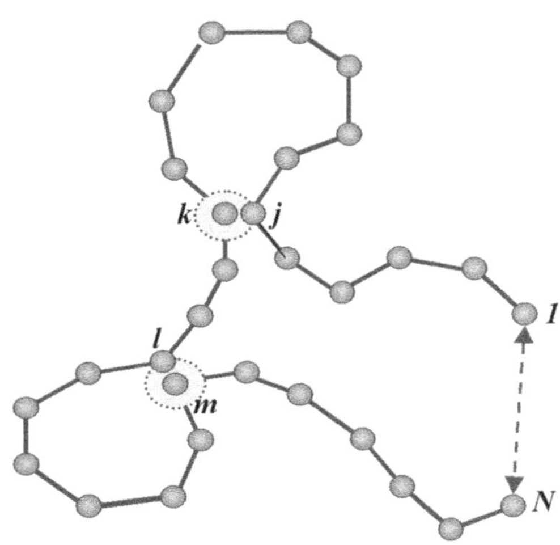

Figure 6.10 A flexible chain can coil back and a repeat unit will come into contact with another bead of the chain (see pairs *j*,*k* and *l*,*m*). For hard-sphere repeat units a volume eight times the bead volume will be excluded.

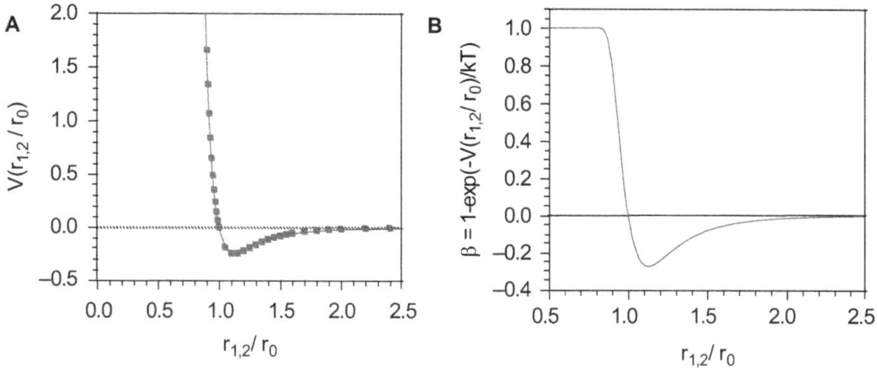

Figure 6.11 A: Lennard-Johns potential of the distance between two beads in a chain. B: the corresponding effect of the excluded volume.

equation as an approximation which turned out to be very useful after adjustment of the numerical parameters:

$$\frac{R_g^2}{R_{g,0}^2} \equiv \alpha^2; \quad \alpha^5 - \alpha^3 = C\beta M^{1/2} \quad \text{or} \quad R_g^2 \rightarrow N^{1+\varepsilon} \cong N^{0.6} \tag{6.22}$$

More precisely, the excluded volume is defined by the integral over an integrand shown in Figure 6.11B, where the excluded volume β of two interacting beads is given by the integral

$$\beta = \int_0^\infty [1 - \exp(-V(r)/kT)]4\pi r^2 \mathrm{d}r \tag{6.23}$$

which has the dimensions of a volume.

Evidently, positive and negative sections contribute to this integral and make the excluded volume smaller than that of a hard sphere which contains no attraction. The excluded volume parameter keeps a significant role in almost all global parameters, *e.g.* radius of gyration, second virial coefficient and intrinsic viscosity. The effect of the excluded volume on these global parameters is difficult to handle mathematically, because the chain statistics becomes a non-Markovian process and arises from long-range interactions. A *nearest* neighbour correlation exists, if a preferred *trans*-configuration is realized.[36] Due to the volume exclusion interaction the perturbation of the chain configuration can occur with any two repeat units if a contact can be realized. The demanding question to be solved is the derivation of the probability distribution for the contact between two repeat units. The problem is quite complex as on each contact a change of the whole chain conformation is induced. For linear chains a reasonable approximate solution of this problem has been possible,[37] but not to this agreeable extent for branched structures. We return to this issue in Section 7.4.

6.4.2 Fractal Behaviour

Experimentalists are often satisfied with approximations which are not yet well confirmed by theory, but it should be good enough to draw sensible conclusions from the fit of experimental data. A first step is the choice of a suitable plot. The molar mass M_w and the corresponding radius of gyration R_g were often measured over two or more decades, and soon a plot of $\log(R_g)$ against $\log(M_w)$ was found to give a suitable presentation. Neglecting the regime of oligomers a linear line could be drawn through the experimental data which can be expressed by a power law equation

$$R_g = KM_w^{\nu} \tag{5.12}$$

with an exponent of $0.5 \leq \nu \leq 1.0$. The exponent varied for different types of chains and with the solution quality of the solvent. The observed exponents were taken as a measure for a good or poor solvent. For flexible linear chains in a θ-solvent (*i.e.* at $A_2 = 0$) an exponent $\nu = 0.5$ was found in agreement with the prediction for unperturbed chains and $\nu \approx 0.6$ for a good solvent.[33–35] For branched chains the assignment to a dissolution quality of the solvent is not yet fully clarified because of the observed behaviour also depends on the extent of branching. An essential step forward was achieved by a minor change for a sensible plot which led to an outstanding new view for a suitable interpretation of structural topologies. Eqn (5.12) could be also written in the inverse form like

$$M_w \propto R_g^{1/\nu} \tag{6.24}$$

which permits a somewhat surprising interpretation: the value $1/\nu = d$ looks like a dimension. This interpretation becomes clear when a similar notation is used for Euclidean bodies like hard spheres, thin discs and thin rods, which are given in the eqn (6.25a–e) which also includes the results for flexible linear chains in the unperturbed and perturbed states:

$$\begin{array}{ll} M \propto R^3 & \text{hard spheres} \\ M \propto R^2 & \text{thin discs} \\ M \propto R_g^2 & \text{unperturbed random coils} \\ M \propto R_g^{1.7} & \text{coils perturbed by excluded volume} \\ M \propto R_g^{1.0} & \text{thin rods} \end{array} \tag{6.25a–e}$$

The examples a, b and e correspond to well-defined Euclidean structures but the examples c and d for random coils are statistically defined. They can be described by a *fractal* dimension and similar behaviour was found for more or less statistically defined *branched* samples with fractal dimensions between $d_f = 2$, well swollen and $d_f = 3$, collapsed structures. It has to be emphasized that the description of the structures by a fractal dimension is a reliable approximation for large objects but for oligomers or short segments significant deviations become noticeable. This is especially true for the mean square radius of gyration which for *small* objects can be described by Flory and Fox's expansion factor[33–35] $\langle R_g^2\rangle / \langle R_{g0}^2\rangle = \alpha^2$ and the equation

$$\alpha^2 = 1 + k\beta N^{1/2}$$

$$\text{or:} \tag{6.26}$$

$$\langle R_g^2\rangle = \langle R_{g0}^2\rangle (1 + K(\beta) N^{1/2})$$

Similar deviations from the fractal conception are obtained with the shape of the particle scattering factor function at small q^2 values, a regime which is universally defined by the mean square radius of gyration. The effect of excluded volume on the shape of the scattering functions will be discussed later separately. Instead we will continue with another parameter that depends on the global (*i.e.* not internal) structure.

6.4.3 Concentration Dependence

In Zimm plots for the $R_\theta(q,c)$ scattering data extrapolations are made for $q^2 \to 0$ with each concentration of the solution. The data show often a distinct concentration dependence which in addition are extrapolated towards $c = 0$ to obtain the reciprocal molar mass $1/M_w$. The question remains, What does the value of $R_{\theta=0}(c)$ mean? To answer the question one has to go back to eqn (6.2) and (6.3), which show that the scattering intensity is dominated by concentration fluctuations which in turn can be expressed by a change of the chemical potential $\Delta\mu$ with the concentration $\partial\Delta\mu/\partial c = \partial^2\Delta G/(\partial c)^2$, but $\Delta\mu$ is proportional to the osmotic pressure which can be expressed in a virial expansion

$$\frac{\Pi}{RT} = \frac{1}{M_n} + A_2 c + A_3 c^2 + \cdots \tag{6.27a}$$

According eqn (6.3) the derivative of eqn (6.27a) is needed, which leads to

$$\frac{\partial\Pi}{RT\,\partial c} = \frac{1}{M_w} + 2A_2 c + 3A_3 c^2 + \cdots = \frac{1}{M_w}\left(1 + 2A_2 M_w c + 3A_3 M_w c^2 + \cdots\right) \tag{6.27b}$$

Note that in the osmotic modulus the *weight* average of the molar mass contributes, whereas for the osmotic pressure the *number* average is effective. For polydisperse linear chains the width of the molar mass distribution is with $M_w/M_n \leq 2$ moderately low, but the distribution of branched samples polymerized in batch reaction can become very broad ($M_w/M_n \propto M_w \to \infty$

when the point of gelation is approached) and for randomly branched samples this has a significant influence on the measured quantities, *e.g.* on the radius of gyration and the shape of the scattering curve.

The virial coefficients A_2, A_3, *etc.* describe the effect of the interaction between two or three *etc.* macromolecules with increasing concentration but at higher concentrations the interaction no longer can be assigned solely to the number of involved particles in the interaction, and therefore it is advisable to keep the thermodynamic notation. To proceed further, two new notations may be introduced:

$$\frac{1}{M_w} + 2A_2c + 3A_3c^2 + \cdots \equiv \frac{1}{M_{\mathrm{app}}(c)} \tag{6.28}$$

and

$$\frac{M_w \partial \Pi}{RT\, \partial c} = \left(1 + 2A_2M_wc + 3A_3M_wc^2 + \cdots\right) = M_w/M_{\mathrm{app}}(c) \tag{6.29}$$

i.e. the complex infinite sum can be described in a compact form of eqn (6.29). The advantage of this new notation is

$$\frac{R_{\theta=0}(c)}{Kc} = \frac{1}{M_{\mathrm{app}}(c)} = \frac{1}{M_w}\left(1 + 2A_2M_wc + 3A_3M_wc^2 + \cdots\right) \tag{6.30}$$

and the whole inter-particle interaction now is defined *via* the measureable $M_w/M_{\mathrm{app}}(c)$. The latter has no dimensions. Interestingly also the concentration dependence can be expressed by one dimensionless quantity. This notation goes back to De Gennes,[37] who presumed that the effect of the higher virial coefficients can be expressed in terms of the second virial coefficients. He denoted

$$A_2M_wc = c/c^* \quad \text{and} \quad c^* = 1/A_2M_w \tag{6.31}$$

which can be used to define an overlap concentration. The overlap concentration is arrived at when the weighed-in concentration just equals the average concentration in a particle. This notation is based on the structure property of the second virial coefficient. With this notation of c/c^* and eqn (6.28) one obtains[37]

$$\frac{M_w}{M_{\mathrm{app}}(c)} = 1 - 2(c/c^*) + 3g_2(c/c^*)^2 + 3g_3(c/c^*)^3 + \cdots \rightarrow \mathrm{const}(c/c^*)^m \tag{6.32}$$

$$\text{with} \quad m = 1/(3\nu - 1) = d_f/(3 - d_f)$$

The arrow means asymptote at $c \gg c^*$. De Gennes derived this scaling function for linear chains and expected universal behaviour for any particle structure. From the fact that strongly beyond the overlap concentration no

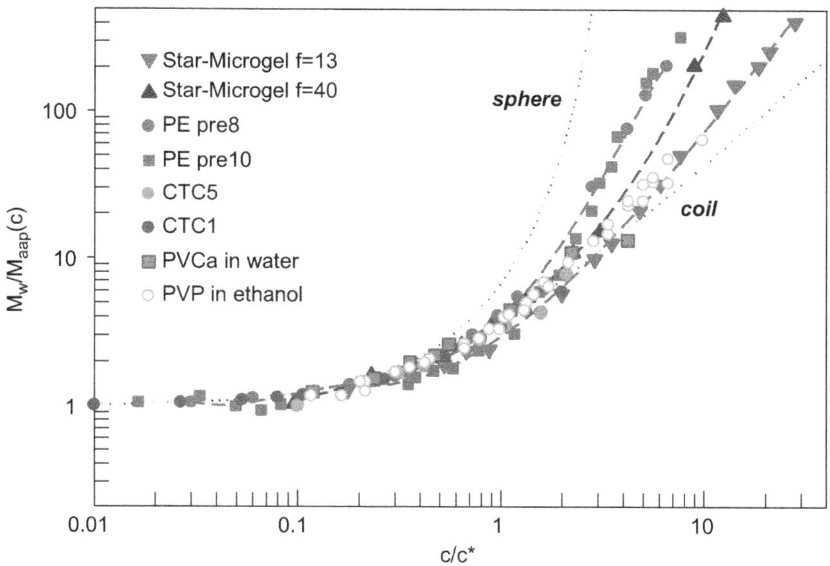

Figure 6.12 Concentration dependence of *inter*molecular interactions demonstrated by plotting $M_w/M_{app}(c)$ as a function of $c/c^* = A_2M_wc$ for some selected examples.

real estimation on the original particle structure can be made he came to this conclusion of a general universality. This appears conclusive if the none scaled concentration is used, but with the parameter $X = c/c^*$ a structure characteristic scaling function clearly distinguishable curves are obtained, which now give us a valuable information on the structure of the dissolved particle. Figure 6.12 shows the results from macromolecules of different architecture or topology,[38] and these are compared with already existing theoretical predictions.[39–41]

Up to close the overlap concentration no significant differences between the various structures can be detected. At higher concentration a clear structure dependence is noticed. This dependence is not detectable only if the non-scaled concentration is used.

6.4.4 Some Properties of the Second Virial Coefficient

Even in dilute solution the effect of the second virial coefficient can be observed with macromolecules. The reason for this effect is recognised from eqn (6.27). The intercept decreases with increasing the molar mass and makes the influence of A_2 increasingly better noticeable despite the fact that A_2 also decreases with increasing M_w. Figure 6.13 shows for some examples a plot of A_2 as a function of molar mass in a log–log presentation.

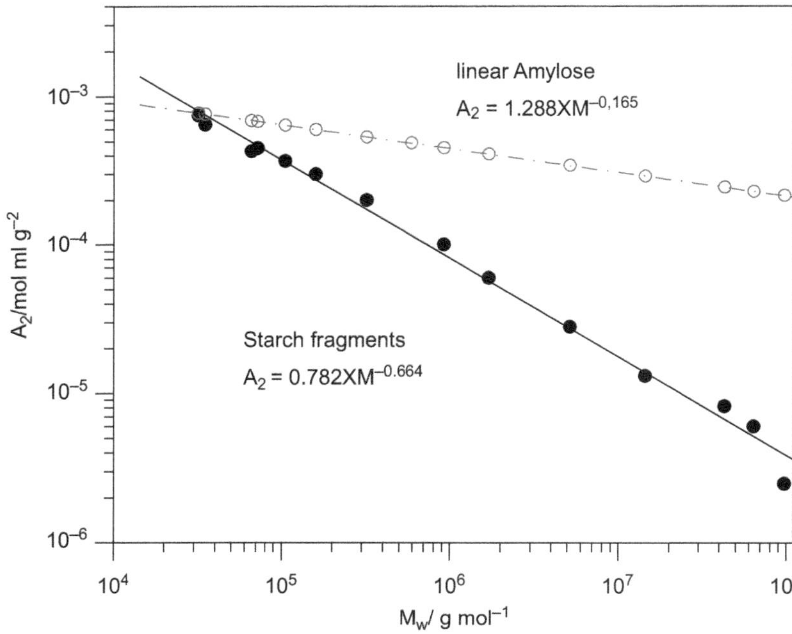

Figure 6.13 Molar mass dependence of the second viral coefficient for linear chains and branched macromolecules.[42]

The curves approach for high molar mass a straight line, and give rise to the suspect that the observed power law is related to fractal behaviour that was already demonstrated with the radius of gyration.

The correlation of A_2 with the volume of a particle or macromolecule was noticed already by van der Waals who used the conception of volume exclusion, now applied to two spheres in contact. He found the now well known relationship

$$A_2 \propto 4 \times \frac{4\pi}{3} \frac{R^3}{M_w^2} \tag{6.33}$$

If the radius of a hard sphere is replaced by the radius of gyration and the exponent ν, derived for R_g, is used the correlation between exponent ν and a_{A2} is

$$a_{A_2} = 3\nu - 2 \tag{6.34}$$

This relationship makes clear that a correlation exists between the various exponents. This correlation has been given the name scaling relationship.

6.4.5 Self-similarity

Systems developing fractal behaviour are mostly considered as self-similar. This impression comes to mind when envisaging electron micrographs from random networks at different magnification where it may be difficult to detect differences in the pictures at various magnifications. In other words the same impression is mediated when the structure is expressed in another scale. One can construct geometric self-similar structures. An example of geometric self-similarity is shown in Figure 6.14.[43]

Another example from nature, is the rosette-like aggregation of liver glycogen.[44] Here the self-similarity is based on spherical sub-structures in an associate from several other particles arranged around it followed by a further assembling to rosette-like structures (Figure 6.15).

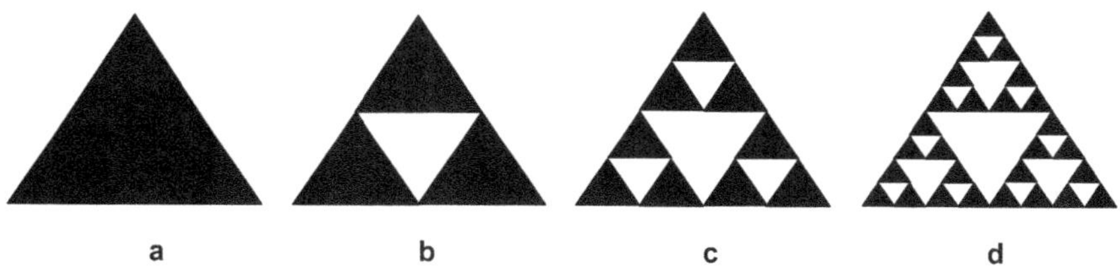

Figure 6.14 An example a self-similar planar geometric structures. The four objects appear to arise from the same triangular topology as repeat units at different magnification.

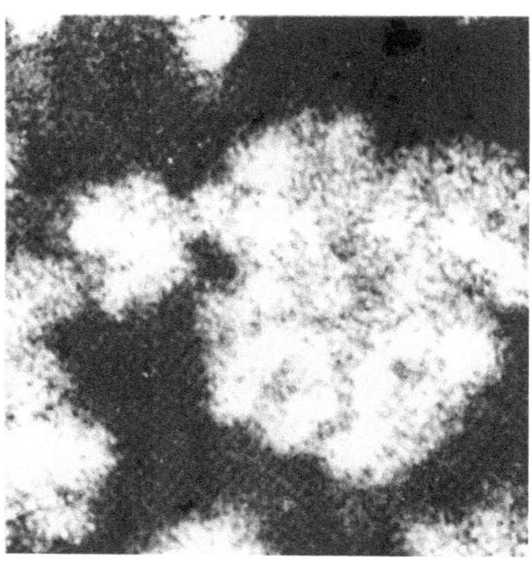

Figure 6.15 A negatively stained electron micrograph of glycogen from rat liver. The small particles at the top correspond to the size of the molecularly highly branched glycogen and are the sub-units of the aggregates of different size but still of the same rosette-like arrangement. The sub-units have a molar mass of $M_w \approx 1.6 \times 10^6$ g mol^{-1} and a radius of gyration $R_g = 2.08$ nm but that of the aggregates is about 312×10^6 g mol^{-1} and a radius of gyration $R_{g\ roset} = 92$ nm.[45]

6.5 Dynamic Light Scattering

6.5.1 Introduction

Dynamic light scattering is at present the most frequently used characterization technique to provide information on the size of macromolecules or supra-molecular structures and compact colloidal particles. With the instruments available on the market measurements are quickly performed with agreeable reproducibility. The instruments are equipped with powerful software such that only a few manipulations are needed to obtain the desired results. The easy handling of the instruments gives a safe feeling of reliability. However, the ease and high reproducibility of measurements is deceptive and can become very misleading if one relies too much on computers and is not sceptical with regard to physical expectations. Mostly, the software of the instruments is based on two significant conditions:

1. The particles or molecules have to be small compared to the wavelength of the light
2. The solution or suspension has to be dilute such that inter-particle interactions can be disregarded

Within these limits a quick result for the diffusion coefficient D is obtained and with the Stokes–Einstein relationship the hydrodynamic radius R_h which gives a valuable impression on the size of the particle. It does not mean that the particle is a hard sphere or even has a spherical shape, and therefore comparison with molecular radii obtained from other techniques gives additional and more detailed information on the shape and conformational properties of the sample under investigation. There are a few highly developed instruments available which allows a check for these conditions. If the dynamic light scattering is recorded in a narrow range of scattering angles or even measured at only one scattering angle a misleading conclusion may be obtained. Apart from the diffusion coefficient these results may be of minor interest in application but they are of great academic value since they give information on the segmental mobility. This knowledge of the internal mobility will probably become interesting also to biochemists and pharmacologists, because it will be essential if one wishes the loading and release of drugs to be optimized.

Static light scattering was introduced to physical chemistry about 70 years ago,[20,22,23] but the corresponding method of dynamic light scattering started much later, more than 20 years with the pioneering work of R. Pecora.[46–49] One may wonder why the apparently easy technique of *dynamic* light scattering became applicable so late, now that equipment has included wonderful electronic designs in the last three decades. This question may arouse suspicion that much greater complexity is hidden underneath the highly developed technique for dynamic light scattering. This suspicion is affirmed when studying the long history before a suitable instrument could be developed. A short review on this history may be helpful to recognize the physical background and it will lead to a deep understanding of the many physical effects which contribute to *dynamic* light scattering.

6.5.2 A Short Outline of the History of the Development of Optics

On a clear night we all are fascinated by the beauty of the firmament with the many stars and the graph representation by the ancient Greeks, filled with the myth of the gods. Scientists never gave up attempting to find an understanding of the universe, and the development of dynamic light scattering is one branch in this effort. It probably started in 1817 with the detection of dark lines in the continuous spectra of stars by Joseph Fraunhofer.[50] He noticed that the same lines are also found in the laboratory; for instance, if the yellow sodium light is placed in front of the bright arc of light from a mercury lamp. This detection immediately induced a closer investigation of spectral lines. After about 15 years Gustav Kirchhoff and his colleague Robert Bunsen[51] found the explanation and assigned the dark lines as the element specific absorption of a cooler gas before a hot light source. The next important step in the understanding of light was made by James Clark Maxwell,[52] who derived the electro-magnetic theory of light, which in 1888 was experimentally proven by Heinrich Hertz,[53,54] who proved light as being an electro-magnetic wave. This theory permitted the prediction of a natural spectral *line-width*. Paralleling to this theoretical development was the design of high-quality instruments such as spectrometers which permitted measurements of the spectral line profile, the so-called power spectrum. The calculated natural line-width turned out to be far too narrow and appeared not to be measurable even with a Fabry–Perot spectrometer. In fact a much broader line-width than expected was measured, which aroused the question, What may be the reason?

This question can be taken as the beginning of studying 'stochastic' processes. A hint was given by the observation that the line-width increased with temperature which gave indications to molecular velocities in a hot gas. At the same time (1842) Doppler[55] published an explanation of the well-known increase of the sound pip if the source of noise is moving towards a person or detector and a decrease if it is moving away. He derived the relationships

$$\omega = \omega_0(1 \pm v/c) \quad \text{and} \quad \lambda = \lambda_0(1 \mp v/c) \qquad (6.35a,b)$$

For a frequency ω and wavelength λ the velocity of the light source may be v and the velocity of light is c. In a gaseous atmosphere the molecules have a high velocity moving erratically in all directions in space with a Maxwell velocity distribution, *i.e.* the motion of the molecules causes fairly broad spectral lines with half line-width that is given by the Doppler shift:

$$\frac{\Delta\omega_{\text{Doppler}}}{2} = \frac{\omega_0}{c}\sqrt{\frac{RT}{M}} \qquad (6.36)$$

where M is the molar mass of the molecules, R the gas constant and T the temperature in Kelvin. The shape of the spectral line is not fully symmetric and because of loss of radiation energy a damping of the oscillation occurs. Taking this fact into account the spectral line has the shape (the power spectrum)

$$I = I_0 \frac{(\gamma/2)^2}{(\omega - \omega_0)^2 + (\gamma/2)^2} \tag{6.37}$$

where γ is the damping factor. Figure 6.16 shows the shape of such a spectral line. The equation is applicable also to small particles in a viscous medium.

The damping was recognized as arising from friction in a viscose medium. This fact caused much interest because, as found by Einstein, Brownian motion of molecules is related to the diffusion coefficient and to the friction of the moving dispersed particles. For the friction he used an equation of Stokes which described this friction of spheres of a radius R_h and finally arrived at the relationship

$$D = \frac{kT}{6\pi\eta_0 R_h} \tag{6.38}$$

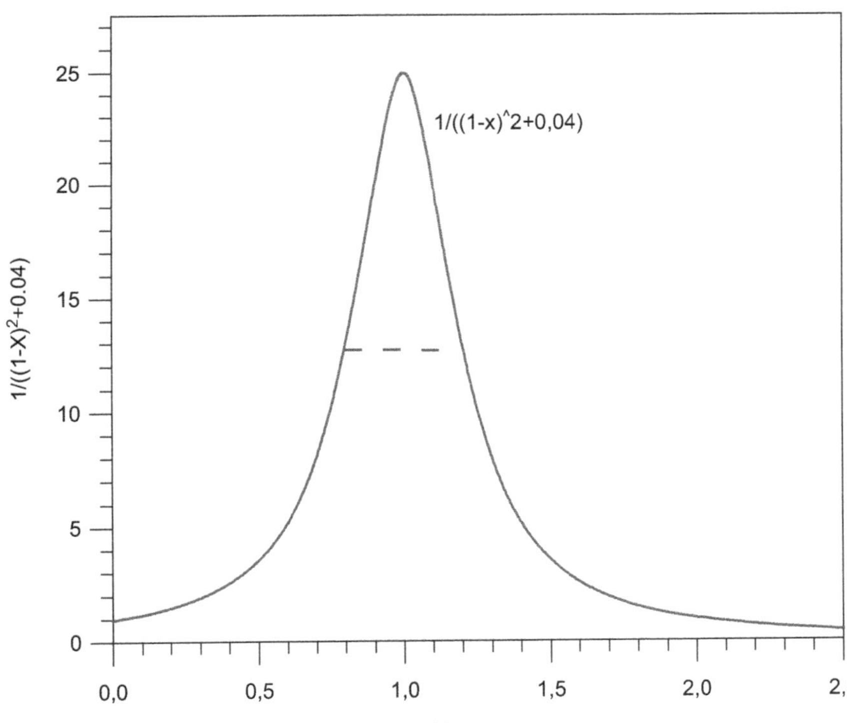

Figure 6.16 Shape of a spectral line which can schematically be expressed with $x = \omega/\omega_0$ and $\Gamma = (\gamma/2\omega_0)$ and a normalization parameter A by the equation represented with $\Gamma^2 = 0.25$.

Einstein was cautious to call R_h as a hydrodynamically equivalent radius of a sphere because it does not mean that R_h must be the radius of a hard sphere, or particle of spherical shape:

$$\frac{I(\Delta\omega)}{I_0} = A\frac{1}{(1-x)^2 + \Gamma^2} \tag{6.39}$$

The high interest in measurements of spectral line-widths originated from the fact that even in the first decade of the 19th century the existence of atoms or molecules of discrete size was considered by many scientists as a doubtful philosophical concept. The desired measurements were not successful for a long time because of the high friction of the viscous medium. The slow Brownian motion of the particles causes a much too low broadening of the spectral line. In addition, the spectral line-width of the light was insufficiently mono-chromatic even when interference filters are used.

A new aspect emerged with the introduction of gas ion-lasers[56,57] whose light is considered to be essentially monochromatic with a very narrow line-width. Even with these light sources the diffusion coefficient could be measured only for fairly low molar mass molecules which have a sufficiently high diffusion coefficient. About the same time a very efficient technique was developed that permitted the desired experiments; this was the design of short time resolving photomultipliers. These two developments led to very successful measurements of *time correlation functions* which turned out to be the Fourier transform of the line-width frequency into the corresponding time domain. This correlation requires some words on the Brownian motion before the technique of dynamic light scattering can be understood. Although the physics of Brownian motion may appear appalling to everyone who has not received a training in physics, it remains imperative to get at least some impression of the main processes. The following section is a simplified representation, which should be understandable to all chemical physicists.

6.5.3 Brownian Motion and its Relation to Stochastic Processes

In 1827 the botanist Robert Brown, using his microscope, watched pollen dust immersed in water. He noticed with surprise a continuous irregular motion of the particles which never ceased to come to a standstill. This irregular motion stimulated Albert Einstein[58] to the following considerations: due to an accelerating force the particle gains a certain initial velocity but this velocity will be slowed down by friction in a viscous medium. After a certain time the motion of the particle must come to a standstill. This can be expressed by the differential equation for the particle motion

$$\frac{\mathrm{d}v}{\mathrm{d}t} + \frac{f}{m} = 0 \quad \text{with the solution} \quad v = v_0 \exp\left(-\frac{f}{m}t\right) \tag{6.40}$$

which indeed causes an exponential slowing down of the velocity. However, this is not observed. Therefore, another force must exist which still keeps the Brownian particle in motion and eqn (6.40) should be completed by an additional term $A(t)$:

$$\frac{\mathrm{d}v}{\mathrm{d}t} = -\frac{f}{m} + A(t) \tag{6.41}$$

$A(t)$ is a fluctuating force which must arise from the thermal energy of the system. Einstein made the following consideration: if some particles in the system get a push they will receive a certain directed velocity. These motions cause a certain increase of the local concentration above the equilibrium concentration. This higher concentration is responded by a local increase of the osmotic pressure, and the corresponding force tries to push the concentration back to equilibrium. This osmotic response must be the force to compensate the frictional force which initiates the diffusion back to equilibrium concentration of the moving particle. This immediately led Einstein to the conclusion that $A(t) = f/m$ and $A(t)$ must arise from the force of the increased osmotic pressure. For the frictional coefficient f Einstein used the Stokes equation $f = 6\pi\eta R_\mathrm{h}$ for hard spheres in a viscous medium and for the osmotic force that stimulates back-diffusion towards equilibrium one has kt/mD, The mass of the particle cancels in this relationship and finally the famous Stokes–Einstein equation $R_\mathrm{h} = \dfrac{kT}{6\pi\eta D}$ is obtained.[58]

It should be mentioned that this derivation only holds if the initial velocity almost immediately loses its original direction and thus comes to a random fluctuation. Then the memory of the initial velocity is lost and velocity (the memory term) can be neglected. At this point it should be mentioned that independently of Einstein, about the same time, Marian von Smoluchowski[59–61] from Krakow University derived detailed description of the Brownian stochastic process which completes Einstein's considerations.

6.5.4 Time Correlation Functions

It is difficult to derive the special feature of such random processes. Only two facts are known for sure which are: (1) on averaging over a sufficiently long time the particles will not have moved from their original position in space, *i.e.* $\langle \Delta r \rangle = 0$, but (2) the mean square displacement $\langle \Delta r^2 \rangle$ is a *positive* quantity and a characteristic parameter for the type of fluctuations. Already Einstein had the idea to cut a longer time into N intermediate time difference, for instance

$$t = \langle t_1 - t_0 \rangle + \langle t_2 - t_1 \rangle + \langle t_3 - t_2 \rangle + \cdots \langle t_N - t_{N-1} \rangle \tag{6.42a}$$

This series of time interval is combined with the length the particle has to move in these time intervals. Assuming that the movements in these time intervals occur at random, the statistics results then can be expressed by a random series of pathlengths of

$$\langle \Delta r \rangle = \langle \Delta r(t_1) \rangle + \langle \Delta r(t_2) \rangle + \langle \Delta r(t_3) \rangle + \cdots \langle \Delta r(t_N) \rangle \tag{6.42b}$$

It is sensible to assume that for all time intervals the $\langle \Delta r(t_j) \rangle$ are, on average, of the same length l. But all the N-processes may be expected to occur independently of each other. The next question is to find the probability distribution to find the particle at a position r after N time intervals. This process corresponds to a random walk and leads to a Gaussian probability distribution

$$W(r) = \frac{1}{\sqrt{2\pi N l^2}} \exp\left(-\frac{\Delta r^N}{2Nl^2}\right) \mathrm{d}r \tag{6.43a}$$

with a mean square displacement

$$\langle \Delta r^2 \rangle = 2Nl^2 \tag{6.43b}$$

Eqn (6.43a) resembles the solution of the differential equation for a common diffusion process of dissolved particles:

$$\frac{\partial c}{\partial t} = D \frac{\partial^2 c}{\partial r^2} \tag{6.44}$$

with the solution

$$W(c, t) = \sqrt{\frac{1}{2\pi D t}} \exp\left(-\frac{\Delta r^2}{4Dt}\right) \tag{6.45a}$$

and mean square displacement

$$\langle \Delta r^2 \rangle = 4Dt \tag{6.45b}$$

Comparing eqn (6.43b) with eqn (6.45b) one obtains

$$Nl^2 = 2Dt \tag{6.46}$$

which discloses the translational diffusion as a random walk.

Statistical independence of each step in a fluctuating system can be expressed by the product of $\langle w(\Delta r_j) \cdot w(\Delta r_k) \rangle = 0$ for $j \neq k$. Such independence is the simplest assumption. More generally, the two intervals can occur with different probabilities, and this leads to a time correlation function (TCF) of[62–64]

$$C(t) = \langle \Delta r(t_j) \cdot \Delta r(t_k) \rangle > 0 \tag{6.47}$$

The steps of length l in the three-dimensional space can occur towards all directions and therefore $\Delta \boldsymbol{r}$ is a vector. For each step a velocity is defined by the ratio $\Delta r(t_j)/(t_j - t_{j-1}) = v_j$ and $C(t)$ in eqn (6.47) becomes a time correlation function of velocities.

6.5.5 Comment on Averages and the Effect of External Forces

The sum over all N values of $\Delta r(t)$ for one selected Brownian particle may be normalized by N steps. Since N is proportional to the elapsed time t this average is a *time* average. On the other hand one can take for each Δr_j the ensemble average. In an ergodic system the ensemble average and the time average are the same. A system becomes non-ergodic when a motion is frozen in, *e.g.* by glass formation. A simple generalization of a random process is a correlation with only the preceding time interval. Such a correlation is a Markov process, but even this simple modification causes trouble in the mathematical treatment and the generalized non-Markov process remains an unsolved problem so far. Still, time correlation functions can be measured in any case, and with this realistic or intuitive interpretations predictions are possible on the basis of physical and chemical properties for the investigated sample. Three special examples may shortly be discussed.

(1) The first is concerned with the effect of the initial velocity in a Brownian process. Now the effect of the inertia force [see eqn (6.40)] has to be taken in account. The differential equation of eqn (6.40) can formally be solved and gives with the frictional coefficient $\zeta = f/m$:[62–64]

$$u(t) \equiv (v(t) - v_0) = \exp(-\zeta t)\int_0^t \exp(\zeta t')A(t')\mathrm{d}t' \tag{6.48}$$

with

$$\langle u \rangle = 0, \quad \text{and} \quad \langle u^2(t) \rangle = \frac{3kT}{m}(1 - \exp(-2\zeta t)) \tag{6.49a,b}$$

Inserting the two averages of eqn (6.49) into the Maxwell velocity distribution leads to the probability distribution for the velocities:

$$W(v, t, v_0) = \left[\frac{m}{2\pi kT(1 - \exp(-2\zeta t))}\right]^{3/2} \exp\left(-\left[\frac{m(v - v_0 \exp(-\zeta t))}{2kT(1 - \exp(-\zeta t))}\right]^2\right) \tag{6.50}$$

This shows that the velocity distribution process corresponds to a dependent Gaussian process. If $\zeta t \gg 1$ the already known Gaussian distribution is obtained. Mostly the friction is very large such that the initial velocity can scarcely be measured.

(2) Another example is the motion of particles which are not hard spheres but are long flexible chains, or are of a branched structure. With such structures the viscosity of the medium is significantly influenced by what is described as the draining effect.[65–67] It is clear that the Brownian motion will have influence on the mobile segments, and since segments are shorter than the diameter of such soft particles, the relaxation modes are faster then that from centre of mass diffusion.

The two modes differ significantly in their relaxation spectrum. Further details will be given when special measurements are discussed.

(3) The third example is of particular interest in pharmacy and biochemistry and includes the perfectly branched dendrimers. These samples (proteins, nucleic acids, and drugs) are ionically charged. Also, in branched polymers in particular, the uniform dendrimers or the polydisperse hyperbranched polymers can be prepared with ionic terminal units. In these

examples the *eletrophoretic* mobility is of essential interest.[68–70] Here the applied electric force is an external force which induces a motion of the particles in direction of the electric field and this is superimposed by the irregular Brownian motion. Special instruments are commercially available and permit a fast and simultaneous determination of the *electrophoretic* mobility and the diffusion coefficient. This possibility will become an essential topic for charged dendrimers and hyperbranched particles, because the electric field may have a somewhat different effect because of the difference in the macroscopic friction to the probable larger friction of the segmental motions inside the particle.

These three issues will be commented upon in some detail. The treatment of example (1) in combination with (2) is highly involved and will only receive superficial comment. First it will be described how, in an experiment, the time correlation function is realized.

6.5.6 The Time Correlation Function in Dynamic Light Scattering

The simplest behaviour is observed with compact particles much smaller than the wavelength of the light ($r \ll \lambda$). Then the particles or molecules can be considered as point-like scattering elements, and only the distance the particles move at random is important. In an experiment a scattering intensity can be recorded in very short intervals. A highly fluctuating scattering intensity is observed. At first sight it appears hopeless to extract from such irregular fluctuations any sensible information.

However, a time correlation function can be constructed (see Figure 6.17). The numbers attached to the various peaks denote the numbers of photons from the equally sized time intervals (here of 0.1 ms). Then the time correlation function (TCF) can be constructed as follows. The scattering intensity at time t_j is multiplied by that at t_{j+k} and this proceeding is repeated about 10^5 times for increasing j but keeping the interval (k–j) constant. A fairly smooth average curve is obtained after summing the products of intensities from the two centres k and j over all j. Eqn (6.51) shows the manoeuvre with a few examples:

$$\langle i(0)i(1)\rangle = C\sum_{j=0}^{N}(i_j i_{j+1})$$

$$\langle i(0)i(2)\rangle = C\sum_{j=0}^{N}(i_j i_{j+2})\,; \tag{6.51}$$

$$\langle i(0)i(k)\rangle = C\sum_{j=0}^{N}(i_j i_{j+k})$$

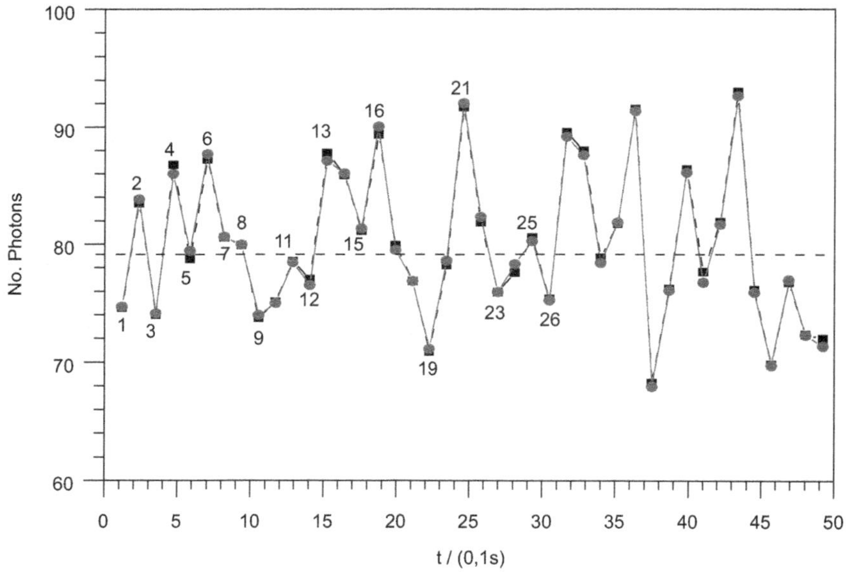

Figure 6.17 Fluctuation of the scattering intensity around the average value (dashed line). The fluctuating data are presented by number of photons collected in the short time intervals of 0.1 ms. The average was taken over 5 ms.

where $t=k\Delta t$ is the delay time. The function $\langle i(t)i(t+k\Delta t)\rangle = G_2(t)$ is the scattering *intensity* time correlation function and is called *homodyne* DLS. The *intensity* TCF has to be distinguished from the *field TCF* $g_1(t) = \langle E(t=0)E^*(t)\rangle / \langle E(t)E^*(t)\rangle$ which is the quantity needed in theory for interpretation. $E(t)$ denotes the amplitude of and $E(t)^*$ the conjugated complex amplitude of an electromagnetic wave, and $\langle E(t)E^*(t)\rangle = i(t)$ is the scattering intensity. Under homodyne conditions the intensity correlation function is related to the field TCF $g_1(t)$ by the Siegert relationship where $g_2(t) = G_2(t)/i^2$ is the normalized intensity correlation function. Note that $G_2(t)$ decays towards a plateau of $\langle i\rangle^2 = G_2(t\to\infty)$. The field correlation function is given by the equation[62]

$$g_1(t) = \sqrt{g_2(t) - 1} \tag{6.52}$$

An example of $g_1(t)$ as a function of the delay time t is given in Figure 6.18 for a monodisperse colloidal sphere in water.

Initially the function decreases like a single exponential but at larger delay times the curves in a plot of $\ln[g_1(t)]$ flattens or decays stronger. The initial slope in a plot of $\ln[g_1(t)]$ as a function of t is called the first cumulant Γ_1 and has dimension $1/s$, and can be taken as a frequency term. The question is, What information can such a correlation function provide? In other words, to which physical parameter the first cumulant is related? Evidently the strong decay in the scattering light means that the correlation becomes

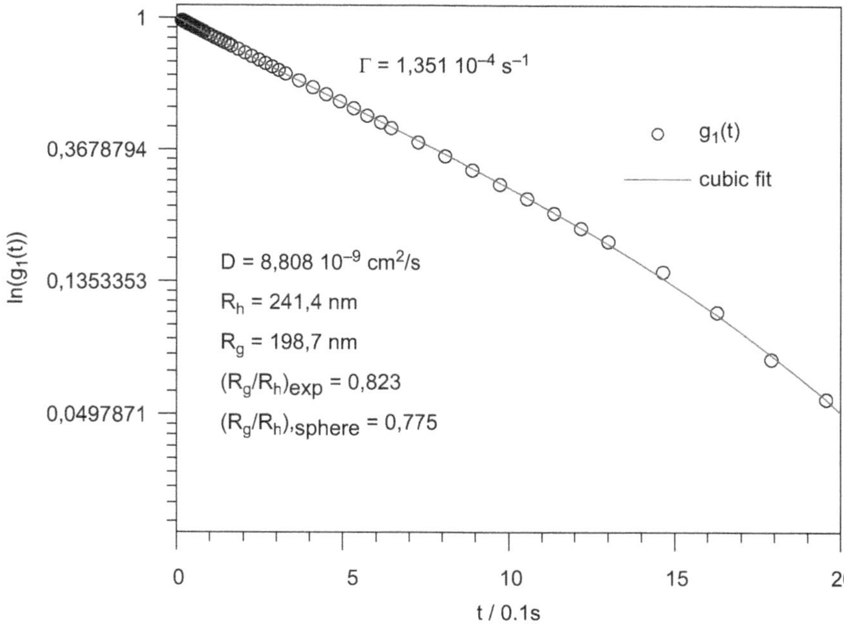

Figure 6.18 The $g_1(t)$ correlation function from spherical latex particle of a sphere radius of $R = 246$ nm.[62]

weaker and weaker as the time proceeds and eventually decays to zero, *i.e.* no correlation at all. We know that a time correlation function is closely related to the Brownian motion and is determined by *translational* diffusion. To check this statement it is useful to consider a Fourier transform of a function $G(r,t)$:

$$F(q,t) = \int_0^\infty e^{iqr} G(r,t) 4\pi r^2 \mathrm{d}r \tag{6.53}$$

It transforms a function $G(r,t)$ in space into a function in a reciprocal space, here denoted by the parameter q of dimension $1/r$. If $G(r,t)$ is the time correlation function in space caused by Brownian motion the Fourier transformed it in a time correlation function. This pair of Fourier transforms is given by the following equations:

$$G(r,t) = \frac{1}{N}\left\langle \sum_{j=1}^{N} r_j(t) - r_j(0) \right\rangle \tag{6.54}$$

$$F(q,t) = \frac{1}{N}\left\langle \sum_{j=1}^{N} e^{iq(r_j(t) - r_j(0))} \right\rangle \tag{6.55}$$

$F(q,t)$ is the time correlation function of the scattered light if $q = (4\pi/\lambda)\sin(\theta/2)$ is chosen, θ is the scattering angle and λ the wavelength of the light in the solvent. In Brownian motion $G(r,t)$ is a probability distribution and, therefore,

$$\frac{\partial^2 F(q)}{\partial q^2} = -\frac{1}{3}\int_0^{\infty} 4\pi r^2 G(r,t)\mathrm{d}r = \frac{1}{3}\langle r^2(t)\rangle \tag{6.56}$$

is 1/3 of the mean square displacement. This mean square displacement can be expressed by the translational diffusion coefficient. In fact, if the differential equation of diffusion

$$\frac{\partial c(x,t)}{\partial i} = D\frac{\partial^2}{\partial t^2}c(x,t) \tag{6.57}$$

is Fourier transformed and with eqn (6.57) one obtains the common differential equation

$$\frac{\mathrm{d}F(q,t)}{\mathrm{d}t} = -Dq^2F(q,t) \tag{6.58}$$

which easily can be integrated:

$$F(q,t) = \exp(-Dq^2t) \tag{6.59}$$

With the meaning of the q-parameter in scattering experiments this function is the dynamic light scattering TCF:

$$F(q,t) = g_1(q,t) \tag{6.60}$$

and the first cumulant is

$$\Gamma_1 = q^2D \tag{6.61}$$

The relationships of eqn (6.54)–(6.61) hold for even more complex systems, provided a differential equation similar to the diffusion type is present. Then, of course, the pre-factor no longer refers to a simple diffusion process. In addition to the motion of the centre of mass the first cumulant contains further parameters which are related to other time dependent structural properties.

Typical parameters in addition arise from the mobility of segments, or from rotational diffusion observed with long rod-like samples. Another influence on Γ_1 arises from the hydrodynamic interaction among the monomer units in a polymeric sample, which is demonstrated in Figure 6.19. The theoretical treatment needed for interpretation of experimental data became a demanding issue to theoreticians. A solution has been possible for a few special structures.[30–32]

6.5.6.1 *Intermediate Summary: Dynamic Light Scattering*

Dynamic light scattering (DLS) is based on a correlation of two scattering intensities which are separated by a certain delay time. The handling of the experimental instruments for DLS is easy and does not require a deeper understanding of the involved theory behind this method. However, before

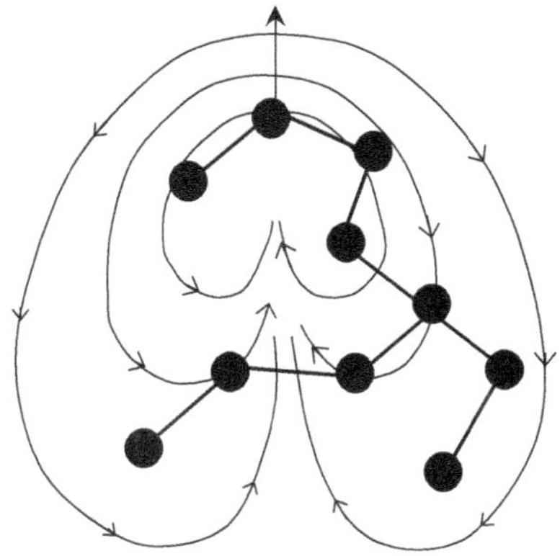

Figure 6.19 Schematic elucidation of the effect of hydrodynamic interaction caused by the motion in an incompressible solvent.

starting an interpretation some facts have to be kept in mind and are summarized below without further arguments:

1. A scheme has been presented to make clear how a time correlation function (TCF) is constructed.
2. Applied to light scattering an exponentially decaying scattering intensity time correlation function TCF, $g_2(t) = \langle i(0)i(t)\rangle/\langle i\rangle^2$ is obtained.
3. This normalized *intensity* correlation function $g_2(t)$ is related to the *field* correlation function $g_1(t)$ by the equation $g_1(t) = (g_2(t) - 1)^{1/2}$ where $g_1(t) = \langle E(0)E^*(t)\rangle/\langle i\rangle$. $E(0)$ is the electric field amplitude at a time zero and $E^*(t)$ the conjugated complex one at a delay time t later.
4. The relaxation parameter Γ_1 (first cumulant) in $g_1(t) = \exp(-\Gamma_1 t)$ is related to the translational diffusion coefficient $\Gamma(q) = Dq^2$, with $q = (4\pi/\lambda)\sin(\theta/2)$, λ the wavelength of the light in the solvent, and θ the scattering angle.
5. In complex systems the TCF is not a single exponential. In these cases a multi-cumulant fit of $\ln[g_1(t)]$ has to be applied [see eqn (6.53)]. From the coefficient Γ_1 the diffusion coefficient $D = \Gamma_1/q^2$ is obtained in the limit of $q \to 0$. For uniform hard spheres a single exponential decay is obtained and permits the determination of the diffusion coefficient at any scattering angle. Problems arise if at the probing angle the static particle scattering intensity has a deep minimum.
6. Time correlation functions are based on irregular Brownian motions. These motions also can affect the internal structure of a large particle. For instance, mobile segments are excited to oscillations. In such cases the ratio $\Gamma/q^2 \equiv D_{\text{app}}(q)$ is q-dependent and requires a careful recording of $g_2(t)$ at different scattering angles. A fit by two or three cumulants is afforded:

$$-\ln(g_1(t)) = \Gamma_1 t + (\Gamma_2/2!)t^2 + (\Gamma_3/3!)t^3 + \cdots \qquad (6.62)$$

in which Γ_1, Γ_2, Γ_3 are the first second and third *etc.* cumulants.[62]

6.5.7 Flexible Particles: Linear and Branched Macromolecules in Solution

The interpretation of experimental data are frequently made under the assumption of hard spheres or compact particles. For such particles the time correlation function is uniquely determined by the differential equation of diffusion as given by eqn (6.44). For soft materials the behaviour is far more complex. Such soft material can consist of flexible linear or branched macromolecules in solution, but it also could be a supra-molecular sample formed by cluster formation of nano-particles. The thermal motion of solvent molecules causes not only the irregular diffusion of the centre of mass but also acts onto the flexibly linked repeat units in a chain and induce oscillations according to their eigen-frequencies. In addition a special interaction takes place. This interaction arises from the fact that two moving spherical particles start to attract each other if they come close together. This attraction arises from streaming field caused by a moving particle in the incompressible solvent. Figure 6.19 illustrates schematically the effect of this *hydrodynamic* interaction.

In such a system the Langevin equation is modified by this additional force:

$$\frac{\mathrm{d}v}{\mathrm{d}t} = -\zeta v + F + A(t) \tag{6.63}$$

The external force $\boldsymbol{F}$ is in this example represented by the Oseen tensor[71,72] of hydrodynamic interaction, and $A(t)$ is the fluctuating force of the Brownian motion. Eqn (6.63) is an example for a Fokker–Planck differential equation.

The bold arrow on top (Figure 6.19) may indicate the push on the particle as the sum of Brownian motion acting on this bead. This unit of the chain pushes solvent away, and because of the solvent incompressibility a back flow takes place indicated by the small arrows. Repeat units in this field of back flow will be drawn towards the centre of the coiled chain and inhibit the flow in this contacted domain. The whole process resembles an attractive interaction. In summary, the velocity between two beads of a chain differs by a certain extent from the velocity of the solvent outside the sample.

On average, the strength of the hydrodynamic interaction decreases as $1/r_{jk}$. Therefore the hydrodynamic interaction has influence mainly on the local structure and for this reason the special architecture of the particle has to be taken into account, as a marginal condition. A complete solution has been possible only for some special cases. One is the viscosity and diffusive behaviour of uniform linear chains in a dilute solution. Without going into detail it is intuitively clear that linear chains can be described as a spring-bead model, the bead represents the monomer repeat unit, and the segment length between such two repeat units can approximately be described by an elastic spring. Due to the Brownian motion, and even more pronounced, due to a shearing force the elastic segments are excited to eigen-value oscillations. These eigen-values could be obtained by diagonalization of the interaction matrix of segment pairs. These normal modes are much faster than the diffusive mobility of the centre of mass. Therefore the effect of these internal modes of motion requires instruments with which the fast vibrations can be measured. Rouse[73] neglected the hydrodynamic interaction but B. H. Zimm[74] succeeded to perform the matrix diagonalization also for linear chains with hydrodynamic interactions.

6.5.8 Small Particles: Proteins

This section does not directly belong to branched systems. On the other hand the many functional end groups in hyperbranched polymers invites use of these functional groups for binding proteins, drugs or special catalysts. Very likely, just single functional groups are not sufficiently effective for this goal, but the efficiency will be improved if the groups are replaced by longer chains to prepare as a shell. This approach will be absolutely necessary if a *hydrophobic* material in an aqueous medium should become encapsulated. For this reason it is helpful to know how such small particles such as proteins behave if not yet bound.

Proteins are fairly compact and small compared to the wavelength used. Table 6.1 gives a list of data from well-known proteins. In most of these

Table 6.1 Molar mass M, diffusion coefficient D, the radii R_h and R_g from some proteins, and the ratio of both radii, which is a valuable quantity that gives a hint whether a global particle or a more elongated one is present.[a]

Protein	M (kg mol^{-1})	$D\times10^{-7}$ (cm^2 s^{-1})	R_h (nm)	R_g (nm)	$\rho=R_g/R_h$
Lysozyme	14 400	118.9	18.14	15.2	0.838
Ribonuclease	17 000	10.4	20.98	14.8	0.705
β-Casein	24 100	6.05	35.87	75.0	2.120
Chymotrypisogen	38 000	7.9	27.09	18.1	0.668
Ovalbumin	45 000	7.3	29.31	24.0	0.819
Bovine serum albumin	66 000	5.93	36.09	29.8	0.826
Hexokinase (yeast)	99 000	6	35.62	24.7	0.693
Lysine tnr ligase	138 000	4.3	49.76	37.5	0.754
γ-Globulin (human serum)	162 000	3.7	57.83	70.0	1.210
Catalase (bovine liver)	225 000	4.1	52.20	39.8	0.763
Fibrinogen	390 000	1.87	115.05	142.0	1.234
Myosin (rabbit muscle)	493 000	1.16	184.48	468.0	2.537
α-2-Myoglobin	820 000	2.41	88.79	63.7	0.717
Sartelite tobacco necrosis virus	2 780 000	2.04	104.91	69.0	0.648
Pyruvate dehydrogenase	3 780 000	1.2	128.33	156.5	0.876
Turnip yellow mosaic virus	4 970 000	1.55	138.06	108.0	0.782
Alfalfa mosaic virus	6 920 000	1.05	203.81	216.0	1.060
Haemocyanin helix pomatia	9 805 000	1.07	200.00	164.0	0.820
Bacteriophage λ (enlarged prehead)	17 000 000	0.69	310.14	285.0	0.9189
Tobacco mosaic virus	39 000 000	0.38	563.16	924.0	1.641

[a]The data for this table were taken from He and Niemeyer.[77]

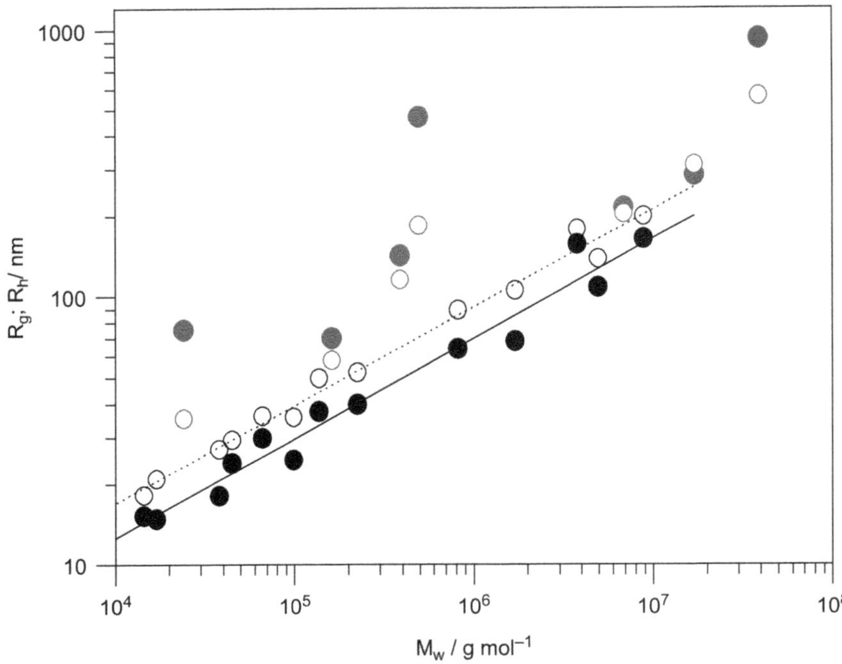

Figure 6.20 Molar mass dependence of the radius of gyration (filled symbols) and of the hydrodynamic radius (open symbols).

examples the normalized static structure factor $P(q) = S(q)/S(q=0)$ shows no detectable angular dependence in *visible* light scattering. This behaviour is always observed when the particle is smaller than the light wavelength in the solvent (*i.e.* $R_g < \lambda/20$). For example, in water the wavelength of the He–Ne laser is $\lambda = 633/1.333$ and therefore the radius of gyration should be $R_g > 15$ nm. In DLS experiments somewhat smaller radii can be measured but below 2 nm problems arise which only in rare cases can be overcome (Figure 6.20).

Most of the examples follow a fairly well defined power law with an exponent of 0.369 ± 0.003 which is slightly higher than 0.333 expected for hard spheres and constant density, *i.e.* which does not change with the distance from the centre of mass and also should contain no porosity. A number of proteins display marked deviations (unfilled symbols) whose origin is made clear in Figure 6.21.

Up to about 50 nm the first cumulant $\Gamma_1 = Dq^2$ shows no further q-dependence. In other words, the particle can be considered as small and close to hard sphere behaviour such that exclusively the translational diffusion coefficient counts for the interference of the particle at $x(t=0)$ and $x(\Delta(t))$ some time later. The data of Table 6.1 are plotted in Figures 6.20 and 6.21 for the radii R_g, R_h and the ratio $\rho = R_g/R_h$.

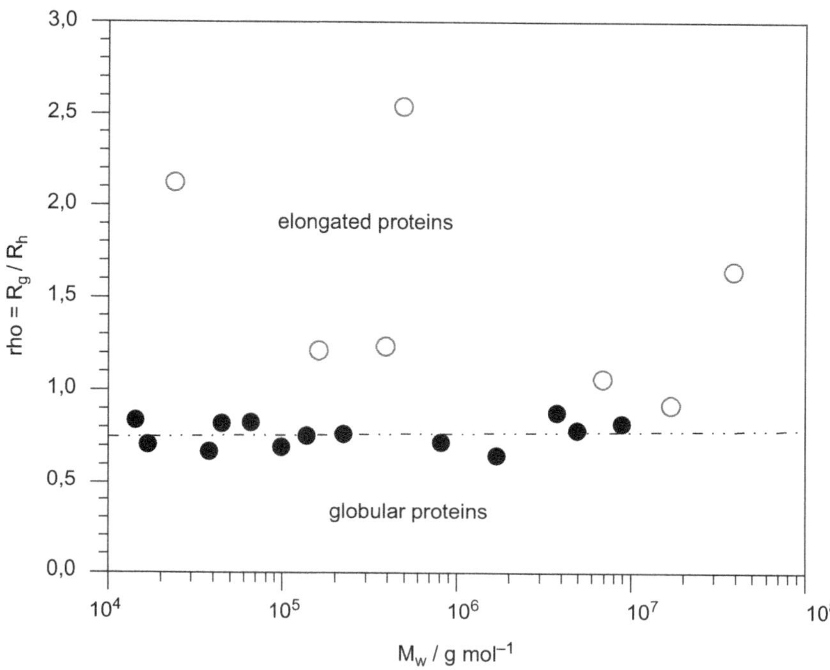

Figure 6.21 Plot of the ρ-parameter of proteins in a wide range of molecular weights. The filled symbols are positioned around a constant value of 0.774 which is the predicted value for compact spheres, and therefore give indication for a globular structure of the proteins. The data of the unfilled symbols can also be interpreted by theory and the high values arise from an elongated structure.[75,76]

6.5.9 Polydispersity and the Effects of Cluster Formation

Proteins often have a strong tendency to aggregate to a dimer or even larger bundles. This fact has been observed with serum albumin, insulin and thrombin activated fibrinogen to mention only a few examples. The aggregates often consist of a defined number of unimers. The time correlation function displays noticeable deviations from a single exponential decay but develop after a fairly fast decay a single weaker but still a single exponential asymptote. In such cases the TCF can satisfactorily be described by a two exponential fit with two different relaxation times or two different diffusion coefficients which give information on the unimer and aggregated sizes. Frequently the two relaxation times differ by three to four decimals. In these cases the fast process decays to a plateau which externs up to large correlation times but then a second decay takes place. The origin of such slow motions can mostly not be disclosed by DLS alone and further experiments have to be carried out with a different method. It is advisable to perform static LS with the same sample and same concentration because the slow motion often indicates a large object which in static LS will display a pronounced angular dependence.

The second now mostly applied method to receive quantitative data on various components in the system is the application of the CONTIN inversion program. In general the DLS time correlation function can be expressed by the integral

$$g_1(q,t) = \int_0^\infty H(\tau)\exp(-t/\tau)\mathrm{d}\tau; \quad \frac{1}{\tau} = Dq^2 \tag{6.64}$$

The integral in eqn (6.64) presents a La Place transformation of the relaxation time distribution. In contrast to Fourier transformation a La Place transformation cannot be analytically transformed. Thus an attempt of inversion has to be made by sensible conjectures established by the Information Theory. The elaborated algorithm works well for continuous single-peaked distributions of $H(\tau)$, but if the distribution consists of two or three well-separated peaks, additional constrains have to be inserted in the conversion program like that of CONTIN. The development of this program is based on the Information Theory and the algorithm has been continuously improved and gave the experimentalist a safe feeling. Still one has to be aware that the derived prediction is not fully determined and a critical judgement of the obtained results remains important.

6.5.10 Electrophoretic/Dynamic Light Scattering

One possibility for such a check for proteins is the further developed electrophoretic dynamic light scattering (ELS). The method was outlined in 1971 by R. Raj, B. R. Ware and W. H. Flygare[68–70] and is based on the method of heterodyne scattering. An excellent description was given by J. P. Wilcoxon and J. M. Schurr.[78] A short description is given below. The idea consists in the study of what can be observed if, vertically to the plane of common scattering, an electric field is applied to a protein. It turns out that nothing special can be observed in the homodyne light scattering, but a very different and informative result is obtained if the heterodyne technique is applied.

Heterodyne means that a small portion of constant (non-scattering) light is added to the scattering field. This condition causes a mixing of the local light with the homodyne scattering field by which the static light (*i.e.* the base line) is much enhanced and only the time-dependent field correlation function is obtained as scattering source.[79] The situation is described by the equation[68–70]

$$\langle i(0)i(t)\rangle \cong I_{\mathrm{local}}^2 + 2I_{\mathrm{local}}\mathrm{Re}(I_1(t)) \tag{6.65}$$

where $I_1(t) = \langle A(0)A^*(t)\rangle$ is the *electric field* TCF.

Only the real part has a physical meaning. The heterodyne correlation function has the advantage that the effect of an external force can be taken into account. Due to this force the scattering particle undergoes not only the

Brownian motion but, in addition, a constant flow occurs. It has been already outlined that the field correlation function can be obtained by the differential equation of a diffusing particle in space. The extended diffusion equation now can easily be obtained by adding the gradient of the external flow to the differential equation. In addition after performing the Fourier transformation this partial differential equation becomes a common differential equation with the solution of

$$\mathrm{Re}[g_1(q,t)] = \mathrm{Re}[\exp(i\boldsymbol{q}\cdot\boldsymbol{v}t)]\exp(-q^2Dt) = \langle N\rangle\cos(\boldsymbol{q}\cdot\boldsymbol{V}t)\exp(-q^2Dt) \tag{6.66}$$

This equation makes it apparent that besides the relaxation due to diffusion now a cosine wave-like time dependence is superimposed. The use of vector notation (bold letters) is required since the direction of scattering vector $\boldsymbol{q}$ will in general form an angle with the direction of the velocity $\boldsymbol{V}$. The scalar product $\boldsymbol{q}\cdot\boldsymbol{V}$ can be written in detail:

$$qV\cos(\varphi) = \frac{4\pi}{\lambda}\sin(\theta/2)V\cos(\varphi) \tag{6.67a}$$

On the other hand the product $\frac{2\pi}{\lambda}V = \Delta\omega$ is a frequency which in the present experimental set-up can be considered as the Doppler shift caused by the moving charged particle due to the external electric force. This makes clear the meaning of the cosine of the angle ϕ between the direction of scattering vector $\boldsymbol{q}$ and the direction of flow velocity $\boldsymbol{V}$. Of course if the electric field causes a flow vertically to the scattering vector no Doppler effect can be observed, the angle ϕ permits *via* the choice of $\cos(\phi)$ to observe the cosine component of the moving velocity. The two angles θ and ϕ may be somewhat confusing, so again, the definition: θ is the scattering angle in the plane of the primary beam and the detector, whereas ϕ is defined as the angle between the flow direction and the direction of the q-vector. The equivalence of

$$q\cdot \mathrm{V} = \Delta\omega_{\mathrm{Doppler\ shift}} \tag{6.67b}$$

permits a two-fold detection of the flow by (1) heterodyne auto-correlation function of the light scattering, *i.e.* the observation how fast the time dependence of the cosine waves passes the detector or (2) the corresponding power spectrum can be recorded and the Doppler shift is detected from the line-width. The first type of experiment was realized by Ware and Flygare and is shown in Figure 6.22, which is a slightly modified copy from the original paper.[68–70]

At zero electric field the common exponential decay to the diffusion is obtained. With increasing field strength (V cm^{-1}) a wave is observed within the same time interval as recorded at zero field strength with increasing number of waves in the same time interval of 2–3 min. From these curves the electrophoretic mobility μ_E and the electric charges of the particle can be obtained. Wilcoxon and Schurr checked the equivalence with the Doppler shift, by carrying out on-line a fast Fourier transformation and recording the

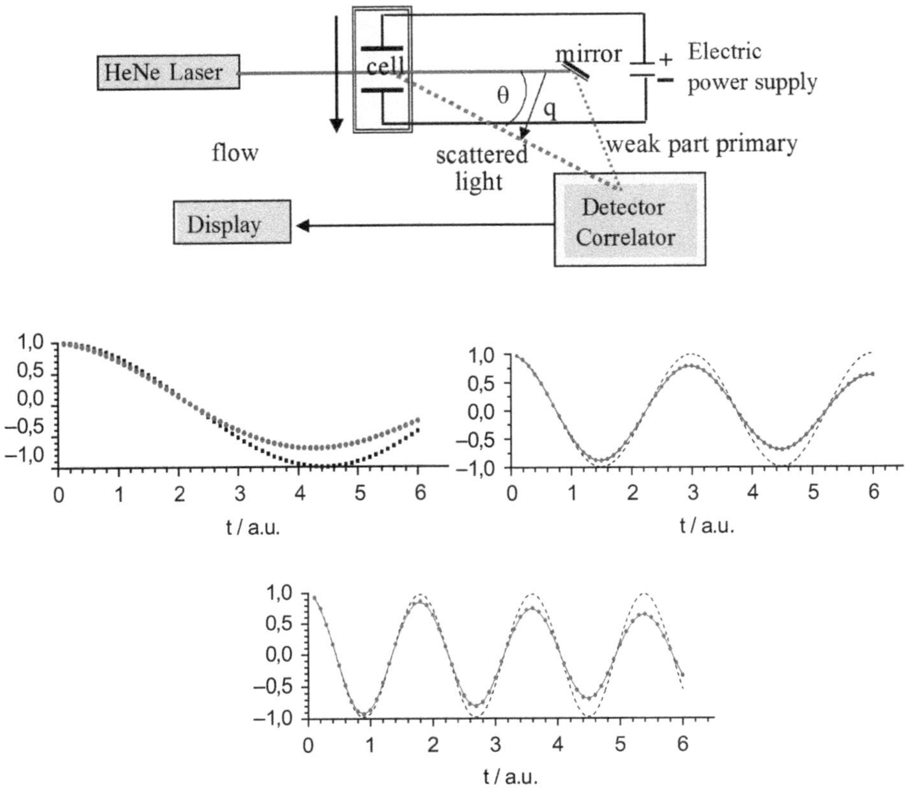

Figure 6.22 Draft of the optical arrangement for measuring electrophoretic mobility *via* dynamic light scattering. θ is the scattering angle and q the corresponding scattering vector. The mixing of a local constant intensity with the fluctuating scattering intensity is arranged in this example by a mirror after attenuation before the two intensities enter the detector and correlator. The applied voltage causes a flow of the charged sample. The response is a weakly attenuated cosine wave (solid lines). At higher voltages a faster flow occurs indicated by more wave cycles within the same time interval. For the three examples, observed within the same time interval, the voltage was increased three and five times. The damping of the amplitudes due to the Brownian fluctuations is mostly fairly weak and can be measured at out switched voltage from the first cumulant and the diffusion coefficient.

occurring power spectrum. The obtained line-widths in their experiments were fairly noisy but still a complete agreement in the electrophoretic mobility was found.

At present three types of ELS spectrometers are on the market. Not many details of the instruments are given in advertisements, but it seems that mostly the TCF is converted by fast Fourier transformation into a power spectrum resulting in a spectral line and the Doppler shift is obtained from the measured line-width broadening.

In the meantime the recording of the Doppler shift could be much improved, such that quick measurements of high quality can be made. The measurements are made at only one fixed angle of ϕ and θ. This means that only small particles of $R_h < 30$ nm can reliably be measured, which is well in the regime of interest with proteins and other globular shapes. The realization of several angles θ, in fact the variation of the angles is restricted by the construction of the instruments which becomes clear from the graph given by Wikipedia[80] which is reproduced here with minor modifications.

6.5.10.1 Summary: Electrophoretic/Dynamic Light Scattering

Mixing of the time dependent scattering light intensity with the local, non-fluctuating light, *e.g.* a weak part of the primary beam, leads to a heterodyne correlation function. This correlation function consists of the squared intensity of the local light source plus the product of the local intensity and the field correlation function. This design of experiments has the advantage that the effect of external forces can be analysed. If the external force is caused by an electric field and the particles are somewhat charged the heterodyne field TCF is modified by a damped cosine wave modulation from which the electrophoretic mobility can be obtained. The damping arises from diffusion of the Brownian particles. The diffusion constant and the related hydrodynamic radius of the charged particle can be measured at zero electric field. Commercially available instruments permit quick measurement of the electrophoretic mobility and estimation of the charges at the particle.

6.5.11 Concentration Dependence

So far the diffusion coefficient at one concentration has been considered. Often no significant concentration dependence is obtained, particularly if small particles are investigated. In such cases the hydrodynamic radius can be directly obtained from the Einstein relationship. Actually, two counteracting interactions are affective at finite concentration. One arises from the concentration dependence of the osmotic compressibility and the other from the increasing friction as the concentration of particles is increased. Diffusion is a process of irreversible thermodynamics and always is effective if the solution is not at a thermodynamic equilibrium and the particles move to re-establish equilibrium. The driving force is the osmotic compressibility:

$$(M/kT)(\partial\Pi/\partial c) = M/M_{\mathrm{app}}(c) = 1 + 2A_2Mc + 3A_3Mc^2 + \cdots \tag{6.68}$$

It can easily be obtained from static light scattering at zero scattering angle:

$$\begin{aligned} \frac{Kc}{R_{\theta=0}} &= \frac{1}{M_{\mathrm{app}}(c)} = \frac{1}{M}\left(\frac{M}{M_{\mathrm{app}}(c)}\right) \\ \frac{1}{M_{\mathrm{app}(c)}} &= \frac{1}{M} + 2A_2 + 3A_3c^2 + \cdots \end{aligned} \tag{6.69}$$

The concentration dependence of the mutual diffusion coefficient then is

$$D_m(c) = \frac{kT}{f(c)} \frac{M}{kT} \frac{\partial \Pi}{\partial c} = \frac{kT}{f(c)} (1 + 2A_2Mc + 3A_3Mc^2 + \cdots) \tag{6.70a}$$

The concentration dependence of the frictional coefficient can be expressed by

$$f(c) = f_o = (1 + k_f c + \cdot \cdot \cdot) \tag{6.70b}$$

combining eqn (6.70a) and (6.70b) one has a concentration dependence which in a first approximation is given by

$$D_m(c) = D_0(1 + k_D(c)) = \frac{kT}{6\pi\eta_0 R_h} (1 + 2A_2c - k_f c) \tag{6.71a}$$

with

$$k_D = 2A_2 - k_f \tag{6.71b}$$

A vanishing concentration dependence of the diffusion coefficient is the result of two different interactions of opposite behaviour. Mostly the thermodynamic effect is stronger than the increasing friction and a positive $k_D > 0$ is observed. For charged proteins the effect can be very strong such that in any case a concentration dependence should be measured. Otherwise too small hydrodynamic radius is obtained. It has to be kept in mind that the Stokes–Einstein relationship holds valid only in the limit of zero concentration, or if no concentration dependence is observed. The behaviour of eqn (6.71b) can be a reason for somewhat confusing effect, which is well known in polymer science. Polymers of different molar mass are often self-similar, *i.e.* a large polymer develops essentially the same behaviour as smaller ones of the same type and differs only by a scaling parameter. Such behaviour is frequently observed also for the diffusion coefficient D_0 but not when data at finite concentration are used. Such an effect is observed for marginally good solvents when the virial coefficient is fairly low but the friction remained effective. In such systems one observes a large positive k_D for high molecular weights but the value decreases continuously for smaller molecular weights, and k_D becomes negative in the range of $M < 5000$ g mol^{-1}. The friction increases stronger than the thermodynamic repulsion between different particles.

A special case is obtained with polymers in a so-called θ-solvent, in which the second virial coefficient is $A_2 = 0$. The particles appear to be in an ideal solution. A similar situation can be created with proteins by adding salt. The particle seems to be unperturbed by interparticle interactions and one may also be inclined to expect no concentration dependence of the diffusion coefficient. On the contrary a very pronounce decrease of D_m is observed since the self-diffusion $D_{self} = kT/(6\pi\eta_0 R_h(1 + k_f c))$ remains effective. A quantitative interpretation of the concentration dependent self-diffusion is not yet fully understood, because with increasing concentration the solvent viscosity can be increased. Certainly also the Brownian motion from an ensemble of obstacles has a marked effect on the increased friction, and for this reason it is extremely more difficult to predict than the non-correlated Brownian motion.

6.5.12 Summary: Dynamic Light Scattering

The easy measurable dynamic light scattering leads to reliable results if one is *sure of probing small particles, i.e.* $R_h \ll \lambda$, the wavelength of the light used. For strong thermodynamic interaction measurements at finite concentration it gives smaller apparent hydrodynamic radii and this amplifies with increasing concentration. The effect is especially pronounced with highly charged samples. *Data obtained from measurements at different concentration have to be extrapolated to* $c = 0$. *Only then can the hydrodynamic radius be determined by the Stokes–Einstein relationship.* The concentration dependence arises from thermodynamic interactions between the particles and the friction caused by the viscosity of the solvent. In dilute solutions the former can be well approximated by the second virial coefficient, *e.g.* measured by static light scattering, but the friction has to be measured separately by other techniques, *e.g.* by sedimentation measurements or by applying a special NMR technique.

For *large particles with* $R_g \approx \lambda$ the first cumulant often develops deviations from the q^2 dependence. This effect is caused by internal motions of the segments and to slower part by a broad molar mass distribution. *Measurements have to be made in a wide angular range. The Stokes–Einstein relationship is applicable only after extrapolation towards the scattering angle* $\theta \to 0$. For weak deviations the apparent diffusion coefficient can be represented by a linear approximation in terms of qR_h. The *coefficient* in this linear approximation *gives indications to elastic motions*. At *large* qR_h values an *asymptotic behaviour* is obtained which is characteristic of *special segmental motions*. For linear and star-branched chains the local dynamics could be solved in theory. Experimental data are largely missing mainly since the required spin-echo measurements by SANS require an exorbitant long recording time.

At present the local motions are considered as less important. However, in order to understand the mechanisms of drug loading and release the local dynamics will become more interesting.

6.6 Effects of Heterodyne Light Scattering

Occasionally with special examples the time correlation does not start at $t = 0$ with the predicted value of unity. This mostly is the case if at large decay times a slow motion appears, which by about four decades in time are slower than the decay observed at short times. Also these slow modes scatter the light very strongly. Both of these observations give indication for the presence of very large structures which dominate the static light scattering despite of a rather low concentration. The immense difference in the time scale between the two components includes the large agglomerates scarcely move in the time interval where the fast motions are recorded, in other words, the large component behaves like a non-fluctuating light source and causes a mixing of the dynamic and static amplitudes of the scattered light which is denominated as heterodyne scattering. This mixing of fluctuating and static light amplitudes was calculated by E. Geissler.[79] In a somewhat simplified version he obtained the equation

$$g_2(t) = 1 + 2R(1 - R)g_1 + R^2 {g_1}^2 \tag{6.72}$$

with

$$g_1(t) = \frac{\langle E(0)E^*(t)\rangle}{\langle E(0)E^*(0)\rangle} \tag{6.73a}$$

$$g_2(t) = \langle i(0)i(t)\rangle = \langle E(0)E^*(0) \times E(t)E^*(t)\rangle \tag{6.73b}$$

and R is the contribution of the homodyne scattering which is found from the intercept of

$$g_2(t=0) - 1 = R^2 + 2(1 - R) \tag{6.74}$$

To find the homodyne scattering curve the quadratic eqn (6.74) has to be solved for R, and after insertion into eqn (6.73) the quadratic equation again has to be solved for $g_1(t)$.

We have to add that a similar problem of heterodyne mixing occurs if very small particles of $R_h \cong 1 \div 2$ nm are studied. In such cases the dynamically scattered light comes into the range of the static scattering intensity of the solvent such that no more than 25% of correlated dynamic scattering insensitivity is obtained. Of course the solvent molecules are in fast thermal motions but the relaxation time is so much faster, that the scattering intensity of the solvent fulfils the condition of static light scattering.

6.7 Appendix: Aspects of Modern Theory of Segmental Dynamics

The development of a suitable theory that also takes into account local motions has been a serious challenge to theoreticians. At the end a fully new conception *via* the response theory was elaborated. The development of this scheme started with the motion of particles in the six-dimensional phase space (three impetus and three Cartesian coordinates) and stepped down to the common Euclidean space by modifying the Liouville operator acting on moving samples in the three-dimensional space. Details of this involved theory are avoided and only the final result will be described and discussed.

The generalized differential equation consists of two parts and is given by the following equation:

$$\frac{\mathrm{d}g_1}{\mathrm{d}t} = -\Gamma g_1(t) + \int_0^t \varphi(u) g_1(t-u)\mathrm{d}u \tag{6A.1}$$

where Γ is called the *frequency* and $\phi(t)$ the *memory* term, and $g_1(t)$ is a time correlation function. In the present case the time correlation function in dynamic light scattering. The integration of eqn (6A.1) is possible though rather involved. It turned out that the contribution from the memory term is rather low for macromolecules and often can be neglected. Without the memory term the differential eqn (6A.1) can formally be solved to

$$g_1(t) = \exp(-\Gamma t) \tag{6A.2}$$

In dynamic light scattering the function $g_1(t)$ is the field time correlation function which may be written as

$$g_1(q,t) \; = \; \rho(q,0)\rho^*(q,t)/(\rho q,0)\rho^*(q,0) \tag{6A.3a}$$

with

$$\rho(q,0) = \sum_{j=1}^{N} \exp(iqR_j(t=0)) \; \rho^*(q,t) = \sum_{k=1}^{N} \exp(-iqR_k(t)) \tag{6A.3b}$$

and

$$g_1(q,t) = \frac{\sum_j^N \sum_k^N \exp(\boldsymbol{q}\,(\boldsymbol{R}_j(0)) - \boldsymbol{R}_k(t))}{\sum_j^N \sum_k^N \exp(\boldsymbol{q}\,(\boldsymbol{R}_j(0)) - \boldsymbol{R}_k(0))} \equiv \frac{\sum_j^N \sum_k^N \exp(\boldsymbol{q}\,(\boldsymbol{R}_{jk}(t)))}{N^2 P(q)} \tag{6A.4}$$

where the asterisk denotes the conjugated complex variable. $P(q)$ is the particle scattering factor for the static light scattering and N^2 denotes the number of pairs of repeat units.

So far, eqn (6A.3) and (6A.4) denote no more than a detailed formulation of the field TCF $g_1(t)$ in dynamic light scattering. To proceed further, a special algorithm for calculating the frequency is required. This is obtained by a modified Liouville operator L acting on a density parameter $\rho(t)$ in a way given by eqn (6A.5):

$$\langle \rho, L\rho^*(t) \rangle \; \equiv \; \int_0^\infty (\rho(0)L\rho^*(0))\psi_0\{\boldsymbol{R}\}\mathrm{d}\{\mathrm{d}\boldsymbol{R}\} \tag{6A.5}$$

where $\psi_0\{\boldsymbol{R}\}$ is the equilibrium distribution of the chain elements and L is the generalized Langevin equation which was first derived by Kirkwood:

$$L = \left[\sum_j^N \sum_k^N (kT)^{-1} \nabla_j U \cdot \boldsymbol{D}_{jk\cdot} \cdot \nabla_k - \nabla_j \cdot \boldsymbol{D}_{jk} \nabla_k \right] \tag{6A.6}$$

This equation [eqn (6A.6)] can be considered as a generalized diffusion differential equation for particles under the constraint of hydrodynamic interactions. The interesting point in eqn (6A.6) is the fact that *via* the action of the L operator only the equilibrium (*i.e.* time independent) distance distributions is required. Omitting all further steps of the derivation one finally obtains

$$\rho.L\rho^* = \sum_{j}^{N}\sum_{K}^{N} \mathbf{q}\cdot\langle \mathbf{D}_{jk}\exp(i\mathbf{q}\cdot\mathbf{R}_{jk})\rangle\cdot\mathbf{q} + \sum \mathbf{q}\cdot\langle\mathbf{D}_{jj}\rangle\cdot\mathbf{q} \tag{6A.7}$$

with

$$\boldsymbol{D}_{jk} = kT(\boldsymbol{T}_{jk} - \zeta^{-1}\boldsymbol{1}\delta_{jk}) \tag{6A.8}$$

The angled brackets mean the average over all pairs of units and $\boldsymbol{1}$ is the unit tensor.

The branched macromolecules are efficiently described by the pearl necklaces model in which the pearls or beads are connected by springs. The velocity of the solvent molecules will have an influence on the motion of the beads. The force acting on the beads is not just a simple vector, but considerably more complex. The solvent velocity $\boldsymbol{u}$ outside the macromolecules differs from that internally of the coil and represents the flow around a bead such that the actual frictional force is given by $F_h = \zeta(u - v)$ where v is the velocity inside the macromolecule and ζ the frictional coefficient. Furthermore, the solvent can be considered as an incompressible liquid which has an important effect.

For instance if a bead B1 in the macromolecular chain is pushed forward by the solvent this will cause a back-flow and if another bead B2 is within this backflow domain B2 will be drawn towards B1. A hydrodynamic interaction is effective which appears to arise from an attractive potential. Clearly, the attraction will become less effective with increasing distances between the two beads. Actually, the effect is more complex because the velocity is a vector and therefore the interaction also depends on the angle which the distance vector forms with the direction of the velocity. This fact can be described by a tensor rather than by a scalar parameter, and is given by the Oseen tensor

$$\boldsymbol{T}_{ij} = \frac{1-\delta_{ij}}{8\pi\eta_0 R_{ij}}\left(1 + \frac{\boldsymbol{R}_{ij}\boldsymbol{R}_{ij}}{R_{ij}^2}\right) \tag{6A.9}$$

with $R_{ij} = R_i - R_j$ and η_0 the solvent viscosity.

This tensor causes considerable difficulties in the treatment of finding an exact solution. Therefore, the tensor is mostly approximated by the average over all orientations:

$$\langle\boldsymbol{T}_{ij}\rangle = \frac{1-\delta_{ij}}{6\pi\eta_0}\left\langle\frac{1}{R_{jk}}\right\rangle \tag{6A.10}$$

which can be calculated if the distance distribution is known. For a Gaussian distribution, $W(R_{jk}) = (3/(2\pi bn))^{1/2}\exp(-(3/2)nb^2)\mathrm{d}n$ the average is with $n = |j - k|$

$$\left\langle \frac{1}{R_{jk}} \right\rangle = \frac{\sqrt{6}}{\sqrt{|j-k|\, b^2}} = \frac{2}{\sqrt{\pi}} \int_0^{\infty} \exp(-nb^2)\mathrm{d}n \qquad (6A.11)$$

It has to be emphasized that this integral representation only holds for unperturbed structures, *i.e.* Gaussian statistics. The introduction of excluded volume interactions is more involved but a solution is still possible.[81]

6.8 Recommended Literature

Flory, P.J. "Principles of Polymer Chemistry", Cornell University Press, Ithaca N.Y. 1953.

Hiemenz, P. C.; Lodge, T.P. "Polymer Chemistry" CRS Press, Taylor & Francis, Boca Raton FL 2007.

Rubinstein, M., Colby, R. "Polymer Physics", Oxfrord University Press, Oxford 2003.

Grosberg, A. Y.: Khokhlov, A. R. "Statistical Physics of Macromolecules" AIP Press, New York, 1994.

See also monographs on Statistical Mechanics:

McQuarrie, D.A. "Statitical Mechanics" Harper and Row NewYork 1976.

Friedman, H.L. "A Course in Statistical Mechanics", Prentice Hall Inc. Englewood Cliffs, NJ 1985.

Yamakawa, H. "Modern Theory of Polymer Solutions" Harper and Row, New York 1971.

Freed, K.F. "Renormalization Group Theory of Macromolecules", John Wiley & Sons, New York, 1987.

Fujita, H. "Polymer Solutions" Elsevier, Amsterdam 1990.

Tanford, C. "Physical Chemistry of Macromolecules", John Wiley & Sons, New York 1967.

References

1. Y. H. Kim and R. Beckerbauer, *Macromolecules*, 1994, **27**, 1968.
2. Y. H. Kim and O. W. Webster, *J. Am. Chem. Soc.*, 1990, **112**, 4592.
3. A. Hult, M. Johannson and E. Malmström, *Adva. Polym. Sci.*, 1999, **143**, 1.
4. F. Galton, *Educ. Times*, 1873, **26**, 17.
5. H. W. Watson, *Educational Times*, 1873, **19**, 115.
6. R. Schneider, PhD, University of Freiburg, 2012.
7. T. C. B. McLeish and S. T. Milner, *Adv. Polym. Sci.*, 1999, **143**, 195.
8. H. Staudinger, *Organische Kolloidchemie*, Vieweg, Verlag, Braunschweig, 1950.

9. H. Staudinger, *Arbeitserinnerungen*, Hüthig, Verlag, Heidelberg, 1961.
10. T. Svedberg and K. O. Pedersen, *The Ultracentrifuge*, Oxford University Press, London, 1940.
11. M. Born, *Optik*, Springer Verlag, Berlin, 1933.
12. H. C. van de Hulst, *Light scattering by small particles*, John Wiley and Sons, New York, 1957.
13. M. Kerker, *The Scattering of Light and Other Electromagnetic Radiation*, Academic Press, San Diego, 1969.
14. W. Burchard, in *Physical Techniques for the Study of Food Biopolymers*, ed. Ross-Murphy, Chapman and Hall, London, 1994.
15. M. B. Huglin, *Light scattering from polymer solutions*, Academic Press Inc., London, 1972.
16. W. Schärtl, *Light Scattering from Polymer Solutions and Nanoparticle Dispersions*, Springer, Berlin Heidelberg, 2007.
17. J. W. Lord Rayleigh, *Proceedings of the Royal Society*, 1914, **A90**, 219.
18. R. Gans and L. Hapel, *Ann. Phys.*, 1909, **29**, 277.
19. R. Gans, *Ann. Phys.*, 1920, **62**, 331.
20. B. H. Zimm, *J. Chem. Phys.*, 1948, **16**, 1093.
21. P. Debye, in *Light Scattering from Dilute Polymer Solutions*, ed. D. McIntyre and F. Gornick, Gordon and Breach Science Publishers, New York, 1964.
22. P. Debye, *J. Appl. Phys.*, 1944, **15**, 338.
23. P. Debye, *J. Phys. Colloid Chem.*, 1947, **51**, 18.
24. W. Kuhn, *Kolloid-Z.*, 1934, **68**, 2.
25. W. Kuhn, *Kolloid-Z.*, 1936, **76**, 258.
26. O. Kratky and G. Porod, *J. Colloid Sci.*, 1949, **4**, 35.
27. K. Saalwächter and W. Burchard, *Macromolecules*, 2001, **34**, 5587.
28. K. Saalwächter, W. Burchard, P. Klüfers, G. Kettenbach, P. Mayer, D. Klemm and S. Dugarmaa, *Macromolecules*, 2000, **33**, 4094.
29. B. H. Zimm and W. H. Stockmayer, *J. Chem. Phys.*, 1949, **17**, 1301.
30. K. F. Freed, *Renormalization group theory of macromolecules*, Wiley, New York, 1987.
31. M. Rubinstein and R. H. Colby, *Polymer Physics*, Oxford University Press, Oxford, 2003.
32. A. Y. Grosberg and A. R. Khokhlov, *Statistical Physics of Macromolecules*, AIP Press, New York, 1994.
33. P. J. Flory and T. G. J. Fox, *J. Am. Chem. Soc.*, 1951, **73**.
34. P. J. Flory, *J. Chem. Phys.*, 1949, **17**, 303.
35. P. J. Flory and T. G. Fox, *J. Polym. Sci.*, 1950, **5**, 745.
36. P. J. Flory, *Statistical Mechanics of Chain Molecules*, Wiley and Sons, New York, 1969.
37. P. G. DeGennes, *Scaling Concepts in Polymer Physics*, Cornell UNoversity Press, Ithaka, London, 1979.
38. W. Burchard, *Results obtained by the Freiburg group of W. Burchard.*
39. N. F. Carnahan and K. E. Starling, *J. Chem. Phys.*, 1969, **51**, 635.
40. M. A. Cotter and D. E. Martire, *J. Chem. Phys.*, 1970, **52**, 1909.

41. T. Ohta and Y. Oono, *Phys. Lett. A*, 1982, **89**, 460.
42. G. Galinsky and W. Burchard, *Macromolecules*, 1995, **28**, 2363.
43. B. B. Mandelbrot, *The Fractal Geometry of Nature*, Freeman & Co., New York, 1983.
44. G. Rosati, *J. Ultrastruct. Res.*, 1967, **18**, 444.
45. W. Burchard, in *Structure of Fibrous Biopolymers*, ed. E. D. T. Atkins and A. Keller, Butterworths, London, 1975.
46. R. Pecora, *J. Chem. Phys.*, 1964, **40**, 1604.
47. R. Pecora, *J. Chem. Phys.*, 1968, **48**, 4126.
48. R. Pecora, *J. Chem. Phys.*, 1968, **49**, 1032.
49. R. Pecora, *J. Chem. Phys.*, 1968, **49**, 1036.
50. J. Fraunhofer, *Ann. Phys.*, 1817, **56**, 264.
51. R. Bunsen and G. Kirchhoff, *Abh. kgl. Akad. Wiss., Berlin*, 1861, 1863.
52. J. C. Maxwell, *A Treatise on Electricity and Magnetism*, Oxford University Press, Oxford, 1873.
53. H. Hertz, *Ann. Phys.*, 1888, **270**, 551.
54. H. Hertz, *Berlin Berichte*, 1888, 297.
55. C. Doppler, *Abhandlungen der königlichen böhmischen Geselschaft der Wisenschaften*, 1843, 465.
56. A. Javan, W. R. Bennett and D. R. Herriott, *Phys. Rev. Lett.*, 1961, **6**, 106.
57. T. H. Maiman, *Nature*, 1960, **187**, 493.
58. A. Einstein, *Ann. Phys.*, 1905, **17**, 549.
59. M. Smoluchowski, *Phys. Z.*, 1916, **17**(557), 585.
60. M. Smoluchowski, *Ann. Phys.*, 1906, **21**, 757.
61. M. Smoluchowski, *Ann. Phys.*, 1908, **25**, 205.
62. B. J. Berne and R. Pecora, *Dynamic Light Scattering,* John Wiley, New York, 1976.
63. S. Chandrasekhar, *Rev. Mod. Phys.*, 1943, **15**, 1.
64. D. A. McQuarrie, *Statistical Mechanics*, Harper & Row, New York, 1976.
65. J. G. Kirkwood and J. Riseman, *J. Chem. Phys.*, 1948, **16**(6), 565.
66. *Macromolecules - John Gamble Kirkwood Collected Works*, ed. J. G. Kirkwood, Gordon and Breach, New York, 1967.
67. H. Yamakawa, *Modern theory of polymer solutions*, Harper & Row, New York, 1971.
68. R. Raj and W. H. Flygare, *Biochemistry*, 1974, **13**, 3336.
69. B. R. Ware and W. H. Flygare, *Chem. Phys. Lett.*, 1971, **12**, 81.
70. B. R. Ware and W. H. Flygare, *J. ColloidInterface Sci.*, 1972, **39**, 670.
71. C. W. Oseen, *Hydrodynamik*, Akad, Verlag Ges., Leipzig, 1927.
72. W. H. Stockmayer, in *Molecular Fluids*, ed. R. Balian and G. Weill, Gordon and Breach Science Publishers, New York, 1976.
73. P. E. Rouse, *J. Chem. Phys.*, 1953, **21**, 1272.
74. B. H. Zimm, *J. Chem. Phys.*, 1956, **24**, 269.
75. T. Coviello, W. Burchard, M. Dentini, V. Crescenzi and V. J. Morris, *J. Polym. Sci., Part B: Polym. Phys.*, 1995, **33**, 1833.
76. T. Coviello, K. Kajiwara, W. Burchard, M. Dentini and V. Crescenzi, *Macromolecules*, 1986, **19**, 2826.

77. L. He and B. Niemeyer, *Biotechnol. Prog.*, 2003, **19**, 544.
78. J. P. Wilcoxon and J. M. Schurr, *J. Chem. Phys.*, 1983, **78**, 3354.
79. E. Geissler, in *Dynamic Light Scattering*, ed. E. Brown, Oxford University Press, Oxford, 1993.
80. en.wikipedia.org.
81. W. Burchard and R. Schweins, manuscript in preparation.

CHAPTER 7

Model Calculation of Different Hyperbranched Structures

7.1 Introduction

The present chapter deals with an algorithm that allows calculation of conformational properties of branched polymers.[1] The derivation is made in a one-to-one correlation to chemical synthesis such that at each step in the algorithm the special chemical reaction is taken into account. This stepwise derivation looks like a cascade of equations and is often denoted as Cascade Theory. In a mean field approximation, which corresponds to our first stage of approach, this cascade can be contracted into one simple relationship, but this is no longer possible if excluded volume interactions are enclosed in this theory and requires a separate outline that is given in Section 7.4. It may be emphasized by this stepwise mimic of reactions the arising polydispersity of the molar mass distribution is automatically included.

The main goal is the prediction of structural properties of single molecules, *i.e.* the degree of polymerization DP_w, radius of gyration R_g, hydrodynamic radius R_h and the derivation of scattering functions for hyperbranched polymers of the AB_2 monomer type and further development to various modified hyperbranched structures. Functional polymers with capabilities to produce specific interactions with other active components are presently in a process of chemical verification, such as reversible binding of drugs or metallic catalysts, and to apply these samples as carriers for directed reactions. This aspect will be enclosed in the theory. Another goal is a variation of the degree of branching (DB) over the whole range from $0\% < DB < 100\%$, a goal desired for use in polymeric or other soft materials.[2,3] Several chemical preparations are already known[4] but a quantitative analysis remained so far fairly intuitive based on chemical reactions.

RSC Polymer Chemistry Series No. 16
Hyperbranched Polymers: Macromolecules in between Deterministic Linear Chains and Dendrimer Structures
By Albena Lederer and Walther Burchard

Published by the Royal Society of Chemistry, www.rsc.org

The following outline has three main sections. One is the theoretical derivation on the basis of the general branching theory which gives insight into the architecture and may allow easy application to predicting new properties. The derived theory will be demonstrated with the already well-explored randomly branched samples and with the Flory type of AB_2 polymers.[5,6] In the second section a few modifications are considered which were already synthesized, and whose properties are under investigation. It follows a detailed outline on attempts how the effect of excluded volume interaction can be introduced to obtain a quantitative agreement with the experimentally observed scattering curves. The third section deals with the extension to more complex systems which partially already were synthesized and whose properties could be predicted. Not all details of the derivation will be given but the various steps of the derivations will be commented in simple term.

The outline of the theory is made mainly for the physical chemists who wish to obtain a quantitative understanding of experimental results. The preparative chemists who are mainly interested in synthesis may skip this first part. The summary at the end of this chapter will provide a deeper insight of the result of their synthesis. This may help to better understand the general discussion of the physical properties.

7.2 Conformation Properties of Branched Polymers in the Light of a General Branching Theory

7.2.1 Some General Remarks

In view of the cascade-like appearing publications on syntheses of modified hyperbranched macromolecules it becomes a demanding task to derive a theoretical scheme that can be used for interpretation of the synthesized physical structure. Probably due to the successful synthesis by Hölter *et al.*[7] the mainstream interest of polymer chemists is presently changing the study of linear and randomly branched polymers to that of hyperbranched samples.[4,8,9] The simplest example, denoted as AB_2 polymerization, was already considered in 1952 and 1953 by P.J. Flory.[5,6] He derived relationships for the molar mass distributions and the corresponding number and weight averages of molar mass by using combinatorial mathematics. He discovered the remarkable fact that, in contrast to randomly branched macromolecules, AB_2 polymers can never reach the critical condition for gel formation.

The mathematical tool of combinatorial mathematics failed to predict conformational properties of polydispersed branched chains, *e.g.* the radius of gyration R_g, the hydrodynamic radius R_h and the angular dependence of scattered light, neutrons and X-rays. These scattering curves give valuable information on the shape, segmental structure and mobility of the particles but a conclusive interpretation of these data remained largely insufficient. The gap for predicting conformational properties was to some extent closed by Dobson and Gordon[10–12] adopting the well-developed general branching theory[13–15] to polymers. Gordon noticed that branching of polymers resembles family trees where, for instance, the grandfather is replaced by the sons, and in the next higher generation the fathers by their sons *etc.*, thus keeping the family name over many generations. Evidently a mathematical algorithm is required in polymer science which mimics the structural facts of polymers with family trees. This mechanism is efficiently presented by the properties of *probability generating functions* (p.g.f.) which is the heart of the branching theory.[16] The main advantage of this mathematical scheme is based on the clear assignment of the reaction from the various functional groups by labels attached to the links formed between a selected functional group in the n-th generation to those in the $n+1$ generation. This labelling of the reactions allows for a unique one-to-one assignment of the various steps in the polymerization reactions. Due to this labelling the scheme enables prediction of properties from structural modifications which recently have been verified by special chemical synthesis (see Chapters 3–5).

Probability generating functions are common in probability theory but are not familiar to chemists. For this reason a compact outline of the properties is given prior to applying the method to branched polymers. The synthesis of complex looking modified AB_2 structures is of great interest in view of novel possibilities to develop soft materials with special tailored properties. Preparative chemists tend to interpret their derived structures more or less intuitively by common sense, but their arguments are often not satisfactorily confirmed by theory and physical experiments.

In the past the Flory–Stockmayer mean field theory[5,6,17,18] was applied to *randomly branched* polymers. The critical phenomenon of gelation could be explained, and the molar mass distribution and the corresponding number- and weight-averages of the degree of polymerization, DP_n and DP_w were derived. The theory was further expanded by Gordon *et al.*[12] to calculations of conformational properties (*i.e.* radius of gyration R_g and hydrodynamic radius R_h) and the scattering behaviour. Clearly the results obtained by a mean field approximation cannot correctly represent the behaviour in good solvents[19] where excluded volume interactions are effective. Nonetheless the mean field theory provides a reliable prediction of the degree of polymerization ($DP_w < 5\times10^3$) and gives valuable insight into the internal architecture. Only at high DPs do the molar mass distribution display significant deviations from that in a poor solvent.[20] To overcome the serious complexity, corrections to the unperturbed structure were introduced using the conception of fractal behaviour,[21,22] but this conception remained unsatisfactorily since only *behaviour* but not the absolute values was predicted in terms of power law dependencies which can hold true only in the limit of large structures. At low DP the predicted power law is not realized, and the magnitude of the fractal effects remained uncertain. A significant improvement was achieved by a conjecture of De Gennes[23] and further development by Freltoft *et al.*[24] and Sinha *et al.*[25] A detailed discussion is given in Section 7.4. The treatment of excluded volume interaction in branched samples is at present a conversely discussed feature. We will mention the various attempts and will point out the still existing gaps in theory. Experimentalists are not competent to decide which of the attempts are better defined. For an experimentalist the agreement between prediction and experimental findings is the final criteria for the applicability of interpretation concepts.

7.2.2 Basic Properties of the General Branching Theory

The present contribution aims at elaborating a theory which efficiently can cope with the complexity of branching processes in polymer science. We develop a scheme which will serve as a reference in order to avoid repetition of the theory in forthcoming papers. The present treatment is confined to the mean field approximation. The influence of excluded volume interaction is considered in Section 7.4.

The mean field approximation to branched particles is presently considered as the description of a phantom because it neglects the presence of the volume of the repeat units in the polymer, but it does not neglect bonds which keep the monomer unit linked together. The history to the prediction of the random coil conformation on this basis still has been very fruitful and remained the basis for theories which intend to describe the effects of the monomers finite volume. We will show that the results from the mean field approach are indispensable to making progress in the development for realistic structure description.

Our contribution starts with a short outline of the properties of probability generating functions. In the ensuing section this scheme is applied to *f*-functional *randomly branched* polymers and the AB_2 polymers as simplest examples. These well-known systems are chosen to allow chemists becoming familiarized with the advanced possibilities of this theory for the description of samples with higher branching complexity. The scheme is also extended to branched co-polymers. The main progress with this branching theory is demonstrated with the derivation of conformational properties, *i.e.* the gyration and hydrodynamic radii and the angular dependence in light, small angle neutron (SANS) and small angle X-ray scattering (SAXS).

7.2.3 Generating Functions of Probable Distributions[16]

7.2.3.1 Definition

Let $h\{p_x\}$ be a probability distribution

$$h\{p_x\}: \quad p_0, p_1, p_2 \cdots p_x \cdots \tag{7.1}$$

which is a *discrete set* of probabilities with $p_x < 1$ and the normalization condition $\sum_{x=0}^{\infty} p_x = 1$. The corresponding *probability generating function* is obtained by multiplying each probability p_x by s^x and summing up all elements:

$$H(p,s) = \sum_{x=0}^{\infty} p_x s^x \tag{7.2}$$

The parameter s can be chosen arbitrarily. It is a label for a unique assignment of a special reaction. This auxiliary parameter can also be considered as a function, but it has to fulfil the condition that this function exists in the interval $0 \leq s \leq 1$ and that this function can be differentiated. In particular, the parameter s can be a function $f(s)$. One advantage of a probability generating function (p.g.f.) is that the discrete set of probabilities is transformed into an analytical function. Setting $s = 1$ a return to the discrete set of probabilities is obtained, and after differentiation at $s = 1$ the average of a quantity over the corresponding distribution is gained. After differentiation at $s = 1$ the average of a quantity over the corresponding distribution is obtained.

7.2.3.2 Moments of Distribution

Repeated differentiation with $s = 1$ leads to the moments of the probability distribution where the first three moments M_0, M_1 and M_2 are of special interest in polymer science:

$$M_0: \quad H(p, s=1) = \sum_{x=0}^{\infty} p_x = 1$$

$$M_1: \quad \frac{dH(p, s=1)}{ds} = \sum_{x=0}^{\infty} x p_x = \langle x \rangle_n \tag{7.3a–c}$$

$$M_2 - M_1: \quad \frac{d^2 H(p, s=1)}{d^2 s} = \sum_{x=0}^{\infty} x(x-1) p_x = \langle x^2 \rangle_w - \langle x \rangle_n$$

M_0 describes the normalization condition of the distribution and M_1 is the expectation value, or the average over a quantity in this distribution. With $x = \mathrm{DP}$ the expectation value corresponds to the *number average* degree of polymerization $\mathrm{DP}_n = \langle x \rangle_n$. Adding M_1 to the second derivation the second moment of the distribution is obtained and with M_2/M_1 the weight average degree of polymerization $\langle x \rangle_w$ and the polydispersity index PDI, of the distribution:

$$\mathrm{DP}_w = \langle x \rangle_w = \frac{M_2}{M_1}; \quad \mathrm{PDI} = \frac{M_2}{M_1^2} = \frac{\langle x \rangle_w}{\langle x \rangle_n} \tag{7.4a, b}$$

Furthermore, the p.g.f. of the *weight distribution* $w\{x\} = x p_x / \langle x \rangle_n$ is derived by differentiation of $H(p,s)$, *i.e.*

$$W(p,s) = s \frac{\partial H(p,s)}{\langle x \rangle_n} \tag{7.5}$$

The *weight average* $\langle x_w \rangle = DP_w = \partial W(p,s)/\partial s|_{s=1}$ is obtained after differentiation. Higher moments can be obtained by further differentiation.

7.2.3.3 Convolution

If two distributions are coupled the resulting distribution is not simply a product of the elements from the two distributions $\{p_a\}$ and $\{p_b\}$ but the probability to find a special DP $= x$ is given by a sum over $p_a p_b$:

$$p^x = \sum_{\xi=0}^{x} p_a^{\xi} p_b^{x-\xi}, \tag{7.6a}$$

which represents all possibilities of short- and long-chain combinations to form a chain of x units in length. The process has to be done for all possible p^x. The process is called convolution of two parental distributions, shortly written as a star-product:

$$h_2\{p\}; = h_1\{p_a\}^* h_1\{p_a\}. \tag{7.6b}$$

In this example the convolution of two equal distributions is chosen for reasons which will become clear when dealing with the reaction of f-functional units of alike reactivity (see eqn (7.8)). Turning to generating functions, the complexity of convolution in eqn (7.6) reduces to a simple product of the parental probability generating functions:

$$H_2(p,s) = H_1(p_a,s) \times H_1(p_b,s) \tag{7.7}$$

An important *example* is the reaction of a unit, carrying f equally reactive functional groups, with an extent of reaction α, *i.e.* the probability α that a group has reacted. A single functional group has only two probabilities, *e.g.* α that a group A has reacted and $1-\alpha$ that it has not. The corresponding generating function is

$$F_A(\alpha,s) = 1 - \alpha + \alpha\, s \tag{7.8}$$

The f-fold convolution of this distribution gives the generating function

$$H_f(\alpha,s) = [F(\alpha,s)]^f = (1 - \alpha + \alpha\, s)^f, \tag{7.9}$$

with $H'(\alpha,s=1) = f\alpha$, the average number of formed bonds. The corresponding members in the *probability distribution* can be found from the coefficients of s^x which in this case are the members in a *Bernoulli* distribution

$$b(k,f,\alpha) = \left(\frac{f\,!}{((f-k)\,!\,k\,!)} \alpha^k (1-\alpha)^{f-k} \right); \quad k = 0,\ 1,\ 2, \cdots f \tag{7.10}$$

This example shows that a complex distribution can be built up by products of simple generating functions. The derivative at $s = 1$ in eqn (7.9) leads to the number of formed *bonds*, *e.g.* for $f = 3$ to 3α bonds, but actually four units are involved in this example. To include the basic unit the eqn (7.9) has to be multiplied by s, and differentiation of $W_f(\alpha,s) \equiv sH_f(\alpha,s)$ then leads to the average:

$$W_f'(\alpha, s=1) = 1 + f\alpha. \tag{7.11}$$

7.2.3.4 *Compound Distribution*

As already mentioned the parameter s in a p.g.f. can be replaced by any analytical function of s. Let $F(s) = (1 - \alpha + \alpha s)^2$ is a generating function for reactions of three functional units in the first generation where one of the three functions was used of being linked to the central unit. With the generating function of $W(\alpha,s)$ for the zero-th generation and $F(\alpha,s)$ for the units in the first generation one has

$$\begin{aligned} W(\alpha, s) &= s(1 - \alpha + \alpha F(s))^3 \\ F(\alpha, s) &= s(1 - \alpha + \alpha s)^2 \end{aligned} \tag{7.12}$$

This process defines the generating function of a *compound* distribution $W(s) = f(F(s))$.

7.2.4 Randomly Branched Polymers

7.2.4.1 *Branched Homopolymers*

We may consider the central unit as the average over all units in the system, *i.e.* on average all monomer units are alike (mean field approximation). It may be taken as the root of a tree similar to the great-grandfather (say the 'Adam' in a family tree) (Figure 7.1).

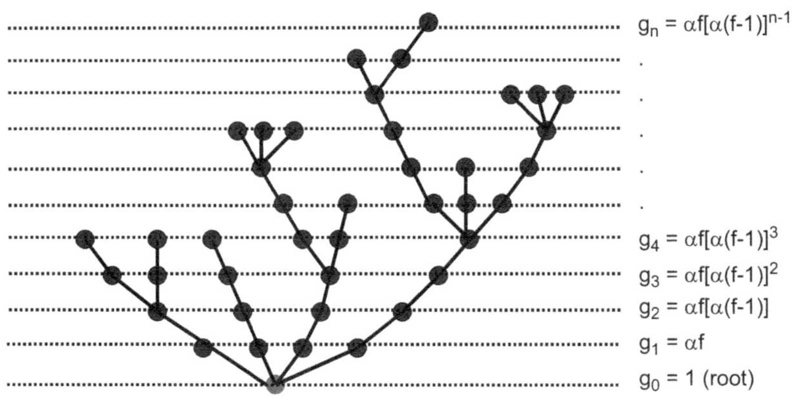

Figure 7.1 Scheme of a rooted tree from a randomly branched polymer of tetrafunctional monomers. Note: this is only one of the four possibilities from which the root may be constructed. The three other consist of (1) three reacted and one not reacted functionalities, (2) two reacted and two unreacted and (3) one reacted and three unreacted functionalities. If all four functionalities have not reacted no tree is formed; in other words, the monomer has not reacted at all. The expression $g_n = \alpha f[\alpha(f-1)]^{n-1}$ denotes the population, *i.e.* the number of units in the n-th generation and α is the probability that a functional group has reacted.

The tree shown in Figure 7.1 refers to a special tree where a monomer unit was chosen at random: The tree in the branching theory corresponds to the *average* over all possible roots from the ensemble of polymers in the system. In a system of f functional units with reactivity α the reactions in the zero-th generation form the first shell or first generation with $f\alpha$ units. One of the f functionalities then has already been used for being liked up to the root and only $f-1$ free functionalities remain for further reaction to form the next generation, *etc.*

Making use of compound distributions we obtain the infinite set of substitution:

$$\begin{aligned} G_0(s) &= sF_0(s) \\ G_1(s) &= F_1(G_0(s)) \\ G_2(s) &= F_2(F_1(G_0(s))) \\ G_n(s) &= F_n(F_{n-1}\cdots F_1(G_0(s)))\,. \end{aligned} \tag{7.13}$$

with :

$$\begin{aligned} F_0(s) &= (1-\alpha+\alpha F_1(s))^f \\ F_n(s) &= (1-\alpha+\alpha F_{n-1}(s))^{f-1} \end{aligned}$$

This cascade indicates the branching process as a Markov process in which the generating functions $F_n(s)$ depend on the preceding generation, and F'_{n-1} becomes the *transition probability* to form the population of the next higher generation. Under mean field conditions the transition probabilities of reaction remains α in all generations. Therefore, one has $F_n(s)=F_1(s)$ for all n. Applying the chain rule of differentiation one obtains

$$G'_n = g_n = sG'_0\prod_{j=1}^{n}F'_j = 1+f\alpha((f-1)\alpha)^{n-1}, \tag{7.14}$$

g_n denotes the average number of units in the n-th generation and the dash denotes differentiation at $s=1$. The sum over all generations leads to the degree of polymerization

$$\mathrm{DP}_w = 1+f\alpha\sum_{n=1}^{\infty}((f-1)\alpha)^{n-1} = 1+\frac{f\alpha}{1-(f-1)\alpha} = \frac{1+\alpha}{1-(f-1)\alpha}. \tag{7.15}$$

This is Stockmayer's equation[17] for f-functional randomly branched chains. The sum in eqn (7.15) converges if $(f-1)\alpha<1$ but diverges at Flory's condition of gelation[5,6,17] $\alpha_{crit}\geq 1/(f-1)$. Eqn (7.15) contains two possibilities for a check of experimental data. In some cases the extent of reaction (probability α) can be determined by spectroscopy, *e.g.* from NMR or other techniques and the predicted weight average degree of polymerization can be checked by static light scattering. Actually the opposite route is preferred, *i.e.* the DP_w is measured and from this the probabilities α is calculated because α is the essential parameter of all other conformational properties.

The derivation of DP_w in two steps can be further simplified and contracted to one single step. This may be shown as follows. Multiplying *each* generating function $F_n(s)$ in eqn (7.13) by s leads to[26–28]

$$\begin{aligned} W(\alpha,s) &= s(1-\alpha+\alpha U_1(s))^f \\ U_1(s) &= s(F_1(U_2(s)))^{f-1} \\ U_2(s) &= s(F_1(U_3(s)))^{f-1} \\ &\cdots\cdots\cdots\cdots \end{aligned} \tag{7.16a}$$

$$\begin{aligned} W' &= DP_w = 1+\alpha U_1' \\ U_1' &= 1+(f-1)\alpha U_2' \\ U_2' &= 1+(f-1)\alpha U_3' \\ &\cdots\cdots\cdots\cdots \end{aligned} \tag{7.16b}$$

The differentiation is always taken at $s=1$ and at this point every generating function reduces to unity. The set of equations can be solved step by step by inserting U_2' in the equation for U_3' *etc.* This cascade of substitutions leads for U_1' to

$$U_1' = \sum_{n=1}^{\infty} [(f-1)\alpha]^{n-1} = \frac{1}{1-[(f-1)\alpha]}, \tag{7.17}$$

multiplication by the denominator leads to the remarkable final result:

$$U_1' = 1 + [(f-1)\alpha]U_1' \tag{7.18}$$

In other words, it is not necessary to go through the infinite cascade of substitution in eqn (7.17) but eqn (7.18) can directly be used and solved for $U_1'(\alpha)$. Inserting the result into the first line of eqn (7.16b) leads to DP_w. Attention has to be paid that a function $U_1(s) = sF_n(U(s))$ was used instead of $G_n(s)$. The additional factor has the effect that, on differentiation, $sG_n(s)$, the population of units from the preceding generation, is added to that in the n-th generation. Thus the summation over all generations g_n can be avoided by using

$$U_1' = 1 + F_1' U_1'; \quad U_1' = \frac{1}{1-F_1'}; \quad DP_w = 1 + \frac{G_0'}{1-F_1'} \tag{7.19a}$$

with

$$\begin{aligned} F_1(s) &= (1-\alpha+\alpha F_1(s))^{f-1} \\ G_0(s) &= s(1-\alpha+\alpha F_1(s))^f \end{aligned} \tag{7.19b}$$

7.2.4.2 Branched Co-polymers

All relationships for random co-polymerization remain similar to those of homopolymers, with the exception that now the scalar transition probabilities have to be replaced by a matrix of transition probabilities. The two monomers A and B may have reactivities α and β for their functionalities A and B units and probabilities $p\alpha$ for an AA linkage. $(1-p)\alpha$ for an AB link, $q\beta$ for a BB link and $(1-q)\beta$ for a BA link. These reactions are described by *two* p.g.f.s, with number fractions n_a and n_b (with $n_a+n_b=1$). Two different types of trees have to be considered, with an A_3 or a B_2 unit as roots. For the branching reaction of A_3 with a linear B_2 monomers the assignments of the probabilities are illustrated in Figure 7.2.

The corresponding probability generating functions are:

- Reaction of the two roots

$$\begin{aligned} U_{A0} &= s(1-\alpha+\alpha p U_{A1}+(1-p)\alpha U_{B1})^3 \\ U_{B0} &= s(1-\beta+(1-q)\beta U_{A1}+q\beta U_{B1})^2 \end{aligned} \tag{7.20}$$

- Transition reactions

$$\begin{aligned} U_{A1} &= s(1-\alpha+p\alpha U_{A2}+(1-p)\alpha U_{B2})^2 \\ U_{B1} &= s(1-\beta+q\beta U_{B2}+(1-q)\beta U_{A2}) \end{aligned} \tag{7.21}$$

and

$$DP_w = n_a U'_{A0} + n_b U'_{B0} \tag{7.22}$$

where the exponents in eqns (7.20) and (7.21) indicate the convolution of the three A and two B functional groups. Using vector/matrix notation the differentiation leads to

$$W'(\alpha,\beta) = n_a U'_{A0'} + n_b U'_{B0} \tag{7.23a}$$

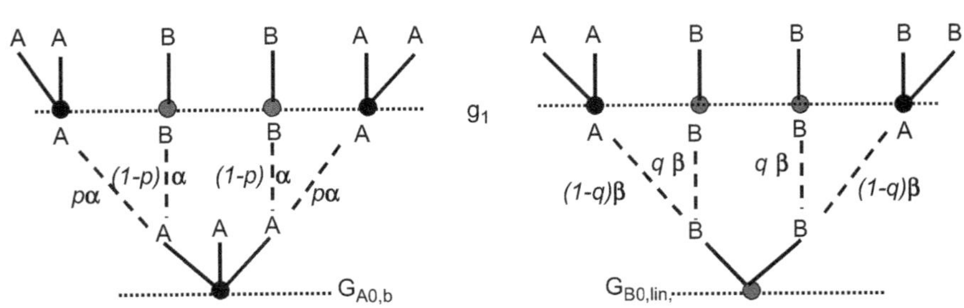

Figure 7.2 Scheme of the reaction between A and B functionalities and the assignment of the various probabilities.

$$U_1' = 1 + TU_2'$$
$$U_2' = 1 + TU_3', \qquad (7.23b)$$
$$U_1' = \sum_1^{\infty} T^{n-1} = (1-T)^{-1}$$

with the matrix of transition probabilities:

$$\mathbf{T} = \begin{pmatrix} 2p\alpha & 2(1-p)\alpha \\ (1-q)\beta & q\beta \end{pmatrix} \qquad (7.24a)$$

and the vector components in the various differentiated generating functions:

$$U_{A0}' = (\, 3p\alpha, \quad 3(1-p)\alpha \,)\begin{pmatrix} U_{A1}' \\ U_{B1}' \end{pmatrix}$$
$$U_{B0}' = (\, 2q\beta, \quad 2(1-q)\beta \,)\begin{pmatrix} U_{A1}' \\ U_{B1}' \end{pmatrix} \qquad (7.24b)$$

To find the degree of polymerization the components of the matrix $\boldsymbol{U}_1'$ in eqn (7.24) has to be determined which would mean inversion of the matrix $(\mathbf{1} - \boldsymbol{T})$. The inversion of a 2×2 matrix may cause no difficulties but if the monomers have more than two functional groups the analytical inversion will become tedious. However, multiplication of both sides of the last line in eqn (7.23b) by $\mathbf{1} - \boldsymbol{T}$ from the left leads to

$$\mathbf{U_1}' = \mathbf{1} + \mathbf{T}\mathbf{U_1}' \qquad (7.25)$$

Written in components, one has to solve the following two equations for U_{A1}' and U_{B1}':

$$U_{A1}' = 1 + 2p\alpha U_{A1}' + 2(1-p)\alpha U_{B1}'$$
$$U_{B1}' = 1 + (1-q)\beta U_{A1}' + q\beta U_{B1'}' \qquad (7.26)$$

Inserting the result into the differentiated eqn (7.20) the weight average degree of polymerization of the two components is obtained and with the number fractions n_a and n_b the DP_w of the total molar mass average.

For application to a special experiment the determination of the four transition probabilities α, β, p, and q is required. This can be done as follows:

1. The four parameters are related to each other by the condition that the number of bonds n_a from A to B units must be the same as that of n_b from units B to A. This leads to $n_a(1-p)\alpha = n_b(1-q)\beta$ and reduces the number of parameters by 1.
2. The two probabilities α and β have to be determined experimentally, *e.g.* by NMR spectroscopy, titration or other techniques.

3. Finally, the last unknown parameter, say p, is obtained from measurements of the molar mass, or DP_w. Special examples will be considered in a forthcoming paper.[1]

7.3 Angular Dependence of Scattered Light

7.3.1 The General Scheme

The use of probability generating functions also permits the derivation of the radius of gyration R_g, the hydrodynamic radius R_h and the angular dependence of the scattered light intensity $P(q)$. The somewhat unexpected fact becomes understandable when the rooted tree, *e.g.* of Figure 7.1 is inspected. One notices that all units in the n-th generation have the same distance from the root. Expressed in number of bond lengths b the corresponding contour length is $nb = l_n$, which connects the unit in the n-th generation with the root. For a random system the segments are formed by random connection of the n bonds and a Gaussian distribution is obtained with the mean square distance of $\langle l_n^2 \rangle = nb^2$, which holds for *unperturbed* chains. The corresponding interference effect of scattered light between the waves from such a segmental chain is

$$\left\langle \frac{\sin(nbq)}{nbq} \right\rangle = \int_0^\infty \left(\frac{3}{2\pi}\right)^{1/2} \exp\left(-\frac{3}{2}\frac{r^2}{\langle l_n^2 \rangle}\right) \frac{\sin(rq)}{rq} 4\pi r^2 \mathrm{d}r$$

or $\qquad$ (7.27)

$$\left\langle \frac{\sin(nbq)}{nbq} \right\rangle = \exp(-nb^2/6) \equiv \phi_n = \phi^n$$

with $q = (4\pi/\lambda)\sin(\theta/2)$ the value of the scattering vector. The sum over all generations corresponds to the sum over all pairs of units in the branched macromolecules and is

$$\frac{R_\theta}{Kc} = M_w \left[1 + \sum_{n=1}^{\infty} g_n \phi^n \right] \tag{7.28}$$

R_θ is the scattering intensity (Rayleigh ratio) at the scattering angle $\theta = 0$, $g_n = G_n'(s=1)$ is the population of units in the n-th generation, K is the contrast factor which depends on the square of the refractive index increment, and c is the weight concentration. Replacing the front factor s by a function $f(s) = s^{\phi_n}$ in all n generations a simple modification of the generating functions in eqn (7.16) is obtained. This leads to the following infinite set of equations,[29,30] for instance with eqn (7.16):

$$\begin{aligned} W(s) &= s\phi^0(1-\alpha + \alpha U_1(s))^f \\ U_1(s) &= s\phi^1(1-\alpha + \alpha U_2(s))^{f-1} \\ U_2(s) &= s\phi^1(1-\alpha + \alpha U_3(s))^{f-1} \\ &\cdots\cdots\cdots\cdots \\ U_n(s) &= s\phi^1(1-\alpha + \alpha U_{n+1}(s))^{f-1} \end{aligned} \tag{7.29a}$$

$$\begin{aligned}
W' &= \phi^0 + f\alpha U_1' \\
U_1' &= \phi^1 + (f-1)\alpha U_2' \\
U_2' &= \phi^2 + (f-1)\alpha U_3^1 \\
&\cdots\cdots\cdots\cdots\cdots\cdots \\
U_j' &= \phi^j + (f-1)\alpha U_{j+1}'
\end{aligned} \tag{7.29b}$$

In contrast to eqn (7.17) the various generations are *not equivalent* to each other, and it seems that one has to perform the tedious set of insertions step by step, *i.e.* U_2' as a function of U_3', U_3' as a function of U_4', *etc.* which finally leads to the sum

$$U_1' = \phi \sum_{n=1}^{\infty} [(f-1)\alpha\phi]^{n-1} = \frac{\phi}{1-(f-1)\alpha\phi}. \tag{7.30}$$

The same result of this equation is also obtained with $\phi = \phi_1$ for all generations:[26–28]

$$U_1' = \phi + (f-1)\alpha\phi U_1' \tag{7.31}$$

which is equivalent to eqn (7.18) with the exception that the probability α is replaced by $\alpha\phi$. The corresponding scattering function is the weight average of $\langle \mathrm{DP} \times P(q) \rangle_w$:

$$\frac{R_\theta}{Kc} = (DP \cdot P(q))_w = U_0'(s=1) = \frac{1+\alpha\phi}{1-(f-1)\alpha\phi}. \tag{7.32}$$

Normalisation by DP_w gives the *z*-average particles scattering factor:

$$P(q) = \frac{(DP \cdot P(q))_q}{DP_w} = \frac{1-(f-1)\alpha}{1+\alpha} \times \frac{1+\alpha\phi}{1-(f-1)\alpha\phi} \tag{7.33}$$

With $\phi = \exp(-b^2q^2/6)$ and $q = (4\pi/\lambda)\sin(\theta/2)$. This result has to be emphasized: There is no need to carry out a separate calculation for the scattering behaviour, but the particle scattering factor $P_z(q)$ is obtained from the equation for DP_w by multiplying the transition probabilities with the function $\phi(q)$. The latter is the interference factor when progressing from one generation to the next higher one, and this is connected to the probability α. The additional factor $\phi \leq 1$ does not violate the convergence. However, we have to emphasize that the simple relationship of eqn (7.30) is restricted to the assumption of unperturbed conformations. It does not hold if excluded volume interactions are effective.

A further comment has to be made on the meaning of *z*-averages. This average type arises from weight distributions in which each element is further weighted by an additional factor. In light scattering the additional weight comes from the coefficients of q^{2n} in the series expansion, which for $n = 1$ is the mean square radius of gyration. This new distribution has to be normalized, which is $\mathrm{DP}_{w.}$. For $q = 0$ the particle scattering factor degenerates to the weight average. For convenience the total particles scattering factor is still denoted as a z-average since all elements with $q^2>0$ fulfil the condition of the z-average, among of them the z-average of the mean square radius of gyration $\langle R_g^2\rangle_z$.

7.3.2 Radius of Gyration

The radius of gyration is given by $R_g \equiv \langle R_g^2\rangle_z^{1/2}$. The mean square radius of gyration is obtained from the coefficient of the first term in the expansion of $P_z(q)$:

$$P_z(q) = 1 - (1/3)\langle R_g^2\rangle_z q^2 + \text{terms of } q^{2n} \tag{7.34}$$

Performing the expansion one obtains

$$R_g^2 = \frac{1}{2}\frac{\alpha f}{(1+\alpha)(1-(f-1)\alpha)}b^2 \tag{7.35}$$

For $b \ll R_g$ the Debye approximation $\varphi = \exp(-b^2q^2/6) \cong 1 - b^2q^2/6$ can be applied to $P_z(q)$[30–32] and leads to

$$P_z(q) \cong \frac{1}{1 + \langle R_g^2\rangle_z q^2/3} \tag{7.36}$$

Surprisingly, this equation for the particle scattering factor shows no direct dependence on the number of functional groups of the monomer and the same angular dependence is obtained for the randomly branched and polydisperse linear chain. All the effects are enclosed in the mean square radius of gyration. We will show in Section 7.4 that this property of the particle scattering factor remains valid over a rather large domain of the q-dependence even when the effect of excluded volume interaction is effective.

7.3.3 Hydrodynamic Radius

The hydrodynamic radius R_h is defined by the equation[33]

$$\frac{1}{R_h} = \frac{1}{2N^2}\sum_{j=1}^{N}\sum_{k=1}^{N}\left\langle \frac{1}{|j-k|b^2}\right\rangle_z. \tag{7.37}$$

No analytical result could be found by solving the sum directly. However, an integral representation exists,[33–35] which corresponds to a continuous chain approximation

$$\left\langle \frac{1}{b|j-k|}\right\rangle = \frac{1}{\pi^{1/2}(|j-k|/6^{1/2})b} = \frac{1}{\pi}\left(\frac{\pi}{nb^2/6}\right)^{1/2} = \frac{1}{\pi}\int_0^\infty \exp(-n(b^2/6)q^2)\mathrm{d}q \tag{7.38}$$

where $|j-k|$ is the number of n units between the two ends j and k but n is also the number of generation. The result has to be summed over all generations. The integrand in eqn (7.39) is just ϕ^n, and the sum over all generations in the integral is $(\mathrm{DP}\times P(q))_w$, and eqn (7.39) leads to[34–36]

$$\frac{1}{R_\mathrm{h}} = \frac{2}{\pi}\int_0^\infty P(q)\mathrm{d}q \tag{7.39}$$

This result is remarkable: Once the particle scattering factor is known in an analytical form *differentiation* of $P(q)$ leads to $\mathrm{d}P(q)/\mathrm{d}q^2 = -R_\mathrm{g}^2/3$, but *integration of* $P_z(q)\mathrm{d}q$ to $(1/R_\mathrm{h})(\pi/2)$. In some cases the integration can be performed analytically, otherwise the integration has to be performed numerically with the experimentally obtained $P_z(q)$. For *f*-functional randomly branched chains in the pre-gel domain one has $P(q) = 1/(1 + R_\mathrm{g}^2 q^2/3)$,[31] and the integration is easily performed. The ratio $\rho = R_\mathrm{g}/R_\mathrm{h}$ is a characteristic branching parameter:[37,38]

$$\frac{1}{R_\mathrm{h}} = \frac{2}{\pi}\left(\frac{\pi}{2}\sqrt{\frac{3}{R_\mathrm{g}^2}}\right) = \frac{\sqrt{3}}{R_\mathrm{g}} \tag{7.40}$$

$$\rho \equiv \frac{R_\mathrm{g}}{R_\mathrm{h}} = \sqrt{3} = 1.73 \tag{7.41}$$

The same results can be obtained for branched samples in a more direct manner that will be outlined at further detail in Section 7.4 which considers excluded volume interactions. Again we have to recall, eqn (7.41) holds for *unperturbed* conformations.

7.3.4 AB_2 Hyperbranched Samples

The polymerization of AB_2 monomers[1] presents in a way a 'hidden' co-polymerization. This fact becomes clear from Figure 7.3 which is drawn to get an overview over the transition probabilities. As with identical *f*-functional monomers, only one type of root can be formed. But the

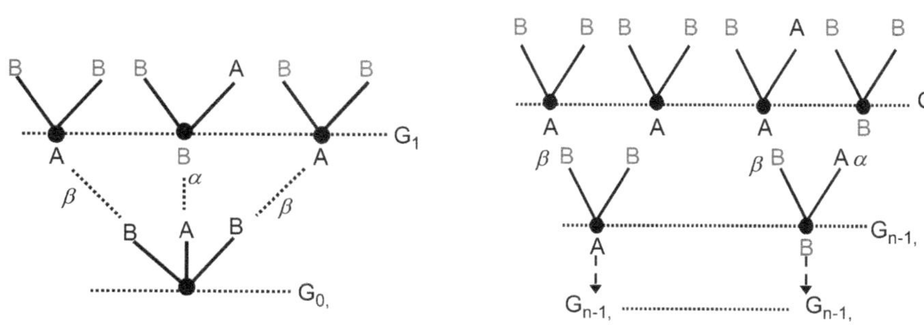

Figure 7.3 Graphs as an overview to find the probability of reaction for the A and B groups to reach the first generation (left graph), and (right) for the transition probabilities by which units in the n-th generation are linked to the $(n-1)$-th generation. Note the constraint $\alpha = 2\beta$.

situation changes with the first generation and the following ones. Now one has to distinguish whether an A-functional group was linked to the root or a B-functional group.

Following the definition of a probability generating function we find for the root of the tree:

$$sG_0(\alpha, \beta, s) = s(1 - \alpha + \alpha F_{1B})(1 - \beta + \beta F_{1A})^2 \tag{7.42}$$

and

$$\begin{aligned} F_{1A} &= s(1 - \beta + \beta F_{1A}(s))^2 \\ F_{1B} &= s(1 - \alpha F_{1B})(1 - \beta + \beta F_{1A}(s)) \end{aligned} \tag{7.43a, b}$$

where the indices in F_{1A} and F_{1B} indicate that the unit in the first generation is linked to the A group of the root or to the B group. With the notation we can set up the generation functions for the first generation and similar for all succeeding generations, where use was made that all generations have the same property which leads to the two eqns (7.43a,b).

7.3.5 Degree of Polymerization and Global Structure Parameters

Differentiation of eqns (7.42) and (7.43a,b) yields

$$\begin{aligned} DP_w &= 1 + \alpha F'_{1B} + 2\beta F'_{1A} \\ F'_{1A} &= 1 + 2\beta F'_{1A} \\ F'_{1B} &= 1 + \alpha F'_{1B} + \beta F'_{1A} \end{aligned} \tag{7.44a}$$

with the solution

$$F'_{1A} = \frac{1}{1-\alpha} \quad \text{and} \quad F'_{1B} = \frac{1}{1-\alpha} + \frac{\beta}{(1-\alpha)^2} \tag{7.44b}$$

and after insertion into the first line of eqn (7.44a) the degree of polymerization is obtained:

$$DP_w = \frac{1+\alpha}{1-\alpha} + \frac{1}{2}\left(\frac{\alpha}{1-\alpha}\right)^2 = \frac{1-\alpha^2/2}{(1-\alpha)^2} = 1 + \frac{2\alpha}{1-\alpha} + \frac{1}{2}\left(\frac{\alpha}{1-\alpha}\right)^2 \tag{7.45a}$$

Eqn (7.45a) is Flory's result[5,6] derived from the weight fraction distribution. It will be noticed that the term $(1+\alpha)/(1-\alpha) = DP_{w,\text{lin}}$ is just the degree of polymerization of a linear chain, and $2\alpha/(1-\alpha)$ is the number of bonds of this chain; therefore, instead of eqn (7.45a) we can write

$$DP_w = DP_{w,\text{lin}} + DP_{w.\text{branched}} = 1 + 2X + \frac{1}{2}X^2; \quad X = \frac{\alpha}{1-\alpha} \tag{7.45b}$$

Replacing α by $\alpha\phi$ and β by $\beta\phi$ and dividing the result by DP_w, the $P_{z,hb}(q)$ is obtained:

$$P_{z,hb} = \frac{1+\alpha\phi}{1-\alpha\phi} + \frac{1}{2}\left(\frac{\alpha\phi}{1-\alpha\phi}\right)^2; \quad \Phi = \exp(-b^2q^2/6) \tag{7.46}$$

The index *hb* denotes hyperbranched. In macromolecules the bond length $b \ll R_g$ and the approximation $\Phi = \exp(-b^2q^2/6) \cong 1 - b^2q^2/6$ can be applied to give[30–32]

$$P_z(q) = \frac{1}{\mathrm{DP}_w}\left[\frac{\mathrm{DP}_{w,\mathrm{lin}}}{1+q^2R^2_{\mathrm{g,lin}}/3} + \frac{\mathrm{DP}_{w,\mathrm{br}}}{(1+q^2R^2_{\mathrm{g,br}}/6)^2}\right] \tag{7.47}$$

$$R^2_{\mathrm{g,lin}} = b^2\frac{\alpha}{1-\alpha^2}; \qquad R^2_{\mathrm{g,br}} = b^2\frac{1}{1-\alpha} \tag{7.48a}$$

and for the mean square *radius of gyration*

$$R^2_{\mathrm{g}} = \frac{\mathrm{DP}_{w,\mathrm{lin}}}{\mathrm{DP}_w}R^2_{\mathrm{g,lin}} + \frac{\mathrm{DP}_{w,\mathrm{br}}}{\mathrm{DP}_w}R^2_{\mathrm{g,br}} \tag{7.48b}$$

7.3.5.1 *Hydrodynamic Radius,* R_h

The derivation for $\langle 1/R_h \rangle$ requires the integration over $P(q)$. With the result of eqn (7.47) according to or the Debye–Zimm approximation[30–32,34,36] the required integrations[36] can be performed to give

$$\frac{1}{R_\mathrm{h}} = \frac{\mathrm{DP}_{w,\mathrm{lin}}}{\mathrm{DP}_w}\frac{\sqrt{3}}{R_{\mathrm{lin}}} + \frac{\mathrm{DP}_{w,\mathrm{br}}}{\mathrm{DP}_w}\frac{\sqrt{3/2}}{R_{\mathrm{g,br}}}. \tag{7.49}$$

7.3.5.2 *Comment*

A closer look at the equation for $F'_{1\mathrm{A}}(\alpha)$ and $F'_{1\mathrm{B}}(\alpha,\beta)$ reveals two essential facts:

1. $1/(1-\alpha)$ is the *number* average units $\mathrm{DP}_{n,\mathrm{lin}}$ of a linear chain which is attached to the root, with probability $2\beta = \alpha$ to a B-functionality.
2. The $F'_{1\mathrm{B}}(\alpha,\beta)$ consists of two terms, one is the $\mathrm{DP}_{n,\mathrm{lin}}$ of the linear chain and the other is the *convolution* of such two chains attached with a probability of $\alpha\beta = 1/2\alpha^2$ to the A-functionality of the root; and the same holds for the link to the A-functionality of a preceding generation. In other words $\alpha\beta = 1/2\alpha^2 = \mathrm{DB}$ is the degree of branching DB.

7.3.6 Summary: Unperturbed Conformations

The global structure parameters, *i.e.* the molar mass M_w and the *z*-average mean square radius of gyration $\langle R^2_\mathrm{g}\rangle_{0z}$ of the unperturbed structure (*i.e.* at

$A_2 = 0$) can easily be predicted from the mean field approximation, because nothing more than the probability of the various functional groups is needed. These data can be determined by spectroscopy or kinetic measurements and are mostly known from the synthesis. The branching theory applied to polymers is an efficient tool to describe these structures of branched samples. Applying the properties of probability-generating functions the correlation between the various probabilities are easy to be handled. For unperturbed polymers the branching theory can be contracted to a simple recipe for the derivation of molar mass and the corresponding conformational properties. The main steps are:

1. The probability-generating functions for the zero-th generation and the first generation have to be set up. The probabilities of reaction and the transition probabilities can be found from the tree presentation for polymers. Differentiation of the generating functions for the zero-th generation leads to DP_w as a function of the first generation.
2. Eqn (7.18) has been solved for the transition probabilities of the first generation and the result has to be inserted into the equation for DP_w.
3. The particle scattering factor $P_z(q)$ is obtained from the equation for DP_w by multiplying the transition probabilities in this equation with the scattering interference function $\phi = \exp(-b^2q^2/6)$.
4. *Differentiation* of $P_z(q)$ leads to the z-average $\langle R_g{}^2 \rangle_z$ and *integration* to the inverse hydrodynamic radius $\langle 1/R_h \rangle_z$, which again is a z-average over the molar mass disribution.

7.4 Effects of Excluded Volume

7.4.1 Introduction: What Does Volume Exclusion Mean?

A linear polymer is considered to consist of a number of monomer repeat units linked by chemical bonds like beads in a pearl necklace. Such repeat units have a finite volume which in a first approximation was neglected and the units were taken as point-like centres. The interest was focussed on the behaviour of chains with $N-1$ bonds linked to form a chain and these bonds were considered to be independent in their motion from each other but under the constraint that they are linked together to neighbours to the left and to the right. The treatment of such phantom chain could be treated with the scheme of a random walk. This assumption immediately led to a random coil in which the end-to-end distance of the chain and also that of shorter segments follow Gaussian statistics. Soon this simple approach was found not to be a satisfactory representative of a real chain, and the finite volume should be taken into account. The motion of the repeat units cannot be fully unperturbed because contacts between such beads should have a noticeable effect. If a bead j comes into contact with bead k a eight times larger than the bead is resulting. The effect of this inhibition is expected to cause an expansion or swelling of the coil. Actually the effective volume exclusion is considerably smaller because of a partially attractive potential that always is present. Thus the effective excluded volume is defined by the following equation

$$\beta = \int_0^{\infty} [\exp(-V(r)/kT) - 1]4\pi r^2 \mathrm{d}r \tag{7.50}$$

where $V(r)$ is the interaction potential as a function of distance r. We refer to eqn (6.22) and Figure 6.10 which illustrates the volume that cannot be entered by two units in contact and the distance dependent potential, for instance if a Lennard-Jones potential is chosen. The corresponding effective excluded volume arises from two parts, a negative contribution causing attraction and a positive contribution due to repulsion at very short distances. A solvent is considered as 'good' if the effect of repulsion strongly exceeds the contribution from the attraction.

Figure 6.10 represents the effect in linear chains where the contact between two segments from a single chain is determined by the *probability to form a ring*, but with a non-covalent bond, because this is the only possible way to make an interaction. The type of interaction as shown in Figure 6.10 is typical of *linear* chains. In contrast, with a branched molecule the majority of interactions arise from *neighboured* chain segments. Ring formation by covalent bonds also takes place and has a significant effect on the reaction, because with each covalently formed ring the number of functional groups is reduced by one and causes in the randomly branched *f*-functional monomers a shift of the gel point towards higher conversion. With branched molecules the question of overcrowding comes up as a new fact in addition to the excluded volume interaction. There are tendencies to treat this phenomenon separately, but actually overcrowding is only a special aspect of excluded volume interaction. Despite the difference in the excluded volume interaction between segments of linear and branched macromolecules all attempts of describing the excluded effect still remains based on the effect of linear chains.

7.4.2 The Flory Expansion Factor

A first mathematical relationship for the coil expansion due to excluded volume interaction was derived 1953 by Flory[5,6,39,40] in terms of an expansion ratio:

$$\alpha^2 \equiv \langle R_\mathrm{N}^2 \rangle / \langle R_{\mathrm{N}\theta}^2 \rangle \tag{7.51}$$

where $\langle R_\mathrm{N}^2 \rangle$ is the mean square end-to-end distance of the chain in a good solvent and $\langle R_{\mathrm{N}\theta}^2 \rangle$ the corresponding value for the same chain in a theta solvent ($A_2 = 0$; unperturbed chain dimensions). Flory assumed a uniform swelling of the coil such that the Gaussian statistics remained valid and derived the equilibrium state by the condition that the swelling force due to repulsive interaction is balanced by the elastic restoring force based on

Gaussian distance distributions. He obtained an analytical expression for the expansion[39,40]

$$\alpha^5 - \alpha^3 = z = c\beta N^{1/2} \tag{7.52}$$

with β the excluded volume of one monomer unit and N the total number of such events. The relationship was much criticized because, in contrast to the assumption of Gaussian chain behaviour, in the end a result was obtained that clearly shows the perturbation of such idealized distributions. Later, de Gennes[23] realized that the effect is caused by the ratio from two distributions in which the volume causes only a second order effect. Despite all reservations experimentalists found the simple equation of Flory's expansion factor very useful to characterize the chain expansion in good solvents, in particular for long chains when the asymptotic behaviour $\alpha^3 \ll \alpha^5$ is considered, which leads to the measurable relationships

$$\alpha^2 \approx AN^{1/5} \quad \text{or} \quad \langle R_\mathrm{N}^2 \rangle \approx b^2 AN^{1.2} \tag{7.53}$$

whereas for the unperturbed theta state $\langle R_{N\theta}^2 \rangle = b^2 N$ is obtained.

Detailed self avoiding simulations on lattices by Domb[41,42] and his coworkers, carried out in the years of 1963 to 1965, proved the reliability of eqn (7.52) with minor corrections for the asymptotic exponent from $\varepsilon = 0.2$ to 0.176. The introduction of a variable exponent ε in eqn (7.53), is sensible because it allows that marginally good solvents with $\varepsilon = 2\nu < 0.176$ could be included in the theory. The magnitude of the exponent $\varepsilon = 0.176$ for good solvent was confirmed somewhat later by the renormalization group (RG) theory with $\nu = 0.588$. The reason for the good agreement of Flory's derivation with the better founded later theories was clarified by de Gennes[23] who noticed that by the assumption of the ideal distribution in both, the expansion and restoring elastic forces the error in Flory's derivation is largely cancelled. Recently, this compensation effect of the used potential was again verified by Rubinstein and Colby.[43]

7.4.3 Size Distribution of Real Chains

The end-to-end distribution is not measurable but measurable structure parameters can still be derived, like the radius of gyration and the hydrodynamic radius without knowing the real distribution which will be shown somewhat later. However, for the derivation of the angular dependence of scattered light or cold neutrons from branched samples in a good solvent the conformational distribution is required. Domb *et al.*[41,42] tried to evaluate this size distribution for uniform linear chains from simulation on the basis of size avoiding random walks (SAW) on different lattices. He obtained, independently of the type of lattices, a distribution

$$W(x) \propto x^l \exp(-Ax^\delta) \tag{7.54a}$$

$$x = r/R_\mathrm{N}\,; \quad \delta = 2/(1-\varepsilon) = 2.43,\ \varepsilon = 0.176 \tag{7.54b}$$

with the exponent ε to describe the mean square end-to-end distance of linear chains or segments the mean square end-to-end distance to be

$$\langle R_N^2 \rangle = b^2 N^{1+\varepsilon}. \qquad (7.55)$$

The value of the exponent $\delta = 2.43$ was generally accepted but the exponent l was difficult to derive and remained for some time a parameter of some vague estimations. This exponent describes the interaction at very short distances. In first applications by Sharp and Bloomfield,[44] Mazur *et al.*[45] and McIntyre *et al.*[46] the exponent l was set $l = \delta$. Utiyama *et al.*[47] were not sure of the equivalence of the two exponents and varied in their calculations the value of l between 2 and 2.43. Further improved derivations showed that even a value of $l = 2$ largely over-estimates the effect of the corresponding volume exclusion since the contacts are not permanent but are strongly fluctuating. An agreeable result was finally obtained by taking into account the *velocity correlation* of the segments in addition to the average position[43,48,49] with the final result

$$W(x) = 0.278x^{0.28}\exp(-1.208x^{2.43}). \qquad (7.56)$$

7.4.3.1 *An Intermediate Summary: Volume Exclusion*

The following facts are understood and can mathematically be described:

1. The swelling of a linear coil in terms of the end to end distance can be expressed by a power law behaviour as a function of the degree of polymerization with an exponent $1 + \varepsilon = 1.176 = 2\nu$, or $\nu = 0.588$, a value that is close to Flory's estimation of $\varepsilon_{\text{Flory}} = 0.2$. For *unperturbed* (not swollen) chains the corresponding exponent is $\nu_0 = 0.500$.
2. This swelling has been derived from a distribution from self-avoiding walk (SAW) simulations on lattices. The normalized distribution was found to display no dependence on the type of lattice used for the simulations. It can be considered as universal such that the results can be used also for flexible chains in solution.
3. One has to keep in mind that the self-avoiding walk was simulated on asserted fixed obstacles whereas in solutions the system is in continuous motion and the chain itself forms the obstacles. Because of the mobility a collision is avoided by a long range correlation due to repulsive interactions. This fact is the reason why the exponent l in the distribution of eqn (7.54) is so much smaller than estimated from static considerations on the size distribution derived from the simulations on a lattice.

7.4.4 The Particle Scattering Factor, *P*(*q*)

The knowledge of the distance distribution is of essential value because it allows for calculations of measurable properties of linear chains, most important for the mean square radius of gyration as a function of the degree of polymerization. In addition, some insight on the segment arrangement inside the coil can be gained. Another important measurable function is the angular dependent scattering intensity which in the normalized version is given by the particle scattering factor:

$$P(q) = \frac{1}{N^2}\sum_j^N \sum_k^N \left\langle \frac{\sin(qr_{jk})}{qr_{jk}} \right\rangle \tag{7.57}$$

with

$$\left\langle \frac{\sin(qr_n)}{qr_n} \right\rangle = \int_0^\infty W(r_n)\frac{\sin(qr)}{qr_n}4\pi r_n^2 \mathrm{d}r_n \tag{7.58}$$

where $r_n = bn$ is the contour length $b(j-k)$ between the two segment ends by j and k with $n = |j-k|$ repeat and the bond length b. The angled brackets $\langle \ldots \rangle$ mean the average of the function $\left\langle \frac{\sin(qr_{jk})}{qr_{jk}} \right\rangle$ which can be formed if the distribution $W(r_{jk})$ is known. For unperturbed conformations this is the Gaussian distribution. This average is easily performed and leads to

$$\left\langle \frac{\sin(qr_n)}{qr_n} \right\rangle = \exp(-b^2 n/6) = \left[\exp(-b^2 q^2/6)\right]^n \equiv \phi^n \tag{7.59}$$

For chains under the influence of excluded volume interactions the distribution of eqn (7.56) is used in following calculations being valid also for segment lengths much shorter than the end-to-end distance of the coil. The derivation of the average in eqn (7.58) with the distribution of eqn (7.56) is principally possible with the distribution of eqn (7.55), which has to be used, and in general the average can be derived. Unfortunately the complexity of the distribution in eqn (7.55) does not allow for the derivation of an analytical equation, but the moments $\langle r^{2j} \rangle$ of the function

$$\left\langle \frac{\sin(qr_n)}{qr_n} \right\rangle = \sum_{j=0}^{\infty} (-1)^j \frac{\langle r^{2j} \rangle}{(2j+1)!} \tag{7.60}$$

can be derived in terms of Γ functions. This was the route which was applied by Sharp and Bloomfield,[44] Mazur *et al.*,[45] McIntyre *et al.*[46] and by Utiyama *et al.*[47] The sum in eqn (7.60) with the alternating components converges very weakly, such that no more than 11 moments could reliably be calculated, and this allows a calculation only for the initial part of the particles scattering factor which probably is already well defined by the mean square radius of gyration. We give here the final equation that was derived by Utiyama *et al.*[47] but which has been adjusted to the $l = 0.28$ exponent in the distribution of eqns (7.54a) and (7.56).

$$P_{\text{uniform coil}}(q) = 2\sum_{j=0}^{\infty} \frac{(-1)^j}{(2j+1)!} \times \frac{\Gamma(a+2j)}{b(1+cj)(2+cj)} (\mathrm{d}x^2)^j \tag{7.61}$$

with the constants: $a = 1.28$, $b = \Gamma(0.5267)$, $c = 1.176$, and $d = 0.782 \times \Gamma(0.267)/\Gamma(0.3691)$.

Mazur *et al.*[45] and McIntyre *et al.*[46] recognized that another route had to be found. They noticed that the main problem arises from the average of $\left\langle \frac{\sin(qbn^{0.588})}{qbn^{0.588}} \right\rangle$ and tried to find an approximation by which the complex sinus function can be transformed into a function which would allow for a solution by integration. We defer a discussion of this complex approximation to a paper that is in preparation, because a very specific argumentation is required, which is beyond the scope of this book. A corresponding check of the theory with very large and uniform coils of linear chains is in progress in cooperation with colleagues from other laboratories.[50]

The impressive results obtained by Mazur *et al.*[45] and McIntyre *et al.*[46] are not much used in applications. Instead the interpretations of scattering data are still made mainly on the basis of unperturbed chains. At first sight such treatment appears senseless but it turned out that the consideration of even unperturbed chains give valuable insight in the arrangement of conformation in linear chains. The reason why this is so, will be shown somewhat later. Before some details of the *unperturbed* chain are considered.

No similar attempts have been made to derive the *perturbed* dimensions of statistically branched macromolecules which encloses the fact of a very broad molar mass distribution. At the moment it appears hopeless to carry on with an approach as was worked out by the treatment given by Mazur *et al.* It appears urgent to look for another approach which would allow a satisfactory fit of experimental scattering curves. Two different approaches will be outlined in the two next sections.

7.4.5 Fractal Approach to Fits of Measured Particle Scattering Factors

Scattering data can often be described by a power law decay when the data are plotted in a $\log(P(q))$ *versus* $\log(qR_g)$ version. Such a power law curves reminds us of fractal behaviour,[24,25] and this gives rise to the question of whether the exponents in such plots can be interpreted as a structure characteristic fractal dimension of the particle. It is a common practice to consider the negative slope in the mentioned plot as $1/d_f$, where d_f is a fractal dimension which is expected to be the mass fractal of the individual macromolecules. This concept is a conjecture and has to be proven to which extent this supposition is valid.

We go back to eqn (7.46), which was derived for unperturbed hyperbranched particles, sometimes called phantom chains. Still the units are linked *via* realistic bonds and due to the finite length of bonds interference of scattered light arises. For this reason already a treatment of unperturbed chains leads to valuable and basic information. The derivation presented by eqns (7.29–7.33) will now be considered under a different aspect.

When, in 1945, Debye derived the now famous equation

$$P_{\text{uniform coil}}(q) = \frac{2}{x^2}[x - 1 + \exp(-x)] \quad x = q^2R^2 \tag{7.62}$$

for the particle scattering factor of uniform linear chains,[30] he compared the effect arising from a single bond with that of the mean square radius of gyration of the sample and realized $q^2b^2/6 \ll q^2R_g^2/3$ such that the approximation $\exp(-b^2q^3/6) \cong 1 - b^2q^3/6$ can safely be applied to the final rather complex exact equation. With this approximation the now well-known Debye equation was obtained.

Appling the Debye approximation[30] to the two components of eqn (7.47) for (polydisperse) hyperbranched samples the following two equations [eqn (7.63) and (7.64)] are obtained for the unperturbed structures:

$$P(q) = \frac{1 + \alpha\Phi}{1 - \alpha\Phi} + \frac{1}{2}\frac{\alpha^2\Phi^2}{(1 - \alpha\Phi)2}; \quad \Phi = \exp(-b^2q^2/6) \cong (1 - b^2q^2/6)$$

which is the same as eqn (7.46),

$$\frac{1 + \alpha\Phi}{1 - \alpha\Phi} \cong \frac{1 + \alpha}{1 - \alpha} \times \frac{1}{1 + q^2R_{g,\text{lin}}^2/3} \tag{7.63}$$

$$\frac{1}{2}\left(\frac{\alpha\Phi}{1 - \alpha\Phi}\right)^2 \cong \frac{1}{2}\left(\frac{\alpha}{1 - \alpha}\right)^2 \times \frac{1}{(1 + q^2R_{g,\text{br}}^2/6)^2} \tag{7.64}$$

The factor $(1 + \alpha)/(1 - \alpha)$ in eqn (7.63) defines the degree of polymerization for a polydisperse *linear* chain[5,6] and the term in eqn (7.64) corresponds to a convolution of two chains and represents the *branched* contribution. The sum of both terms in eqns (7.63) and (7.64) gives the degree of polymerization of an AB_2 hyperbranched polymer.[39,40] The somewhat unexpected contribution of linear chains has the simple explanation of an outer chain that is anchored to the branched core. The *length* of the chain *between two branching points* has on average the same length of the outer chain.

The q-dependencies of the two equations are of special interest, because the first one represents the Zimm particle scattering factor for linear polydispersed Gaussian chains[31] and the second one represents the Debye–Bueche equation[51] which was derived to describe the scattering behaviour of 'heterogeneous' systems by a radial distribution

$$g_{\text{DB}}(r) = \exp(-r/\xi_{\text{DB}}) \tag{7.65}$$

with a correlation length that turned out to be related to the mean square radius of gyration

$$\xi_{\text{DB}}^2 = R_{g,\text{DB}}^2/6 \tag{7.66}$$

Eqn (7.65) is just the Fourier transform of the Debye–Bueche particle scattering factor:[51]

$$P_{\mathrm{DB}}(q) = P_{\mathrm{branch}}(q) = \frac{1}{(1 + q^2 R_g^2/6)^2} \tag{7.67}$$

The Fourier transform of the Zimm equation [eqn (7.63)] is also well known and leads to the corresponding radial distribution

$$g_{\mathrm{lin\ coil}}(r) = \exp(-r/\xi_{\mathrm{lin}})/r \tag{7.68}$$

At this point the question arises, which shape a particle scattering factor would have if a radial distribution function is chosen with properties between those of a Zimm and the Debye–Bueche form. Before showing this the radial distribution of polydisperse, rigid, infinitely thin chain may be mentioned. This function can be obtained from the Fourier transformation of the particle scattering factor that was derived in a very different approach by Goldstein[52] for thin polydisperse rods with a most probable molar mass distribution. He obtained

$$P_{\mathrm{rod}}(q) = \frac{\arctan(qR_{\mathrm{g,rod}})}{qR_{\mathrm{g,rod}}} \tag{7.69}$$

with the corresponding radial distribution

$$g_{\mathrm{inter}}(r) = \exp(-r/\xi_{\mathrm{rod}})/r^2 \tag{7.70}$$

Now we know the radial distribution and the corresponding particle scattering factors for objects of dimensions of d_1, d_2 and d_3. One can try to make an interpolation between the already known radial distributions. This was done for the first time by De Gennes[23] in 1979 who regretfully neglected the exponential factor defined by a correlation length which takes care that all chains have a finite length. In the percolation theory the fact of finite chain length had to be introduced artificially as a cut-off function, but in a statistical system the exponential decay is a more natural choice. (It may be recalled that such a correlation function also occurred in Debye's derivation of the particle scattering factor for uniform linear chains.) In all three examples the exponential factor is an essential part of the radial distribution. De Gennes' idea may be completed by adding the corresponding exponential correlation function which is

$$g(r) = \exp(-r/\xi)/r^{3-d} \tag{7.71}$$

with the Fourier transform

$$P(q) = \frac{\sin[(d-1)\arctan(q\xi)]}{(d-1)q\xi[1 + (q\xi)^2]^{(d-1)/2}} \tag{7.72}$$

This Fourier transform was performed by Sinha *et al.*[25] in 1984 and led him exactly to eqn (7.72). The Fourier transformation left open how the correlation length is defined, but this question is easily solved by the expansion of eqn (7.72) in terms of q^2 whose coefficient has to be $1/3\langle R_g^2\rangle_z$ with the result

$$\xi^2 = R_g^2 \frac{2}{d_f(d_f+1)} \tag{7.73}$$

The d_f may be interpreted as the fractal dimension of the particle.

Some general properties of eqn (7.73) are demonstrated by the three graphs of Figures 7.4–7.6.

Interestingly, the negative slopes do not agree with the fractal dimension which is clearly shown in Figure 7.6. Evidently the asymptotic exponents do not in all cases represent the expected fractal dimension of the particle. If the fractal dimension is $d_{\text{fractal}} > 2.2$ a larger value is obtained from the negative slope in the log–log presentation. This fact was already noticed by Porod[53] who tried to explain the scattering behaviour of polydisperse hard spheres, and he interpreted this higher exponent, larger than 3, being caused by the formation of a well-defined smooth surface. The same negative slope of −4 was also obtained in the derivation of the Debye–Bueche[51] function. The negative slope of −4 is observed with colloidal or other densely packed structures, and now is denoted as the Porod regime. In all these cases the samples are three-dimensional, but mostly disordered. It is

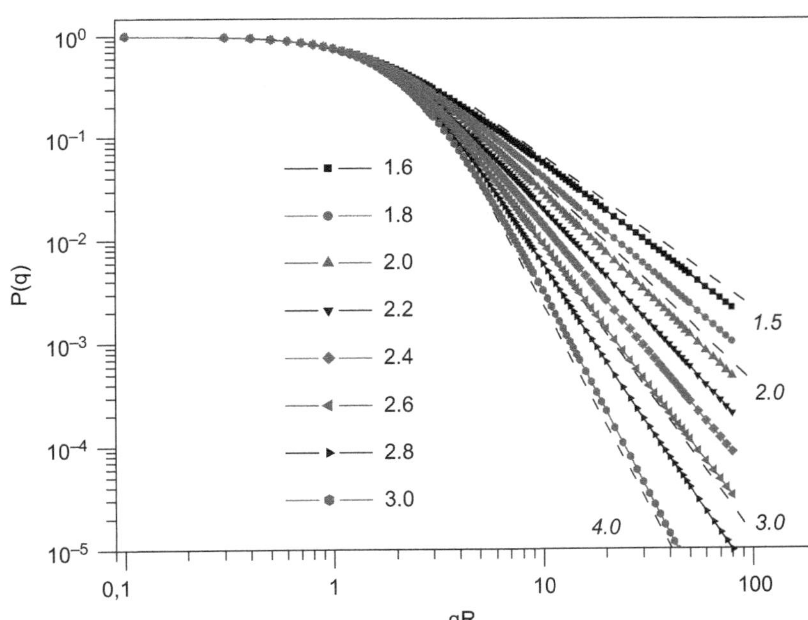

Figure 7.4 The particle scattering factor $P(q)$ as a function of qR_g for a set of different fractal dimensions indicated by the numbers to the left of the graphs. The plot was made in a log–log presentation. The asymptotic power laws are indicated by the dashed lines and by numbers in italics.

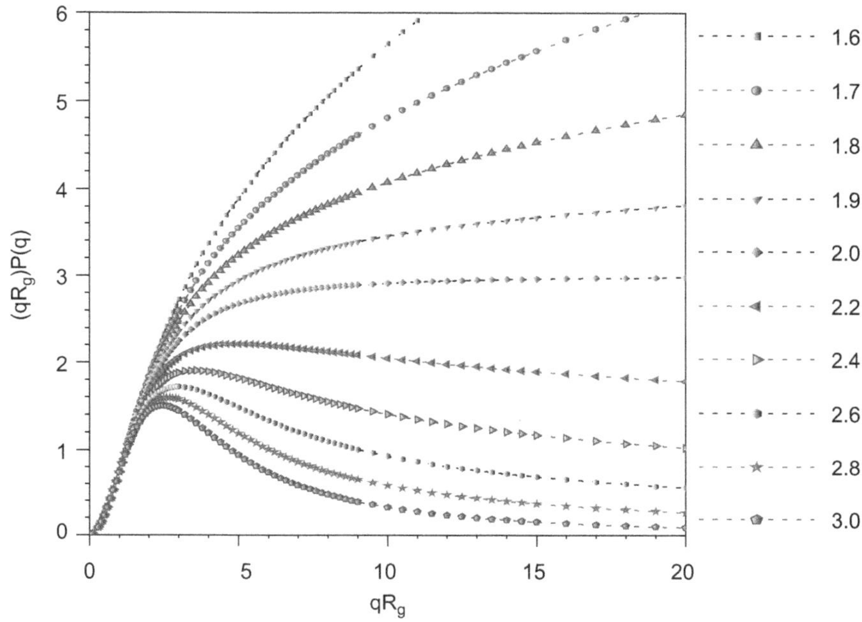

Figure 7.5 The same curves as shown in Figure 7.4 now presented as normalized Kratky plots. The numbers to the right of the figure denote the fractal dimension. The predicted plateau at $d_f = 2$ for polydisperse flexible chains is clearly verified, but for other structures characteristic deviation in the asymptotic regime of large qR_g are obtained.

of interest that, for a correlation function of $d = 3$, a negative slope of -4 is predicted. At present it is not fully clear whether solely the surface is causing this behaviour and how much a large polydispersity contributes to this behaviour and may cause an even larger negative slope than -4.

This assumption is confirmed by the linear chain as another example. It is well known that the uniform linear chain approaches in the Kratky plot a plateau of the height $P(q)q^2R_g^2 \to 2$ for $qR_g > 4$, and according to Zimm's equation[31] for polydisperse coiled linear chains the plateau should reach the value 3. Both predictions were experimentally realized for linear chains in a θ-solvent. The plateau height of 3 is also nicely reproduced by the equation of Freltoft *et al.*[24] and Sinha *et al.*[25] In summary one should be aware that to the value of the negative slope not only the fractal dimension of the particle contributes but also the width of the molar mass distribution may have a significant contribution. We return to this point in Chapter 8 where experimental data are discussed.

Aside from this fact the use of a Kratky plot has another important effect. As shown in Figure 7.4 the scattering intensity decreases dramatically at large qR_g such that important local structure properties are not detectable. After multiplication of $P(q)$ by $(qR_g)^2$ a strong magnification of the asymptotic regime is obtained which now is related to fairly short distances *i.e.* to

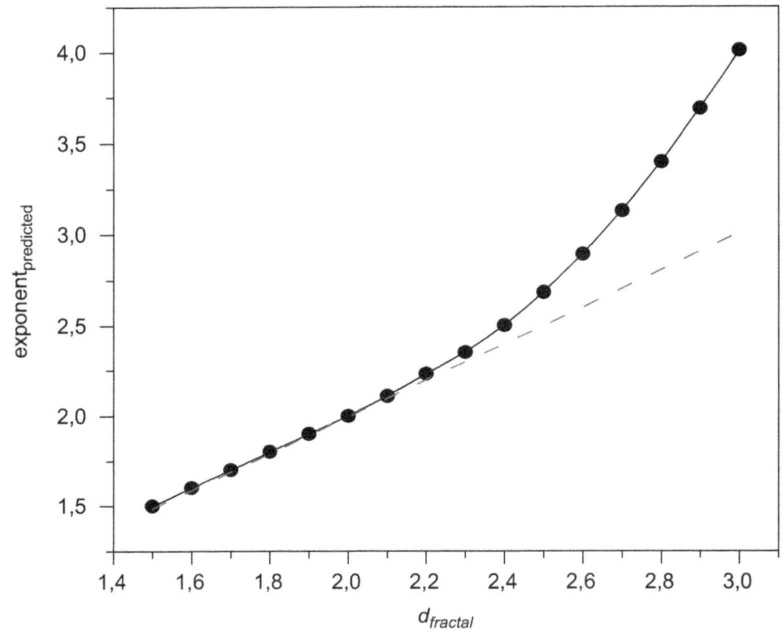

Figure 7.6 Predicted exponents from the asymptote of $P(q)$ at large qR_g-values (unfilled circles) compared with the fractal assumption for disordered systems.

behaviour of local structures. This fact is impressively presented with the predicted scattering behaviour of fractal objects as shown in Figure 7.5. This plot also clearly shows that up to values of $qR_g \leq 3$ no specific fractal dimension can be discerned. All effects of excluded volume interaction are enclosed in the behaviour of the mean square radius of gyration. This fact clarifies why in common experiments mostly the Debye equation is used and still a very reliable conclusion could be drawn. In fact, synthetic polymers scarcely can be prepared with radii of gyration larger than 100 nm. Large structures up to 1 μm are frequently achieved with natural polymers since the enzymatic synthesis is far more selective than it is possible by synthetic means. The situation drastically changes when branched structures are probed by small angle neutron scattering (SANS) where values of up to about of $qR_g = 40$ can be obtained.

7.4.6 The Radius of Gyration and Hydrodynamic Radius of Branched Macromolecules in Good Solvents

For convenience we repeat a few facts known on the influence of excluded volume interactions. The prediction of scattering curves from uniform linear chains *with* excluded volume interaction has a long history. Probably the first idea for treating this problem was given by Sharp and Bloomfield[44] but it was mainly Mazur *et al.*[45] and McIntyre *et al.*[46] who in a clear manner worked out a scheme that soon was applied by Utiyama and his co-workers[47] for a fit of actual measurements. The basis of the Mazur scheme is the self-avoiding walk (SAW) simulation on lattices by Domb *et al.*[41,42] who finally suggested a reliable end-to-end distance distribution as already was given:

$$W(x) = 0.278x^l \exp(-1.208x^{2.43})$$

which is eqn (7.56). A reasonably good fit could be obtained from the initial part of eqn (7.56) but a complete description of the actual curve could not be obtained, even for uniform linear chains. So far known to us the numerical calculations of the Utiyama group[47] were not repeated with the now accepted exponent of $l = 0.28$. The equation for the particle scattering factor $P(q)$ as given by eqn (7.56) now contains the correction. In a second paper, Mazur *et al.*[45] and McIntyre *et al.*[46] suggested an approximation for $P(q)$ for larger q-values which we will not comment upon here.

The Mazur and McIntyre procedure[45,46] can, in principle, be incorporated into the scheme of the branching theory which was used before to derive $P(q)$ of polydisperse but *unperturbed* linear or branched chains. The derivation of the number of units in the n-th generation remains not affected by the excluded interaction, but for all distances an essential change is caused. With the transition probabilities the probability generation functions for all generations can be set up also for perturbed conformations. For convenience some points of the mathematical scheme will be repeated with simple model of f-functional monomers, because we wish to make clear which changes have to be made in order to include the excluded volume interactions. For the first and all other generations we have for the generating functions the cascade of compound equations

$$\begin{aligned} F_1(\alpha, s) &= s(1 - \alpha + \alpha F_2(\alpha, s))^{f-1} \\ F_2(\alpha, s) &= s(1 - \alpha + \alpha F_3(\alpha, s))^{f-1} \\ &\textit{etc.} \end{aligned} \tag{7.74}$$

In order to treat conformational properties a weighting function ϕ_n has to be introduced as an exponent, *i.e.* the effect of light interference that arises from the step length between two adjacent units has to be multiplied as an exponent to as s^{φ_n} with the generating function for the n-th generation by $\varphi_n(q)$, where ϕ_n is a function of the scattering vector $\boldsymbol{q} = (4\pi/\lambda)\sin(\theta/2)$ by which eqn (7.74) is modified to

$$\begin{aligned} F_1(\alpha, s) &= s^{\varphi_1}(1 - \alpha + \alpha F_2(\alpha, s))^{f-1} \\ F_2(\alpha, s) &= s^{\varphi_2}(1 - \alpha + \alpha F_3(\alpha, s))^{f-1} \\ &\textit{etc.} \end{aligned} \tag{7.75}$$

Differentiation at $s=1$ leads to

$$\begin{aligned} &F_1'(\alpha)=\Phi_1+(f-1)\alpha F_1'(a)\\ &F_2'(\alpha)=\Phi_2+(f-1)\alpha F_2'(a)\\ &etc.\\ &F_1'(\alpha)=\Phi_1+(f-1)\alpha\Phi_2+(f-1)^2\alpha^2\Phi_3+\cdots((f-1)\alpha)^{n-1}\Phi_n \end{aligned} \tag{7.76}$$

with $((f-1)\alpha)^{n-1}$ as the number of units in the $(n-1)$-th generation and finally to the sum over all generations:

$$G'(\alpha,F_1'(s=1))=\alpha f\sum_{n=1}^{\infty}[(f-1)\alpha]^{n-1}\phi_n \tag{7.77}$$

For $\phi_n=1$ for all generations the sum in eqn (7.74) leads to the degree of polymerization DP_w but not for all ϕ_n parameters is this the case. Characteristic parameters are obtained if the sum of distances over all generations come into play for instance in the derivation of the mean square radius of gyration and the hydrodynamic radius. These two parameters are easily derived for the unperturbed conformation, but also can be obtained in the same scheme for the perturbed conformation by excluded volume interactions.

7.4.6.1 Unperturbed Conformations

7.4.6.1.1 Radius of Gyration, $\phi_n=nb^2$. The *mean square radius of gyration* of an unperturbed, randomly branched sample is given by the equations

$$\langle R_g^2\rangle=\frac{1}{2N^2}\sum_{j=1}^{N}\sum_{k}^{N}\langle r_{jk}^2\rangle=\frac{1}{2N_{w,\text{bonds}}}\sum_{n=1}^{\infty}b^2n\alpha f((f-1)\alpha)^{n-1} \tag{7.78}$$

The first expression represents general definition, and the second term is the corresponding relationship if the number of bonds in the various generations is used. The total number of bonds is just DP_w-1, which for randomly branched samples is

$$\langle N\rangle_{w,\text{bonds}}=\frac{af}{1-(f-1)\alpha}. \tag{7.79}$$

The infinite sum over second term in eqn (7.78) is easily performed:

$$\sum_{n=1}^{\infty}b^2nf\alpha((f-1)\alpha)^{n-1}=b^2\frac{\alpha f}{1-(a(f-1))^2}$$

and with eqn (7.79) the result is

$$\langle R_g^2\rangle_{z,f}=b^2\frac{1}{1-(f-1)\alpha} \tag{7.80}$$

Eqn (7.80) agrees with the result obtained from the expansion of the particle scattering factor if expressed in bonds rather instead of degree of polymerization.

7.4.6.1.2 Hydrodynamic Radius, $\phi_n = \dfrac{1}{b\sqrt{n}}$

$$\left\langle \frac{1}{R_\mathrm{h}} \right\rangle_{z,f} = \frac{1}{bN_{w,\mathrm{bonds}}} \sum_{j=1}^{\infty} \frac{\alpha f(\alpha(f-1))^{n-1}}{n^{1/2}} \cong \frac{\alpha f}{bN_{w,\mathrm{bonds}}} \int_0^{\infty} \left[\frac{\exp(-(1-\alpha)(f-1))}{n^{1/2}} \right] \mathrm{d}n \tag{7.81}$$

$$\int_0^{\infty} \left[\frac{\exp(-(1-\alpha)(f-1))}{n^{1/2}} \right] \mathrm{d}n = \sqrt{\frac{\pi}{b^2[1-\alpha(f-1)]}} \tag{7.82}$$

$$\left\langle \frac{1}{R_\mathrm{h}} \right\rangle = \frac{1}{b} \sqrt{\pi[1-\alpha(f-1)]} \tag{7.83}$$

The approximation of the sum by the integral is justified since $1-\alpha(f-1) \ll 1$.

The ratio of the two radii is:

$$\rho_0 = \langle R_\mathrm{g}^2 \rangle^{1/2} \langle 1/R_\mathrm{h} \rangle = \sqrt{\pi} = 1.772 \tag{7.84}$$

which is close to $\sqrt{3} = 1.732$, obtained from the integral derivation of the particle scattering factor for unperturbed structure.

7.4.6.2 Perturbed Conformations by Excluded Volume Interactions

7.4.6.2.1 Radius of Gyration, $\Phi_n = \langle r_n^2 \rangle = b^2 n^{2\nu}$; $2\nu = 1 + \varepsilon$. The summation over all generations cannot be performed analytically and had to be approximated by an integral:

$$\begin{aligned} &\langle R_\mathrm{g}^2 \rangle = b^2 \alpha f \int_0^{\infty} \exp -[1-\alpha(f-1)] n^{2\nu} \mathrm{d}n / N_\mathrm{bonds}\,; \\ &N_\mathrm{bonds} = \alpha f / (1-\alpha(f-1)) \quad \langle R_\mathrm{g}^2 \rangle = b^2 \frac{\Gamma(2\nu+1)}{[1-\alpha(f-1)]^{2\nu}} \end{aligned} \tag{7.85}$$

7.4.6.2.2 Hydrodynamic Radius, $\phi_n = \left\langle \dfrac{1}{r_n} \right\rangle = \dfrac{1}{bn^\nu}$

$$\left\langle \frac{1}{R_\mathrm{h}} \right\rangle N_\mathrm{bonds} = (\alpha f / b) \int_0^{\infty} \exp(-[1-\alpha(f-1) n^{-\nu} \mathrm{d}n]) = \frac{\Gamma(1-\nu)}{b[1-(\alpha(f-1))]^{1-\nu}} \tag{7.86}$$

$$\left\langle \frac{1}{R_\mathrm{h}} \right\rangle = \frac{\Gamma(1-\nu)}{b} [1-\alpha(f-1)]^{\nu} \tag{7.87}$$

$$\langle R_g^2 \rangle^{1/2} \langle 1/R_h \rangle \equiv \rho = \Gamma(1-\nu)(\Gamma(2\nu+1))^{1.2} \tag{7.88}$$

where $\Gamma(x)$ is the gamma function.[54]

It is interesting to compare the branching parameter $\rho_0 = R_{g0}/R_{h0}$ for the unperturbed chain with $\rho = R_g/R_h$ of chains with excluded volume interaction:

$$\nu = 0.5 \quad \rho_0 = \sqrt{\pi} = 1.772\,; \qquad \nu = 0.588 \quad \rho = 2.246 \tag{7.89a, b}$$

$$\rho/\rho_0 = 1.267. \tag{7.90}$$

Thus the ratio of the radius of gyration to the hydrodynamic radius is, in a good solvent, about 27% larger than that for the unperturbed chains in fair agreement with experimental results, which is similar to the value of $\rho/\rho_0 = 1.234$ obtained by Akcasu *et al.*[55]

7.4.6.3 *Angular (or q)-dependence of Scattered Light, Neutrons or X-rays*

The scheme as developed by Mazur *et al.*[45] and McIntyre *et al.*[46] can also be introduced into the branching theory. The scheme had to be slightly modified but an analytical evaluation of the occurring integrals could not be solved analytically. We arrive in the last step at the same type of problems which Mazur tried to circumvent by a reasonable approximation which could be integrated. The work is still in progress and the results will be published some time later elsewhere.[50]

7.5 Summary

This chapter has dealt mainly with the theoretical background with the goal of gaining better insight on the structure of branched samples from fits of experimental curves. For *unperturbed* structures branching theory can efficiently be used to obtain simple expressions for quantitative interpretation. The main advantage arises from the fact that this theory builds up the structure formation in a unique one-to-one relationship, where *instead of kinetic constants* the corresponding *probabilities of reaction* are used, similar as outlined by Flory. This approach also includes the effects of broad molar mass distributions but deals essentially with the averages from these distributions. Two simple examples are discussed in detail: the randomly branching A_f polymerization and the AB_2 hyperbranching process. These examples make clear how the meanwhile chemically prepared modification of hyperbranching can be treated.

The theory was further developed to include *excluded volume interactions.* The presented branching theory allows for the derivation of the global parameters M_w, R_g and R_h which for branched samples has not been possible before.

Results obtained by scattering with visible light, cold neutrons and X-ray enclose many additional effects. The angular dependence of the scattered light can strictly be described for low $qR_g < 1.5$ values, but because of a slow convergence in the exact theory the whole q dependence could not yet be fully derived. The search for a suitable approximation is on-going.

Excellent *fits of experimental scattering curves* can be obtained by an equation on the basis of fractal properties with an equation derived by Freltoft *et al.*[24] and Sinha *et al.*[25] We show that this approximation is much better anchored to a basic theory than generally suspected.

References

1. W. Burchard and R. Schweins, manuscript in preparation.
2. A. Khalyavina, F. Schallausky, H. Komber, M. Al Samman, W. Radke and A. Lederer, *Macromolecules*, 2010, **43**, 3268.
3. A. Lederer, T. Hartmann and H. Komber, *Macromol. Rapid Commun.*, 2012, **33**, 1440.
4. A. Hult, M. Johannson and E. Malmström, *Adv. Polym. Sci.*, 1999, **143**, 1.
5. P. J. Flory, *J. Am. Chem. Soc.*, 1952, **74**, 2718.
6. P. J. Flory, *Principles in Polymer Chemistry*, Cornell University Press, Ithaca, New York, 1953.
7. D. Hölter, A. Burgath and H. Frey, *Acta Polym.*, 1997, **48**, 30.
8. Y. H. Kim and O. W. Webster, *J. Am. Chem. Soc.*, 1990, **112**, 4592.
9. Y. H. Kim and O. W. Webster, *Macromolecules*, 1992, **25**, 5561.
10. G. R. Dobson and M. Gordon, *J. Chem. Phys.*, 1964, **41**, 2389.
11. G. R. Dobson and M. Gordon, *J. Chem. Phys.*, 1965, **43**, 705.
12. M. Gordon, *Proc. R. Soc. London, Ser. A*, 1962, **268**, 240.
13. F. Galton, *Educational Times*, 1873, **26**, 17.
14. H. W. Watson, *Educational Times*, 1873, **19**, 115.
15. H. W. Watson and F. Galton, *Journal of the Antropological Institute of Great Britain and Ireland*, 1874, **4**, 138.
16. W. Feller, *An Introduction to Probability Theory and Its Applications*, John Wiley and Sons, New York, 1957, ch. 11&12.
17. W. H. Stockmayer, *J. Chem. Phys.*, 1943, **11**, 45.
18. W. H. Stockmayer, *J. Chem. Phys.*, 1944, **12**, 125.
19. *See Textbooks in Polymer Science.*
20. M. Weissmüller and W. Burchard, *Polym. Int.*, 1997, **44**, 380.
21. B. B. Mandelbrot, *The Fractal Geometry of Nature*, Freeman & Co., New York, 1983.
22. D. Stauffer, *Introduction to Percolation Theory*, Taylor and Francis, London, 1985.
23. P. G. DeGennes, *Scaling Concepts in Polymer Physics* Cornell University Press, Ithaka, London, 1979.
24. T. Freltoft, J. K. Kjems and S. K. Sinha, *Phys. Rev. B: Condens. Matter Mater. Phys.*, 1986, **33**, 269.

25. S. K. Sinha, T. Freltoft and J. Kjems, in *Kinetics of Aggregation and Gelation*, ed. D. P. Landau, Springer Publishing Company, 1984, pp. 87.
26. M. Daoud and J. E. Martin, in *The Fractal Aproach to Heterogeneous Chemsitry*, ed. D. Avnir, John Wiley & Sons, Chichester, 1989.
27. I. J. Good, *Math. Proc. Cambridge Philos. Soc.*, 1955, **51**, 240.
28. I. J. Good, *Math. Proc. Cambridge Philos. Soc.*, 1960, **56**, 367.
29. K. Kajiwara, W. Burchard and M. Gordon, *Br. Polym. J.*, 1970, **2**, 110.
30. P. Debye, Rubber Reserve Company, Technical Report no. **637**, 1945.
31. B. H. Zimm, *J. Chem. Phys.*, 1948, **16**, 1099.
32. B. H. Zimm, R. S. Stein and P. Doty, *Polym. Bull.*, 1945, **1**, 90.
33. H. Yamakawa, *Modern theory of polymer solutions*, Harper & Row, New York, 1971.
34. K. Kajiwara, personal communication, 1969.
35. W. Burchard, in *Light Scattering from Polymers*, Springer Berlin Heidelberg, 1983, ch. 1, vol. 48, p. 1.
36. I. S. Gradshteyn and I. M. Ryzhik, *Tables of Intergrals, Series and Products*, Academic Press, New York, 1965.
37. W. Burchard, in *Branched Polymers II*, ed. J. Roovers, 1999, vol. 143, p. 113.
38. W. Burchard, M. Schmidt and W. H. Stockmayer, *Macromolecules*, 1980, **13**, 1265.
39. P. J. Flory, *J. Am. Chem. Soc.*, 1941, **63**, 3096.
40. P. J. Flory, *J. Am. Chem. Soc.*, 1941, **63**, 3083.
41. C. Domb, *J. Chem. Phys.*, 1963, **38**, 2957.
42. C. Domb, J. Gillis and G. Wilmers, *Proceedings of the Physical Society of London*, 1965, **85**, 625.
43. M. Rubinstein and R. H. Colby, *Polymer Physics*, Oxford University Press, Oxford, 2003.
44. P. A. Sharp and V. A. Bloomfield, *J. Chem. Phys.*, 1968, **49**, 4564.
45. J. Mazur, D. McIntyre and A. M. Wims, *J. Chem. Phys.*, 1968, **49**, 2896.
46. D. McIntyre, J. Mazur and A. M. Wims, *J. Chem. Phys.*, 1968, **49**, 2887.
47. H. Utiyama, Y. Tsunashi and M. Kurata, *J. Chem. Phys.*, 1971, **55**, 3133.
48. A. R. Khokhlov and A. Y. Grosberg, *Statistical Physics of Macromolecules*, AIP Press, New York.
49. J. des Cloizeaux and G. Jannik, *Polymers in Solution. Their Modelling and Structure*, Clarendon Press, Oxfrod, 1990.
50. W. Burchard and P. Lu, Manuscript in preparation.
51. P. Debye and A. M. Bueche, *J. Appl. Phys.*, 1949, **20**, 518.
52. M. Goldstein, *J. Chem. Phys.*, 1953, **21**, 1255.
53. G. Porod, *Kolloid Z. Z. Polym.*, 1951, **124**, 83.
54. M. Abramowitz and I. Stegun, *Handbook of Mathematical Functions*, Martino Fine Books, Mansfield Centre (USA), 2014.
55. A. Z. Akcasu, M. Benmouna and C. C. Han, *Polymer*, 1980, **21**, 866.

CHAPTER 8

Scaling Conception and Interpretation of Structural Parameters and Scattering Experiments

8.1 Molecular Properties in a Global Scale

8.1.1 General Comments

Macromolecules in solution and in melts are soft materials. This statement indicates that not only is the viscosity of the solution much higher than that of the solvent, but also the individual macromolecules are deformable and show significantly deviating properties from that of hard spheres. Certainly the softness in melts facilitates processing of the materials. Thus, a comprehensive investigation of the molecular properties is an essential prerequisite in the field of polymer science. In scattering experiments, mainly by SANS, one can study the shape and the internal structure of the individual molecules. This becomes possible since by changing the scattering angle from small to large values much smaller distances inside the particle can be probed. This is accomplished by the scattering vector $q=(4\pi/\lambda)\sin(\theta/2)$ which is of inverse dimension $1/d$ $(1/\lambda)$. At small scattering angles the whole particle is seen and probed by the experiment and at large angles the segmental structure. It is sensible to discuss these two different scales separately. In the end we will show that these two regimes are two aspects of a more general conception denoted as fractal behaviour. We start this discussion with the so-called global properties, because some of these

RSC Polymer Chemistry Series No. 16
Hyperbranched Polymers: Macromolecules in between Deterministic Linear Chains and Dendrimer Structures
By Albena Lederer and Walther Burchard

Published by the Royal Society of Chemistry, www.rsc.org

structural parameters which are obtained from scattering experiments can also be measured by conventional techniques.

8.1.2 Molecular Parameters

In Table 8.1 techniques are listed by which global parameters can be measured. With global we mean structural parameters like the molar mass M_w; the corresponding radius of gyration R_g, hydrodynamic radius R_h, the intrinsic viscosity $[\eta]$ and the second virial coefficient A_2 by which the particles are probed as an integer equivalent sphere.

The parameters Φ and Ψ depend on details of the internal structure and cannot be measured directly, but from combination of at least two other global structural parameters some information on the internal structure can be obtained. From the physical behaviour Φ is related to coil *penetration* of linear or branched particles by the *solvent*, and Ψ results from segment penetration from two particles into contact.

Another way of looking at the properties is to describe the measured quantities by power laws in terms of molar mass:

$$\begin{aligned} [\eta] &= K_\eta M^{a_\eta} \\ D &= K_D M^{a_D} \\ A_2 &= K_{A_2} M^{a_{A_2}} \\ R_g &= K M^{\nu} \end{aligned} \tag{8.2}$$

Evidently these four relationships are closely related *via* molecular radii which may differ in their magnitude but should show similar molar mass

Table 8.1 Molecular parameters of hard sphere particles which can be obtained by measurements of osmotic pressure, static and dynamic light scattering and viscometry.

Molar mass	M_n; M_w		Osmometry, static LS
Intrinsic viscosity	$[\eta] = \lim_{c\to 0} \dfrac{\eta(c) - \eta_{\text{solvent}}}{\eta(c)\,c} = \Phi \dfrac{R_g^3}{M_w}$	(8.1a)	Capillary viscometry
Diffusion coefficient	$D = \dfrac{kT}{6\pi\eta_{\text{solvent}} R_h}$	(8.1b)	Boundary spreading Dynamic LS
Second virial coefficient	$A_2 = 4\pi^{3/2} N_A \dfrac{R_g^3}{M_w^2} \Psi$	(8.1c)	Osmotic pressure, static LS
Radius of Gyration	$R_g \equiv \langle R_g^2 \rangle^{1/2}$	(8.1d)	Osmotic pressure Static LS, SANS

dependencies. With this conjecture the following correlation between the exponents is obtained:

$$a_\eta = 3\nu - 1$$

$$a_D = -\nu \tag{8.3}$$

$$a_{A_2} = 3\nu - 2$$

The exponent ν describes the molar mass dependence of the radius of gyration $R_g = KM^\nu$. The relationships have been known for a long time and can easily be derived from the set of eqn (8.1a–d). The relationships of eqn (8.3) are denoted as *scaling relationships*,[1] a notation which requires further comments given below when the principle of scaling is introduced.

The molecular radii in eqn (8.1) are indicated by special subscripts which indicate that these radii should have the value, but actually differ significantly. The question arises what the reason is for this fact.

A critical check of the physical background reveals that only the mean square radius of gyration is a universal parameter, which without any assumption on the structure is uniquely defined by the interference of scattered light. The radius of gyration is the *decisive* parameter for all other structural parameters, and the deviations of the other radii from R_g are affected by additional physical phenomena. There are three main issues which have to be taken into account:

- The statistical nature of structure formation
- The response to friction caused by the viscosity of the solvent of dispersing medium
- The internal mobility of the segments in a particles, due to irregular, Brownian motions, and a possible deformation of the particle shape by external forces, *e.g.* by shear.

8.1.2.1 Monodisperse Colloidal Particles

Colloids are mostly of spherical shape with a smooth surface and homogeneous density, and they cannot be deformed by an external force. For polymers spherical particles are *idealized* structures, and the radii indicated in eqn (8.4a–c) should have the same value for all particles in the ensemble.

$$R_h = \frac{kT}{6\pi\eta D} = 0.7746 R_{sphere} \quad \text{Einstein}$$

$$R_\eta = \left(\frac{3[\eta]M}{10\pi N_A}\right)^{1/3} \quad \text{Einstein} \tag{8.4a–c}$$

$$R_{A_2} = \left(\frac{3A_2M^2}{16\pi N_A}\right)^{1/3} \quad \text{van der Waals}$$

For hard spheres, agreement with these relationships was experimentally verified.

8.1.2.2 Polymers and Soft Spherical Particles

Polymers in solution possess highly fringed surfaces and possess no constant radial density but the density finally decays exponentially if a certain correlation length is exceeded and the correlation length should be related to the radius of gyration, which can directly be measured as the root mean square radius of gyration $R_g = \langle R_g^2 \rangle^{1/2}$ where the angle bracket means the average over all possible arrangements of the segments in the particle. To find this radius of gyration from experiments by one of the relationships in eqn (8.1a–c) one requires the introduction of two new parameters, *i.e.* Φ and Ψ. The one is determined by a specifically hydrodynamic interaction,[2,3] the other by the statistics of repulsion between the repeat units from two polymers in contact.[4–6] As already mentioned, in contrast to hard spheres the polymer particles have fringed outskirts such that a partial interpenetration occurs and if two particles come into contact the clouds of segments partially interpenetrate. The statistical treatment of this problem is complex and for some special structures[7,8] the Ψ-value has been derived, *e.g.* for random coils of uniform linear chains and for star-branched macromolecules up to about 10–12 arms, and of course for $\Psi_{\text{hard sphere}}$. The corresponding value for hard spheres can be easily calculated from the van der Waals equation replacing the sphere radius by the corresponding radius of gyration.

8.1.2.3 The Ψ-parameter

The value of this quantity can be obtained from measurements of the second virial coefficient and the molar mass. Figure 8.1 and Tables 8.2 and 8.3 show some results from different types of macromolecules.

The Ψ-parameter depends for small macromolecules on the molar mass but approaches an asymptotic value Ψ^* at large dimensions.[7,8] So far it is not clear whether the quoted data for the branched samples represent already the asymptote. Nonetheless the data display a sensible trend, and increase with samples of higher segmental density. At present the Ψ interpenetration function has not evoked much interest, but it will become an *important* quantity for *hyperbranched* and *modified hyperbranched materials*, because of absent entanglement with these macromolecules which can take place if at least a certain depth of segmental interpenetration is possible.[9,10] For dendrimers no penetration at all can be expected but in addition some special repulsive interactions may cause an even larger value than of $\Psi_{\text{hard sphere}} = 1.62$, expected for spheres.

8.1.2.4 The Viscosity Draining Parameter, Φ

Because of the well-defined smooth surface of hard spheres only the friction between solvent molecules has to be taken into account, *i.e.* between an

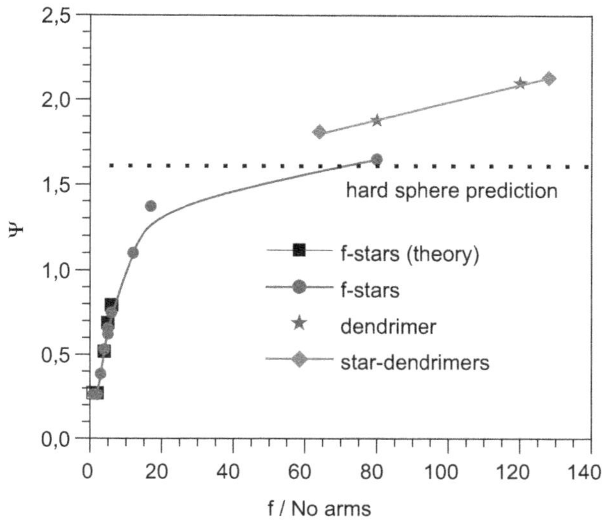

Figure 8.1 Inter-penetration value for $\Psi(f)^*$ from linear and star-branched chains and of some hyperbranched samples. The data are given in Table 8.2. The circles represent experimental data and the squares correspond to theoretical predictions.[8]

Table 8.2 Collection of some values of the inter-penetration function Ψ^*.[11]

Polymer structure	f	Ψ^* (exp.)	Ψ^*(theor.)	Reference
Linear chains	1	0.26	0.269	Yamakawa[12]
Linear chain	2	0.26	0.269	Norisuye[13]
Star	3	—	0.384	Douglas[14]
Star	4	0.53	0.517	Bywater[11]
Flexible rings	5	0.62	0.687	Roovers[15–18]
Star	6	0.75	0.793	Burchard[19–21]
Star	12	1.1	—	Burchard[19–21]
Star	17	1.81		Burchard[19–21]
Star dendrimer	64	1.81	—	Roovers[15]
Dendrimer	80	1.89	—	Rietveld[22]
Star dendrimer	128	2.13	—	Roovers[15]
Star microgel	S10	1.24	—	Lang[23]
Star microgel	S20	2.03	—	Lang[23]
Hard sphere	—	—	1.61	Yamakawa[12]

adsorbed solvent layer and the surrounding solvent molecules. This problem was solved by Einstein[27,28] and he found the now well-known relationship, as quoted in eqn (8.4b). The situation becomes more complex if the outskirts of the polymer particle resemble dangling fringes such that the solvent can partially flow through the interior of the particles. The simplest assumption that can be made is that each monomer repeat unit is considered as a little bead and would contribute to the friction, as was mentioned for a colloidal sphere. This approximation is denoted as the *free draining* case.

Table 8.3 Some data of the inter-penetration function Ψ from hyperbranched starch fragments[24,25] and from AB_2 hyperbranched polymers together with four pseudo-dendrimers of the same core (below).[21,26]

$M_w \times 10^{-3}$ (g mol^{-1})	Ψ	Φ (mol^{-1})	$A_2M_w/[\eta]$	$\rho = R_g/R_h$	d_{app} (g cm^3)
66	0.050	0.38	2.95	1.84	0.0399
72	0.079	0.51	2.19	1.69	0.0577
105	0.075	0.43	2.22	1.77	0.0523
150	0.115	0.66	1.95	1.55	0.0541
320	0.110	0.67	1.89	1.44	0.0384
920	0.140	0.81	2.09	1.34	0.0322
1700	0.120	0.75	2.37	1.41	0.0255
5200	0.170	1.09	2.33	1.35	0.0252
14 500	0.140	0.92	2.15	1.35	0.067
43 000	0.200	0.99	2.25	1.20	0.125

G	M_w g mol^{-1}	Ψ	Φ 10^{-24} mol^{-1}	$A_2M_w/[\eta]$	$\rho = R_g/R_h$	d_{app} g cm^{-3}
1	23 300	0.0756	0.54	1.88	1.53	0.007
2	53 000	0.230	1.26	2.45	1.10	0.018
3	101 000	0.337	2.29	1.98	1.04	0.038
4	116 200	0.316	4.70	0.90	0.80	0.078
5	186 500	0.458	7.06	0.85	0.76	0.125

In this model the assumption is included that the flow velocity at each of the units is the same for all beads. It was Kirkwood and Riseman[29] who pointed out that the flow velocity of the solvent around the beads differs from the macroscopic flow velocity because of hydrodynamic interactions which becomes effective for short distances of surrounding other beads in the coil. Hence, the velocity at positions of the units is $v + \Delta v$ where v is the macroscopic flow velocity, and Δv refers to the effect of hydrodynamic interaction. As outlined already in Chapter 6 this hydrodynamic interaction causes an attraction between neighboured chains but decreases on average like $1/r$ where r denotes the distance between two beads. As expected, the mathematical treatment becomes involved because now not only the distance distribution between the various beads has to be taken but also the hydrodynamic interaction in addition to the free draining effect.

The problem could be satisfactorily solved for samples undergoing Brownian motion, which is the basis for diffusion and dynamic light scattering. Unfortunately, in addition, the intrinsic viscosity arises from applied shear forces. No analytical solution was possible despite many efforts, and the occurring integral equation had to be solved by sensible approximations. The best result so far was obtained by Kirkwood *et al.*,[3] further revised by Auer and Gardner[2] for uniform linear chains who expressed the effect of draining by a function of XF(X) with $X = (6/\pi)^{1/2} n^{1/2}$ where n is the degree of polymerization. The function of XF(X) is listed in the book by Kurata *et al.*[12] Figure 8.2 shows the dependence of this draining function (curve 1) as a function of the degree of polymerization for uniform linear chains compared with experimental data from polydisperse branched samples.

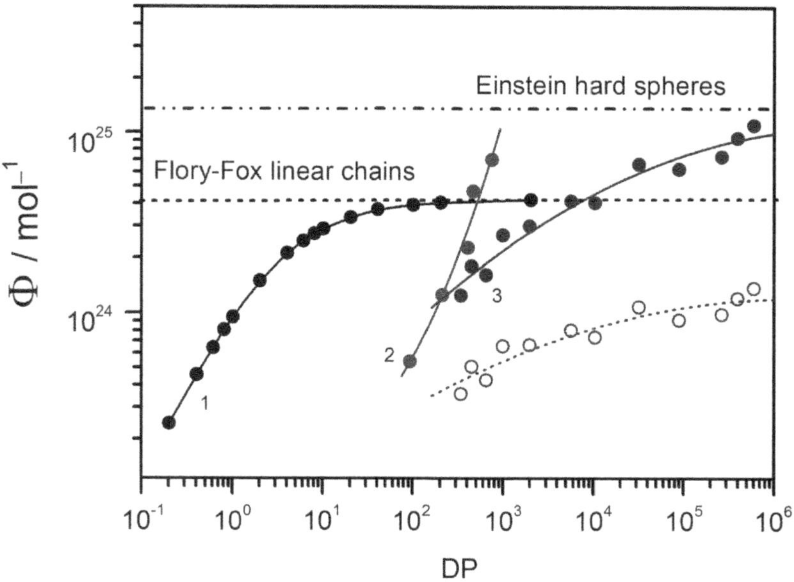

Figure 8.2 Viscosity draining parameter as a function of degree of polymerization from two different examples in comparison to prediction by theory. Curve 1 and the symbols on it represent the theoretical predictions by Kirkwood *et al.*[3,29] Curve 2 and the symbols on it show the behaviour of pseudo-dendrimers of one to four generations.[21] Curve 3 and the symbols on it represent experimental data from starch fragments obtained by a controlled degradation of native potato starch in ice-water.[24,25] The open circles refer to the data corrected for uniform (fractionated) samples.

The draining arises from hydrodynamic interactions which for linear chains approaches a constant value of $\Phi_{\text{Flory}} = 4.22 \times 10^{24}$ mol^{-1}. A hard sphere makes draining impossible; the corresponding value is $\Phi_{\text{hard sphere}} = 13.57 \times 10^{24}$ mol^{-1} that can be derived from Einstein's equation [eqn (8.4b)].[27,28] Curve 3 in Figure 8.2 were obtained with starch fragments corresponding to AB_1B_2 hyperbranched polymers. The filled symbols refer to the unfractionated samples and the open ones refer to the same samples after correction for the broad size distributions. In the AB_1B_2 model the functional group B_1 of the C_4OH group of the glucose unit has a higher reactivity than the B_2 group, corresponding to the C_6CH_2OH group of the glucose, which causes the branching. The degree of branching is $DB = 2\alpha^2 p_1 p_2$ and is expected to be considerably smaller than the $DP = 0.08$ of native starch. α is the probability that the glucose had reacted and αp_1 and αp_2 are the probabilities of reaction of the B_1 group and B_2 groups. It might be useful to recall that the degree of polymerization is determined *via* the probability α and is given by the equation

$$DP_w = 1 + 2X + 2p_1 p_2 X^2 \quad \text{with} \quad X = \alpha/(1-\alpha) \tag{8.5}$$

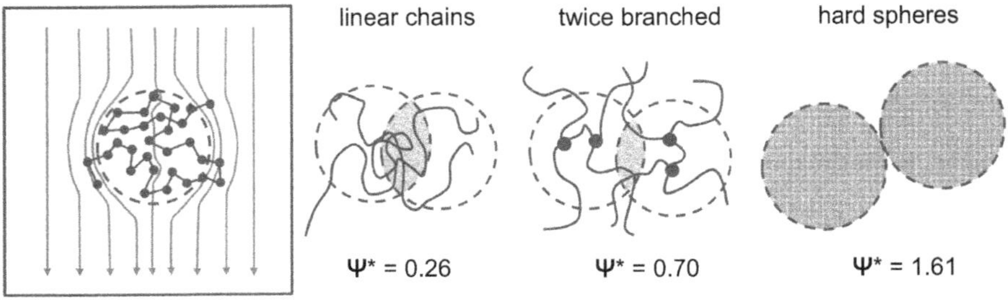

Figure 8.3 The *left* cartoon illustrates how the flow pattern is perturbed by the presence of a macromolecule and that the solvent molecules can partially flow through the lower density part in the outskirts of a coil. The *right* cartoon demonstrates the hindered penetration of segment clouds if two coils meet ($\Psi \approx 0.26$) compared to a much stronger restriction with two branched particles ($\Psi \approx 0.70$) and the case of two hard spheres where no penetration at all is possible ($\Psi \approx 1.61$).[31]

Flory and Fox[30] came to the conclusion that for most linear polymer chains the degree of polymerization is higher than 100 repeat units, and therefore they considered the asymptotic draining parameter as a reasonable representative for linear chains, independent of polydispersity. This limiting draining is often taken as a universal law, applicable also to hyperbranched samples and otherwise structurally constructed chains. However, curves 2 and 3 in Figure 8.2 clearly demonstrate significant deviations from the Φ_{Flory} parameter. Apparently the draining effect strongly depends on the segment density. Furthermore the data obtained from the hyperbranched starch fragments display a marked influence of the polydispersity, but the value for hard spheres was never exceeded. Corresponding data from synthetic hyperbranched polymers are not yet available, because of the small size of these macromolecules which did not permit determination of the radius of gyration.

The scheme of Figure 8.3 may be instructive to understand the difference between draining and segment interpenetration. A remark on this effect was made in Chapter 5. The right cartoon makes it clear that the segment clouds of two linear chains can fairly well interpenetrate. With increasing branching such penetration becomes hampered by dangling chains, and the penetration is reduced. Finally for hard spheres no penetration can be possible.

8.1.3 Generalized Ratios of Molecular Parameters

The following outline is in some aspects a repeat of discussion presented already in Chapter 5, but it encloses a number of other examples, which will disclose a much larger complexity than expected. A similar consideration as shown in this section was made previously by Oono *et al.*,[32–34] Freed[8] and in a review by Douglas *et al.*[14] The last authors used preferentially the comprehensive data from star-branched macromolecules from the study by Roovers. We will not use their data which are focused on the contraction values and will confine our outline mainly to our own results which are represented in a somewhat different manner.

8.1.3.1 The Ratio $A_2M_w/[\eta]$

It will be noticed [see Chapter 5 and eqn (8.4b and c)] that the intrinsic viscosity and the effect of the inter-particle interaction can be expressed by a specific volume/mass, *i.e.*

$$[\eta] = \text{vol.}(\eta)/\text{mass}$$
$$A_2M_w = \text{vol.(Interaction)}/\text{mass} \qquad (8.6a, b)$$

$[\eta]M_w$ responds with an effective volume of the particle to a shear in solution, whereas $A_2M_w^2$ represents the excluded volume for two interacting particles in contact. The two relationships in eqn (8.6a,b) have the same dimensions, *i.e.* mL g^{-1}. It is of interest to know to which extent these two specific volumes differ from each other. This question can be investigated in two ways: (1) by using the expected values for hard spheres or (2) by using the more specific relationships for partially drained particles. This leads to the two equations

$$(1) \quad \frac{A_2M_w}{[\eta]} = 1.6\frac{\text{vol.(Interaction)}}{\text{vol.}(\eta)} \text{ hard sphere approximation} \qquad (8.7)$$

$$(2) \quad \frac{A_2M_w}{[\eta]} = 4\pi^{3/2}\frac{\Phi}{N_A\Psi} \text{ drained penetrating particles} \qquad (8.8)$$

Both equations are equivalent but eqn (8.8) is more specific since the volumes in eqn (8.7) are not real but are *apparent* or effective volumes.

The graphs in Figures 8.4–8.6 show some experimental results from various polymer structures compared with theoretical predictions. They are similar but differ for the *star-branched* molecules. In the first case the *total* molar mass was used whereas in Figure 8.4 the molar mass of the *arms* are taken. In both cases an increase of the $A_2M_w/[\eta]$ ratio with the molar mass is obtained for the star molecules. The $A_2M_w/[\eta]$ ratios from *star molecules* display in both graphs in Figures 8.4 and 8.5 roughly the same behaviour, but if the overall molar mass is used instead of those from the arms a larger spread of error is obtained. The effect of the number of arms is not clearly discerned.

The *hyperbranched* starch fragments (Figure 8.4), obtained by a controlled acid degradation at temperatures slightly below 0 °C, display no molar mass dependence for the not fractionated samples. These fragments have a fairly large polydispersity index (PDI) which approximately increases like $\text{PDI} \propto \text{DP}_n$. The molar mass distribution of these starch fragments is well known and was reported by Flory[35] and Erlander and French.[36]

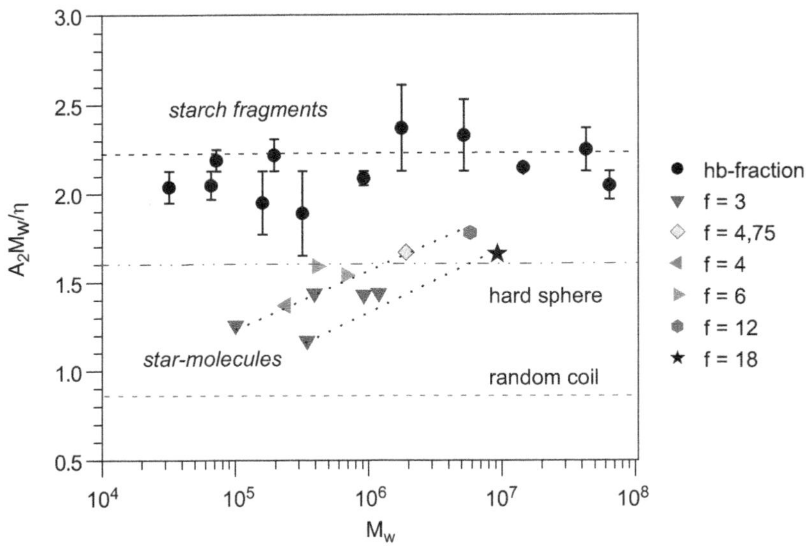

Figure 8.4 The dimensionless ratio of $A_2M_w/[\eta]$ as a function of molar mass M_w for hyperbranched starch fragments[24,25] (filled circles), and from various star-branched chains[15–18] (other symbols) compared to the theoretical predictions for hard spheres (upper dashed line) and flexible linear chain.[8,35]

The polydispersity of cross-linked chains is with PDI $\approx M_w$ exorbitantly large. A sensible correction for such broad distribution has been possible, and attempts have been made in special cases.[24,25,37,38] We will not discuss these attempts but only mention that complex and significantly different behaviour was obtained. With regards of the intrinsic viscosity uncertainty remains mainly based on the unexplored effect of draining in cross-linked and hyperbranched samples.

The curves in Figure 8.5 resemble in a way star-branched macromolecules, but the effect is considerably more pronounced. Remarkable is the difference in behaviour of the samples from the pre-gel state compared to those from the sol fraction (extracted from already formed networks). In the pre-gel state (Figure 8.6) a linear increase is obtained, much above that of uniform hard spheres, whereas the increase of the data from the post-gel domain is slightly bent but remarkably stronger. A careful investigation of the intrinsic viscosity as a function of the extent of cross-links gave a hint to the reasons for this observed difference. With increasing conversion the intrinsic viscosity of the sample from the gel domain is lower than that from the samples of the same degree of polymerization in the pre-gel domain, which indicates *ring formation* of the cross-linked chains in the constraining environment of a network. Figure 8.7 shows the effect.

Tentatively, the following conclusions can be drawn from the $A_2M_w/[\eta]$ ratio. Evidently the theories developed for uniform linear chains are *not*

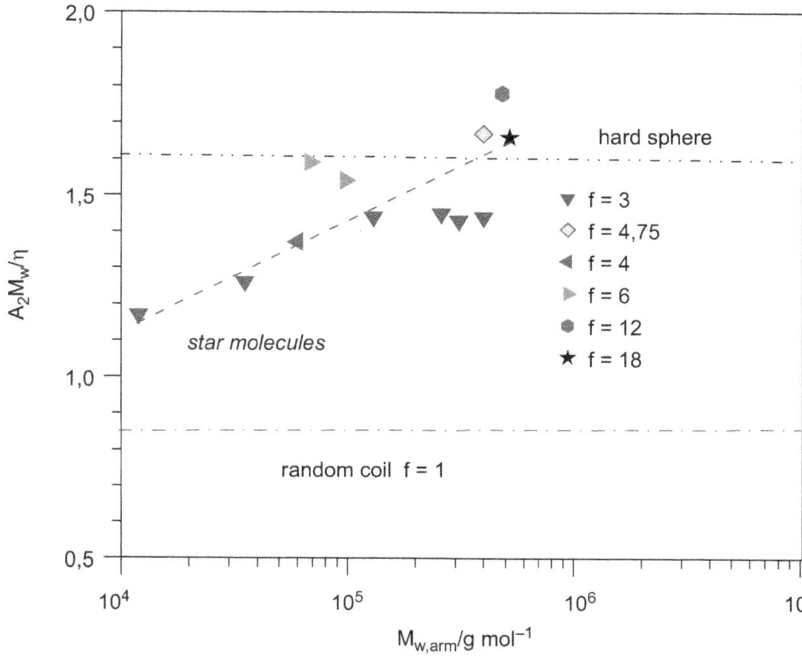

Figure 8.5 Plot of $A_2M_w/[\eta]$ for the same star molecules as shown in Figure 8.4 but presented as a function of molar mass of the arms.

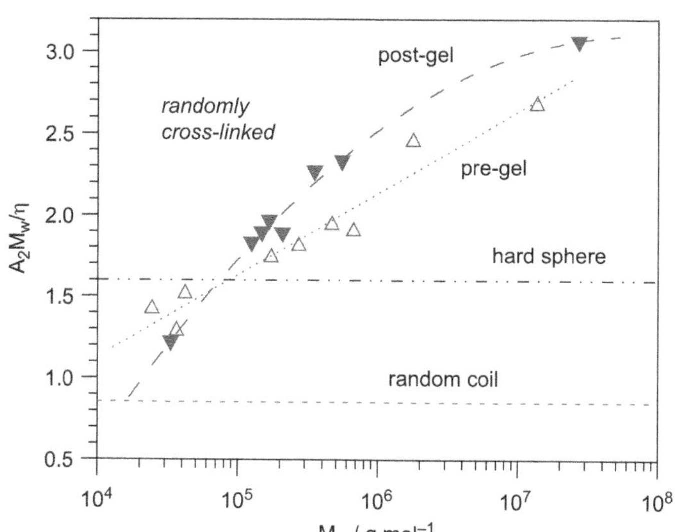

Figure 8.6 The same type of plot for randomly cross-linked polyester chains (A_2B_4 type) for samples before gelation took place (pre-gel domain) and samples from sol fractions extracted from already formed gels (post-gel domain).[39]

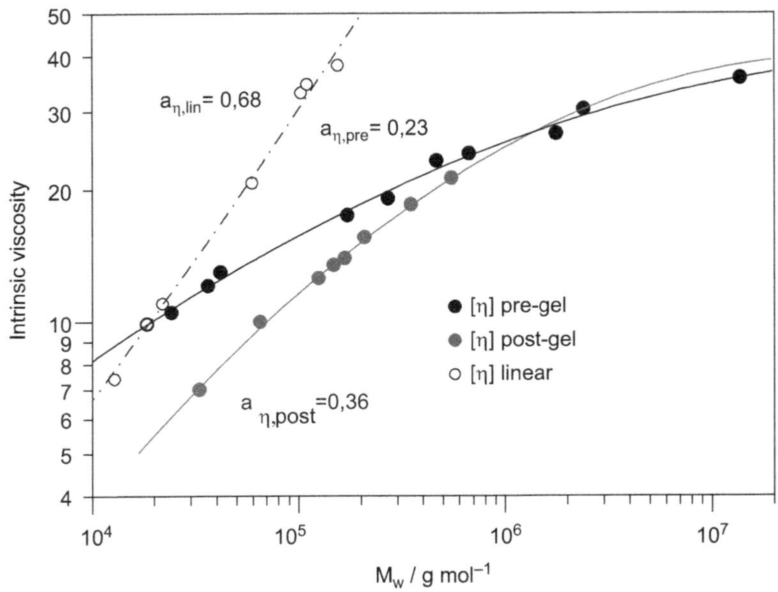

Figure 8.7 Molar mass dependence of the intrinsic viscosity of samples from the pre-gel and the post-gel states.[39] At the gel point the two curves meet at one same point. With increasing conversion the segment density of the particles from the gel domain becomes significantly larger than that from the pre-gel state. Such condensing is produced by intra-molecular reactions or ring formation.[40–42]

universally valid and cannot be applied without corrections. A significant effect is caused by the segment density and the width of the molar mass distribution. The hydrodynamic interaction and the interpenetration of segment clouds from interacting particles requires a more detailed re-consideration in theories with suggestions for feasible experiments. At present we can only conclude that the impediment of the segment penetration is stronger than that of solvent draining, which appears sensible. Figure 8.3 shows a cartoon of how the hydrodynamic radius changes with branching, demonstrated with a linear coil and in cross-linked micro-gel of the same radius of gyration. Obviously the penetration of segment clouds from two interacting particles is hindered even stronger, and a ratio $A_2M_w/[\eta]>1$ can be expected.

Intermediate Summary: the Ratio $A_2M_w/[\eta]$. A ratio $A_2M_w/[\eta]\approx 0.8$ corresponds to the prediction for linear chains. In any case of branching $A_2M_w/[\eta]$ exceeds the value of 1. The effect is most pronounced with randomly cross-linked chains. In nearly all cases the mentioned ratio approaches for small macromolecules ($M_w<1.5\times10^3$ g mol^{-1}) the value $A_2M_w/[\eta]=1$ as expected for the free draining case and free penetration of segment clouds. An exception in behaviour seems to be effective with the hyperbranched samples from starch fragments. One possible explanation arises from the large size of the anhydrous-glucose repeat units combined with a higher chain stiffness, *e.g.* cellulose derivatives. The attempts of interpretation are still based on presumptions, and the data published so far give not yet sufficient insight into the structure of the polymer particles. The known relationships should cautiously be used for interpretation. The influence of the size distribution or the PDI has not yet been sufficiently examined. At present no reliable theory is available on the depth of segment clouds interpenetration as a function of the branching density.

There will be no simple concept regarding how this issue can be tackled. Still, the second virial coefficient can be measured from the concentration dependence of the osmotic pressure, or the osmotic compressibility, obtained from the concentration dependence of scattered light. This is possible even for small particles which show no angular dependence of the scattered light, and measurement of $[\eta]$ causes no significant problems. This point is important since so far most synthetic hyperbranched polymers have radii smaller than 10 nm and these small sizes strongly restrict the characterization possibilities of such macromolecules. The $A_2M_w/[\eta]$ ratio is fairly accurately measurable and will become a valuable quantity for an inexpensive characterization, and will be useful for further progress in theory.

8.1.3.2 The Ratio $\rho = R_g/R_h$

Both radii are well defined and can be measured without knowing anything about the polymer conformation. The radius of gyration R_g is obtained from *static* light scattering or small angle neutron scattering, and the hydrodynamic radius R_h by *dynamic* light scattering, both from the initial slope of the scattering curves as a function of q^2:

$$\left\langle R_g^2 \right\rangle = -3 \times \frac{dR_\theta(q^2)}{dq^2} \frac{1}{R_\theta(q=0)} = -3 \times \frac{\text{slope}(q^2)}{R_\theta(q=0)} \tag{8.9}$$

$$\frac{\Gamma_1(q^2)}{q^2} = D = \frac{kT}{6\pi\eta} \left\langle \frac{1}{R_h} \right\rangle \tag{8.10}$$

where

$$\Gamma_1 = -\left. \frac{d\ln(g_1(t,q))}{dt} \right|_{\text{at } t=0} \tag{8.11}$$

It may be worth mentioning that $\langle R_g^2 \rangle \equiv R_g^2$ is the 2nd moment M_2 of the segment distribution around the centre of mass and $\left\langle \frac{1}{R_h} \right\rangle \equiv \frac{1}{R_h}$ - is the 1st moment M_{-1}. The first moment $M_{-1} = \langle R \rangle = 0$ vanishes because after averaging over all orientations of the radial vector the number of positive direction are compensated by the number of opposite direction. It is well known that a distribution can exhaustively be described by their moments though the convergence may be slow, but even if only the first two are known valuable information on the structure can be obtained. For details see Chapter 7. Therefore, the dimensionless ratio

$$\left\langle R_g^2 \right\rangle^{1/2} \left\langle \frac{1}{R_h} \right\rangle \equiv \frac{R_g}{R_h} \equiv \rho \tag{8.12}$$

is a characteristic quantity of the segment distribution which is indicative for special classes of polymers. The ratio was found to be a valuable parameter that sensitively depends on the degree of branching (see Chapter 5). As shown in Chapter 7 this ratio can be calculated for polydisperse unperturbed linear chains and branched particles, and even for particles in a good solvent when excluded volume interactions are effective. This can be done with the branching theory that is briefly outlined in Chapter 7. This branching theory requires no more than the extent of reaction of the individual functional groups on a repeat unit (*i.e.* the probabilities of reaction). The effect of the excluded volume on the branching parameter ρ has been calculated so far only for the randomly branched structures and was found to increase by about 28% from 1.772 for the unperturbed samples to 2.28, a value which approximately agrees with experimental results from randomly branched samples. The cartoon of Figure 8.8 demonstrates with two examples the skeletal effect on this parameter.

The ρ-value is *not* a universal quantity but depends on details of the branched skeleton, the degree of polymerization and the hydrodynamic response called draining. Figures 8.9 and 8.10 make this statement apparent with three examples: the randomly cross-linked polyesters, from the pre-gel state, *i.e.* before gelation took place, and the others were members from the sol fraction from the gels, *i.e.* under-critically cross-linked samples which could be extracted by a solvent from pieces of ruptured gel.

The ρ-values of the cross-linked samples from the pre-gel domain display a weak molar mass dependence in agreement with prediction for unperturbed chains. In contrast the ρ-value from the post-gel domain increases significantly with increasing network formation but finally decreases again. This fact can be explained as follows. At nearly full conversion of cross-linkers essentially only the linear samples (which carry no cross-linker) remain not bound to the network, and for this reason the ρ-parameter decreases again to a value of a polydisperse linear chain. Furthermore, not all of these chains are open chains but to a considerable content ring-formation has taken place. This conclusion was confirmed by the measurements of the intrinsic viscosity (see Figure 8.7).

The other example refers to the starch fragments and from poly-inimers which actually are hyperbranched samples.[25,44–50] Experimentally a considerably different behaviour was obtained for poly-inimers or arborescent

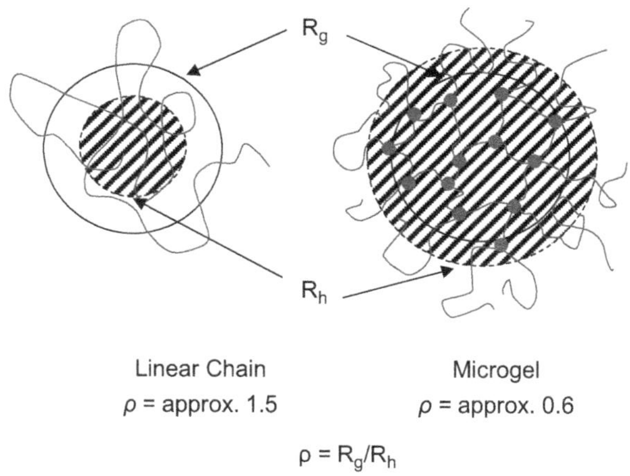

Figure 8.8 Schematic presentation of the effect of draining and of particle–particle inter-penetration demonstrated with a linear coil and a branched microgel. The penetration of the solvent is inhibited by the segment density which is much higher in the microgel than in a linear coil. The hydrodynamic radius increases in comparison with the radius of gyration.

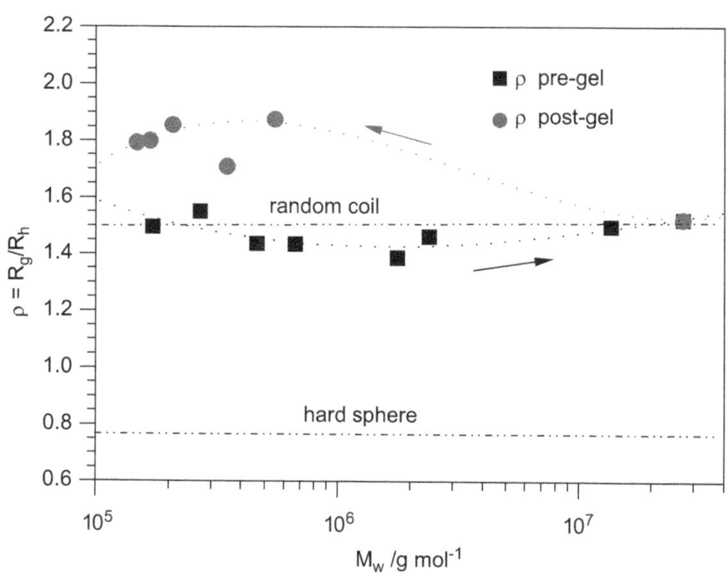

Figure 8.9 Molar mass dependence of the $\rho = R_g/R_h$ ratio for a cross-linked polyester. The squares refer to the samples before gelation occurs (pre-gel domain) and the circles denote to results from the extracted soluble polymers from the already formed network (post-gel domain). The arrows indicate increasing amount of cross-linker per linear chains.[43] The dotted lines correspond to the calculations for unperturbed dimensions.

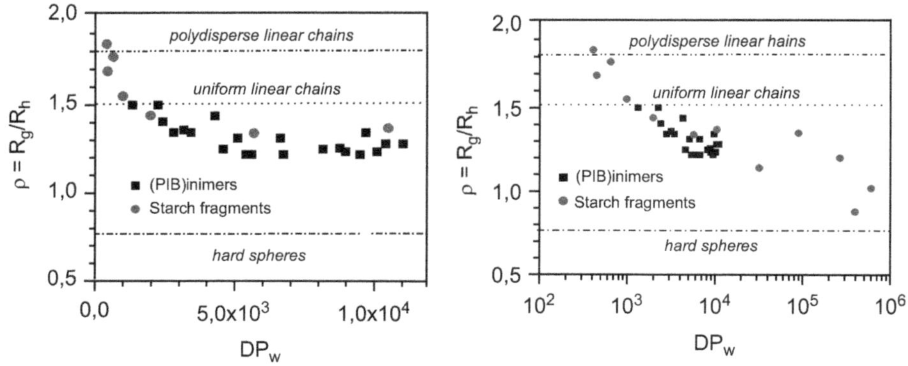

Figure 8.10 Dependence of the ρ-parameter as a function of the degree of polymerization, from starch fragments (circles)[24,25] and from poly(isobutylene) (PIB)-inimers[44–46] (squares); Left graph: linear plot in a limited DP_w range, right graph: log–log presentation over a much larger range of DP_w.

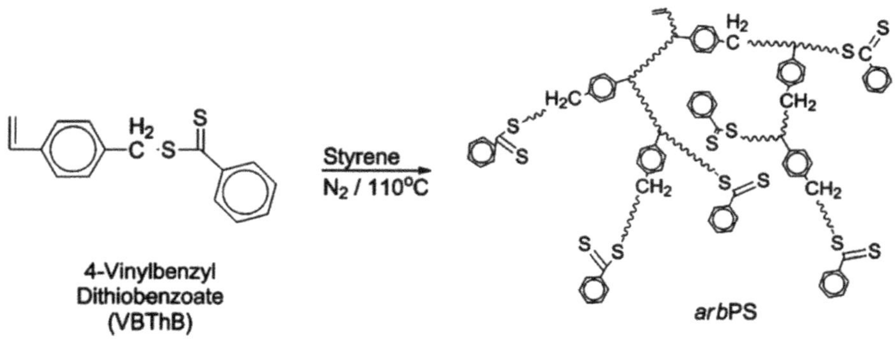

Figure 8.11 Scheme of the one-pot synthesis of *arb*PS.
Reprinted with permission from ref. 46. Copyright © 2011 Wiley Periodicals, Inc.

polymers and the starch fragments. Inimers are monomers with polymerizable double bonds which in addition carry a functional group that acts as an initiator.

The structure of arborescent polymers was synthesized for the first time by Gauthier *et al.*[51–53] and has found considerable interest for synthesis of other examples. Paulo and Puskas,[44] Heidenreich and Puskas,[45] and Puskas *et al.*[46] prepared, besides the PIB-inimer,[44–46] also an arborescent polystyrene[45,46] macromolecule by radical polymerization using the RAFT technique with 4-vinylbenzyl dithiobenzoate. The architecture formed is demonstrated in Figure 8.11.

Remarkably up to a limit of a $DP_w = 13\,000$ the chemically very different polymers of starch fragments and of the PIB-inimers display essentially the same behaviour. The behaviour at much higher DP_w remains insufficiently clear because of experimental errors and less number of data points.

Despite these uncertainties it seems that asymptotically an almost constant value appears and is approached at a plateau that corresponds to a prediction for AB_2-type polymers in a good solvent. Paulo and Puskas,[44] Heidenreich and Puskas,[45] and Puskas *et al.*[46] determined the ρ-parameters also for the *arb*PS samples but for no clear reasons the experimental data scattered much stronger around a well-defined curve. A decrease at low DP_w seems to occur, but at higher values for DP the data points scatter around the curve (for further information see the original publications by Paulo and Puskas,[44] Heidenreich and Puskas,[45] and Puskas *et al.*[46]).

Star Molecules. It is of interest to compare the ρ-parameter from the inimers with those from star molecules which are shown in Figure 8.12. The special decrease of this parameter may be helpful for orientation on which architecture a sample of investigation has to be assigned in cases

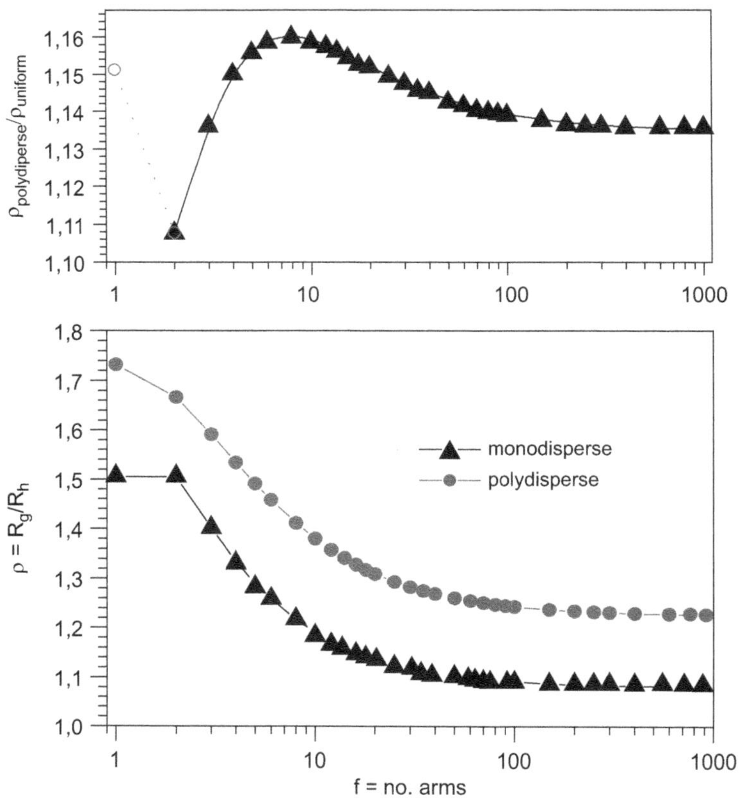

Figure 8.12 Change of the ρ-parameter with the number of arms of a star molecule with uniform arm length (triangles) and polydisperse arms of PDI = 2. The upper diagram shows the ratio of these two ρ-parameters. No arm length dependence was observed.[54]

where no special details on the synthesis are known. The equations derived for unperturbed star molecules are[54]

$$\rho_{\text{uniform}} = \left(\frac{3f-2}{f\pi}\right)^{1/2} \frac{8}{3} \frac{(2-f) + 2^{1/2}(f-1)}{f} \tag{8.13}$$

$$\rho_{\text{polydisperse}} = \left(\frac{6f}{f+1}\right)^{1/2} \frac{f+3}{2(f+1)} \tag{8.14}$$

No molar mass dependence was found for linear polystyrene in a θ-solvent and this may be valid also for the star molecules.

The influence of excluded volume interaction has yet been solved only for randomly branched polymer architecture (see Chapter 7), but for uniform linear chains Akcasu *et al.*[55] derived the equation

$$\rho(\nu) = \frac{5.066}{(1-\nu)(2-\nu)[3\pi(1+\nu)(2+\nu)]^{1/2}} \tag{8.15}$$

where the exponent ν is given by the equation $R_g = bn^{\nu}$ which can vary between $0.5 \leq \nu \leq 0.588$. The lower value refers to the unperturbed chain in a θ-solvent, the upper to good solvent behaviour. The value of $\rho_0 = 1.504$ for linear chain in a θ-solvent is well known. A check of eqn (8.15) for $\nu = 0.5$ leads to $\nu_{\text{Alcasu, Benmouna}} = 1.136$ and a better agreement with experimental data may arise if the factor 5.006 is replaced by 6.654. For $\nu = 0.588$ (good solvent) a value of $\rho = 1.856$ is obtained and for the ratio $\rho/\rho_0 = 1.23$. A slightly higher value of $\rho/\rho_0 = 1.27$ was obtained from branching theory [eqn (7.90)] for polydiserse linear chains.

8.1.4 Contraction Parameters Related to Branching

This topic was discussed in an earlier chapter but we take up this issue again under a somewhat different aspect. Branched samples may be envisaged as being formed, for instance, by cutting a rod-like wire, say into six parts of the same length, and might be soldered together by one of the two ends, hence in this case a star-like structure is obtained. Immediately it will be recognized that such structures have a significantly smaller dimension than the linear one although the weight of these two examples remains the same. Based on this simple model one can expect smaller sizes with increasing degree of branching. This is the beginning of the development of special branching parameters. This contraction parameter is obtained by comparing a special structure parameters of the branched macromolecule with those of the linear one of the same molar mass. As a first example Zimm and Stockmayer[56] compared the mean square radius of gyration of branched and linear polystyrene. They introduced the parameter

$$g = \left.\frac{R^2_{\text{g,br}}}{R^2_{\text{g,lin}}}\right|_{\text{same } M_w} \tag{8.16}$$

This g-ratio is smaller than 1 and is recently denoted as *contraction* parameter. To avoid a misunderstanding this name does not mean a contraction of the chain dimensions but solely indicates that a branched macromolecule has a smaller radius of gyration than the corresponding linear chain of the same molar mass. The contraction is more pronounced the higher the degree of branching is. The interpretation of the experimental data became an important issue. *Via* living anionic polymerization almost perfectly uniform linear chains can be prepared, but branched samples usually have very broad size distributions. To overcome this problem Zimm and Stockmayer[56] derived equations for both the polydisperse materials and the fractionated polymers of nearly uniform structure. Despite some ambiguity the derived equations turned out to be very useful in application to estimate the number of cross-links obtained from fractions by precipitation techniques, or now more advanced, by size exclusion chromatography. The equation of the contraction parameters for uniform *fractions* of A_3 and A_4 randomly polymerized monomers are given by eqn (8.17) and (8.18):

$$g_{3ZS} = \left[\left(1+\frac{n}{7}\right)^{1/2} + \frac{4n}{9\pi}\right]^{1/2} \qquad (8.17, 2.2)$$

$$g_{4ZS} = \left[\left(1+\frac{n}{6}\right)^{1/2} + \frac{4n}{3\pi}\right]^{1/2} \qquad (8.18, 2.3)$$

where n is the number of branching units. If a spacer of m repeat units is included the degree of polymerization is obtained:

$$\mathrm{DP}(m) = m \times n \qquad (8.19)$$

where m is the number of linear repeat units between two branching points. Eqn (8.17) and (8.18) were calculated for uniform size (no mass distribution). For the AB_2 model as representative for hyperbranched polymers the calculations were performed with the aid of the branching theory as described in Chapter 7. The corresponding equation is

$$g_{\mathrm{AB}_2} = \frac{6(1+2B_w)^{1/2}}{\left[1+(1+2B_w)^{1/2}\right]^2} \qquad (8.20)$$

Here B_w denotes the number of branching points in the polydisperse system. In experiments molar masses are determined, and in this case B_w includes the molar mass of the segment molar mass connected to two branching points. Eqn (8.20) was applied to characterize the branching of starch fragments, which were obtained by a controlled acid degradation.[57] Figure 8.13 shows the result.[24,25]

The experimental curve in the log–log plot can be shifted towards the $g(AB_2)$ curve by dividing the M_w of the fragments by a factor of about $M = 6000$. As shown in Figure 8.13 a very satisfactory agreement with the

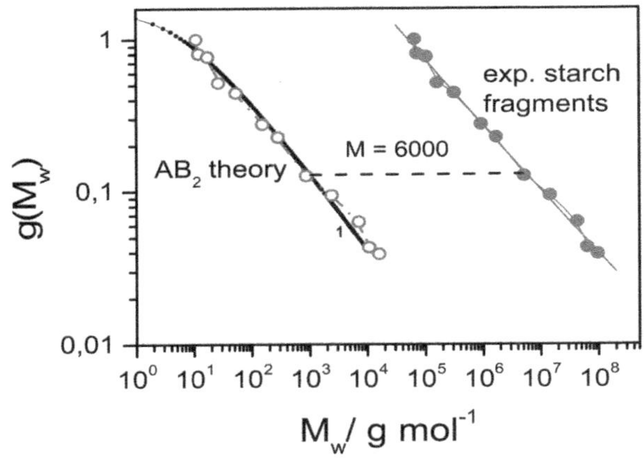

Figure 8.13 Contraction parameter from 12 starch fragments[24,25] as a function of the M_w (filled symbols) compared with the predicted behaviour of eqn (8.19) with $M=1$ (curve 1). A shift of the experimental curve by a factor $M=6000$ leads to a very satisfying agreement with the theoretical curve (open symbols). The shift factor M corresponds to a chain length of 37 repeat units for segments connecting two branching points.

theory is obtained. Dividing M of the connecting segments by the molar mass of the anhydro-glucose, $M_0=162$, a $DP_w=37$ for the connecting segments is obtained which lies well in the region of debranched starches.

A direct application of the Zimm–Stockmayer equation [eqn (8.17)] to size exclusion chromatography (SEC) appears not possible in most cases because the fractions are often too small in size such that the radii of gyration could not be measured in-line. This was not possible in the experiments where three-arm star molecules were cross-linked by diisocyanate, although the molar mass of the arm was $M_{\text{arm}}=6.3\times10^4=583\ M_0$. On the other hand, from non-fractionated samples of high molar mass the radius of gyration and the intrinsic viscosity could be measured which led to a correlation between the contraction factor $g=R^2_{g,\text{br}}/R^2_{g,\text{lin}}$ based on the mean square radius of gyration and $g'=[\eta]_{\text{br}}/[\eta]_{\text{lin}}$ based on the intrinsic viscosity. An exponent $\varepsilon\equiv0.6$ in eqn (8.21) was estimated by Kurata *et al.*[58] who approximated the correlation by a power law:

$$g_\eta \approx g^{\varepsilon}_{R_g} \qquad (8.21, 5.14)$$

where

$$g_\eta = \frac{[\eta]_{\text{br}}}{[\eta]_{\text{lin}}} = g' \qquad (8.22, 2.8)$$

Although the value of the exponent ε is still a point of controversy, Weissmüller used the value 0.6 as suggested by Kurata *et al.*[58] and converted the SEC curve from cross-linked three-arm star molecules, observed with in-line detectors for static LS and viscosity, from g' to g. He also measured

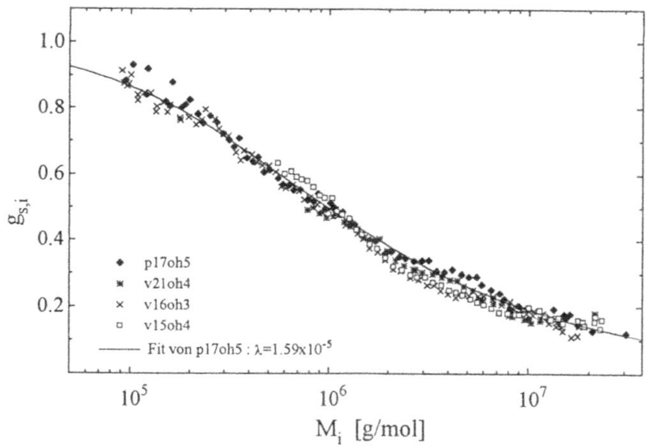

Figure 8.14 Molar mass dependence of the contraction factors for fractions of OH-terminated three-arm stars which were end-linked *via* different amounts of diisocyanate. Three samples from the pre-gel domain was taken and one also from the sol fraction (indicated by p) obtained by elution of an already formed gel. The solid line denotes the average fit with eqn (8.16). The fit was made with $\lambda = n\, M_0$ where M_0 is the molar mass of the three-arm stars and n is the number of branching points. Reproduced with permission from ref. 59. Copyright © 1997 WILEY-VCH Verlag GmbH & Co. KGaA.

the corresponding SEC diagrams from linear PS chains. For the fit he used eqn (8.17) for g_{A_3} (uniform fractions). The result from four different cross-linked samples is shown in Figure 8.14.[59]

The contraction parameter based on the intrinsic viscosity was introduced by Stockmayer and Fixman[60] and was already defined in eqn (8.22). A further contraction factor can be defined with the second virial coefficients from the ratio of the branched samples and the linear ones at the same molar mass:

$$g_{A_2} = \frac{A_{2,\text{br}}}{A_{2,\text{lin}}} \tag{8.23}$$

The ratio is closely related to that of g_η because both are based on the volumes of the dissolved particle, but they often differ significantly.

The reason has already been discussed and is caused by the solvent draining in g_η and the segment interpenetration of two interacting chains in g_{A_2}. The contraction factor of eqn (8.22) and (8.23) was used by Galinsky and Burchard.[24,25] Figure 8.15 shows the results as an example for the hyperbranched starch fragments.

The reason for the different behaviour was extensively discussed in a paper by Douglas *et al.*[14] and compared with recent results derived by the Renormalization Group Theory.[8] Both of these theories and recent experiments make clear that the generalized ratios are *not universal*[10] parameters but depend on the molecular architecture and the molar mass dependence.

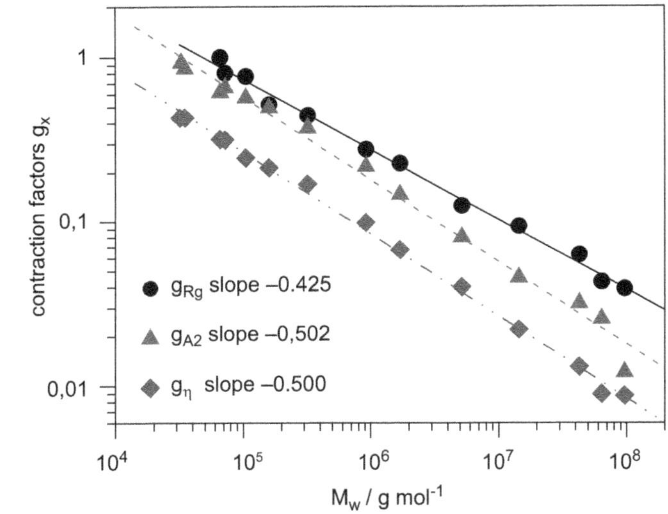

Figure 8.15 Molar mass dependence of three types of contraction factors presented here with the examples of the hyperbranched starch fragments.[24,25] The ratio was formed with the polydisperse fragments and uniform linear amylose chains, prepared by enzymatic synthesis. The aqueous 0.5 M NaOH solvent is a thermodynamically good solvent for these polysaccharides.

Results obtained with one type of branched polymers cannot thoughtlessly be used for interpretation of data from other topologies.

At present the correlation between the various contraction factors is not yet fully explored, and interpretation can be made only empirically from research projects where as many as possible other molecular parameters have to be measured. This demand was not much followed in the literature, and often only the intrinsic viscosity is measured or now, more recently, also the hydrodynamic radius, which can be recorded in-line by size exclusion chromatography. Despite the uncertainty in the mentioned correlation between the various parameters it remains clear that, in a way, all are related to the radius of gyration R_g. As already mentioned: *the radius of gyration is the key parameter in all relationships for conformational characterization.* It is the only parameter which uniquely can be measured, by static light scattering or SANS, without knowing anything of the sample, even if the molar mass or the refractive index increment is unknown.

It is informative to derive the molecular equivalent hard sphere radii from the various measurable quantities and to plot the various apparent radii against the degree of polymerization or the molar mass. This was done in the past with some of the available data. The following two figures show such plots. Regretfully, the second viral coefficient is often not systematically measured, probably because this parameter is of minor interest to preparative chemists. In the study of branched polymers the interaction parameter compared to the intrinsic viscosity is of particular interest because

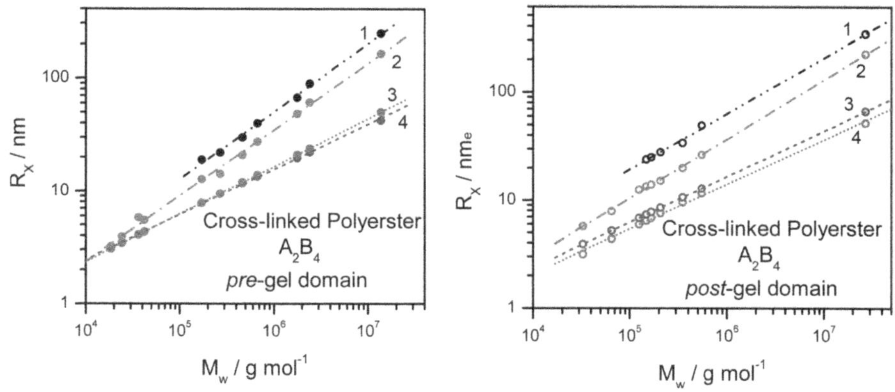

Figure 8.16 Molar mass dependencies, derived from four different types of measurements of the cross-linked anhydride cured epoxides in the pre-gel and the post-gel domain,[39] and the end-linked PS arm stars. In both panels: curve 1, R_g; curve 2, R_h; curve 3, R_η; curve 4, R_{A_2}.

the intrinsic viscosity is measurable on-line by SEC but not the second virial coefficient. To date the quantitative interpretation of the curves in Figure 8.16 is not possible as the draining and penetration parameters for branched samples are not yet been predicted by theory.

Possibly, by comparing the behaviour from several other structures, a conjecture can be developed which physical effect is dominating. This will be helpful in the derivation of simplified models to set up a sensible theory that will allow for a more specific interpretation of experimental data.

8.1.5 Molar Mass Distribution

In this section we come to a serious problem which so far has been vaguely left open to discussion. The question is what influence the molar mass distribution has on the experimental data. We know that the PDI can be extremely large, for instance with randomly branched samples where PDI $\propto DP_w$, and the gel-point DP_w diverges to infinity and thus also the PDI.

The number average DP_n remains very low from which we realize that a large amount of molecules have only the size of oligomers. Unavoidably in the process of precipitation and 'purification' to remove undesired chemical concomitants, a large amount of low molar mass molecules is lost, and a much lower PDI is obtained. Still, in some experiments a value up to 800 for the PDI has been possible to measure. Figure 8.17 presents the results from three examples.

The effect of polydispersity is less pronounced with hyperbranched samples where a PDI $\rightarrow DP_n/2$ is obtained $(DP_z/DP_w \rightarrow 3)$ where the arrows indicate asymptotic behaviour. The change of the PDI with the molar mass for

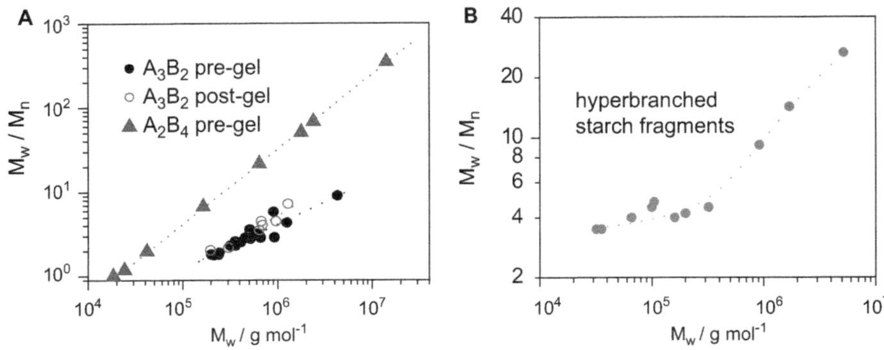

Figure 8.17 Increase of the measured PDI $= M_w/M_n$ as a function of M_w. (A) triangles, cross-linked polyester from the pre-gel domain;[39] filled circles, end-linked PS three-arm stars from the pre-gel regime, and open circles, from the post-gel state.[37,43,59] (B) Starch fragments as an example for hyperbranched polymers.[24,25]

the hyperbranched starch fragments is shown in Figure 8.17B. Evidently the curve has two branches: a weak increase not much larger than DPI = 2.4 and a much stronger increase for larger samples. The first part reminds of linear chain the second one to branched samples. This indeed is a typical characteristic of AB_2 polymers which show the behaviour of linear and branched chains. The linear contribution dominates for small particles but the branching contribution grows quadratically with the monomer conversion α whereas the linear contribution grows only linearly with α, and this leads to the characteristic branching feature, *i.e.*

$$\mathrm{DP}_w = 1 + \frac{2\alpha}{1-\alpha} + \frac{1}{2}\left(\frac{\alpha}{1-\alpha}\right)^2; \quad \mathrm{DP}_n = \frac{1}{1-\alpha};$$
$$\mathrm{DP}_w/\mathrm{DP}_n \cong 2\alpha + \frac{\alpha^2}{2(1-\alpha)} \tag{7.45}$$

To avoid confusion, the linear chains are not free chains but are tethered to the body of the branched core; they represent the outer chains. The outer chains and the linear sections between two branching points have on average the same length. Thus some similarity should exist between star-branched macromolecules with a random length distribution of the arms and the hyperbranched macromolecules. This feature becomes apparent from experiments and the branching theory as outlined in Chapter 7.[54,61]

By use of size exclusion chromatography (SEC) the characterization of branched materials can be made also with broad sized distribution, and the various parameters can be detected in-line with detectors for the refractive index, viscosity and dynamic light scattering. With this design fractions of more or less uniform molar mass are detected which permits determination of the molar mass, and the intrinsic viscosity, but the radii of gyration are often much smaller than the wave length of the light used.

8.1.6 Pending Problems to be Solved

At that point problems come into play because the obtained results from SEC measurements do not agree with the data from the broad distribution. For a quantitative characterization a suitable theoretical support is required. The weight average molar mass M_w of the non-fractionated samples is easily derived by the cascade-like branching theory as outlined in Chapter 7, and also the mean square radius of gyration can be calculated, even for particles with excluded volume interaction in a good solvent.[61]

One can use the corresponding radii derived from the intrinsic viscosity which is less influenced by a broad size distribution expressed in terms of equivalent hard spheres. However, then the fact of draining is disregarded. The draining effect is taken into account by the Flory–Fox equation for the intrinsic viscosity of uniform linear chains. There is a tendency to assume universality of the Flory draining parameter Φ_{Flory} in the hope that correct results are obtained also for branched samples. Recently we demonstrated with several published data[10] that the draining parameter Φ varies for branched samples with molar mass and branching architecture. It remains unclear to what extent the draining and other properties are affected by the special topology. A detailed discussion on this subject has been given in Chapter 5. Much activity is presently invested to gain better insight into the behaviour of different topologies[62–69] deriving analytical equations for the properties of uniform fractions, or otherwise by simulations. Both approaches could not yet be checked by us, but it evokes keen interest in our groups in particular, in view of applicability to better interpretation of actual experiments.

Because of the low size of the hyperbranched-samples the problem of finding the correct radius of gyration remains for the moment largely being unsolved. It is advisable to include non-fractionated samples in the investigation of uniform samples, which are obtained from SEC in combination with in-line detection for the viscosity and dynamic light scattering; and the molar mass dependencies of these data should be compared. It has become common to apply the concept of fractal behaviour which at present is loaded with serious doubts.

8.1.7 The Fractal Approach

The scheme of fractal behaviour was briefly discussed in Section 7.4. Here we will give a few other details, which may help to judge observations from measurements other than scattering experiments.

It is important to keep in mind that the concept of fractals is applicable only to a disordered system.[70] The scheme was discovered by strange results from measurements. For instance an increased accuracy in measuring the border length between two countries led more and more to longer lengths and apparently diverged to infinity which clearly is a senseless result. Of course, divergence occurs if in an overdone test the length of each stone or sand grain is measured. Evidently the border between countries cannot sensibly be described by a well-defined one-dimensional line. To obtain a sensible answer one has to allow for a certain width or strip of land, and the average length within this strip has to be taken. Interestingly if the width is made continuously broader a shorter length is obtained which converges to a reproducible value. In other words the border cannot be by described by a one-dimensional line, but it has a dimension somewhat between a linear and a planar dimension. Within a certain width the lines fluctuate at random. This meandering can, on average, be described by a power law with an exponent that describes the average length of the line in terms of a fractal dimension d_f. Further systematic studies led to the conclusion that such a power law is always observed with *disordered* systems and can be well described by a *fractal* dimension. All polymers in solution are disordered systems and this is with regards to two properties:

1. Let us take one single macromolecule, with a structure like a pearl necklace. The various beads are linked together but still have largely free mobility. This leads to a random coil or to a modified coil due to the excluded volume interactions. Clearly such a coil is a disordered structure. For uniform chain length the disorder is somewhat restricted by the condition of uniform length. Thus a uniform chain cannot be described solely by a mass fractal but must contain this restriction. This constraint of uniform length can be taken into account by adding a 'cut-off' function which for a disordered length must have the form of an exponentially decaying function, a point that was emphasized by Debye and Bueche,[71] mathematically further specified by Ross[72] and was pointed out by Porod.[73] This cut-off function must also contain a *correlation parameter* which would correspond to the average size of the macromolecule, *e.g.* the mean square radius of gyration.
2. The strict constraint related to uniform size is somewhat released if a polydisperse system is considered. A correlation length has still to be introduced which takes into account that even in a random type of polymerization no chain can become infinite in length, but this is already included by the molar mass distribution. Chemical reactions are largely described by kinetics, *i.e.* by statistical processes. If the statistical nature of the polymerization kinetics is kept in mind, one may wonder whether the arising molar mass distribution may also have the property of a fractal system. This would mean that the disorder of the non-fractionated samples is caused by the segmental mobility and in addition by the irregularity of the molar mass distribution. This point was disclosed in a profound study by Stauffer *et al.*[74,75] on aggregation and branching processes, applying the concept of percolation with simulations on lattices of various types.

The twice disordered system should lead to a more generalized view of fractal behaviour. This was confirmed by some recent experiments. To give an example, for the description of the conformation of uniform linear chains one has to calculate the average positions of all pairs in the chain under the constraint that no segment can be longer than the contour length of the polymer chain. In the free radical reaction a special segment length is determined by the probability how often the radical has formed a bond. A probability α is always smaller than unity and therefore after n consecutive reaction on an active chain the probability of n consecutive reactions goes to zero since $\alpha^n \to 0$ for $n \to \infty$. In other words, the constraint for a certain length is automatically included by the extent of reaction α, because for $\alpha \cong 1$ one has $\alpha^n \cong \exp[-(1-\alpha)n]$ which is the required cut-off function for *linear* chains. For *random* branching of *f*-functional monomers Stockmayer derived the equation for the molar mass distribution:

$$w(x) = \frac{(1-\alpha)^2}{\alpha} fx \frac{\exp[-(1-\beta/\beta_c)]}{2\pi^{1/2}(f-1)^{5/2}x^{5/2}} \to Cx^{-5/2}; \quad \beta = \alpha(1-\alpha)^{f-2} \tag{8.24}$$

In the vicinity of the point of gelation (indicated by the subscript *c*) the distribution displays fractal behaviour with a power of −1.5. An important feature is the exponential decay which fulfils the required property of a cut-off function.

Before discussing the distribution obtained by Stauffer from percolation simulations, it is useful to go back to the question of how polymer chemists make use of the fractal concept. We have already shown that the hydrodynamic radius R_h, the intrinsic viscosity $[\eta]$ and the second virial coefficient A_2 can be expressed by the radius of gyration R_g. Based on the scaling relationships eqn (8.3) we gave a list how the various exponents in the log–log presentation as a function of the molar mass are related. An excellent agreement for linear polystyrene in a good solvent is obtained, and this scaling prediction is reasonably well fulfilled also with other linear chains.

The question remains of how good these scaling laws are applicable to branched samples. Figure 8.18A shows the molar mass dependence of the

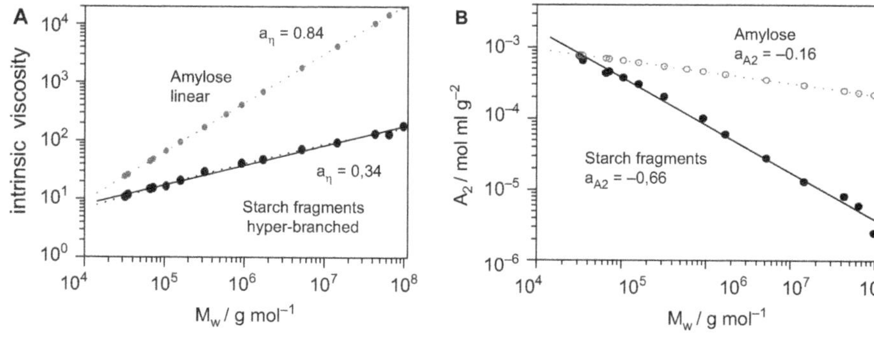

Figure 8.18 (A) Molar mass dependence of the hyperbranched starch fragments in 0.5 N NaOH and of the liner amylose in the same solvent. (B) The same plot for the molar mass dependence of the second virial coefficient in the same solvent.

intrinsic viscosity of the hyperbranched starch fragments and Figure 8.18B that of the second virial coefficient in comparison to linear amylose. In order to get a more comprehensive impression the corresponding data from the randomly cross-linked polyesters in DMF are shown in Figure 8.19A and B.

The first impression is that the measurements illustrate behaviour as expected: the exponent of the intrinsic viscosity of the branched samples is considerably lower than for the corresponding linear ones and the same is found for the exponents from the second virial coefficients. A somewhat disturbing impression is obtained if the scaling relationships are checked. This is shown in the Tables 8.4 and 8.5. The curves in Figures 8.18 and 8.19 are repeated here because we wish to demonstrate the experimental accuracy from which the exponents were obtained.

In these tables the following notations have been used: the exponent ν describes the power in the molar mass dependence of the radius of gyration; the dash in ν' denotes the calculated one using the scaling conception and $1/\nu = d_f$ or $1/\nu' = d_f$ are the resulting fractal dimensions of the particle. The exponent ν_x corresponds to the exponents obtained in the various kinds of experiments.

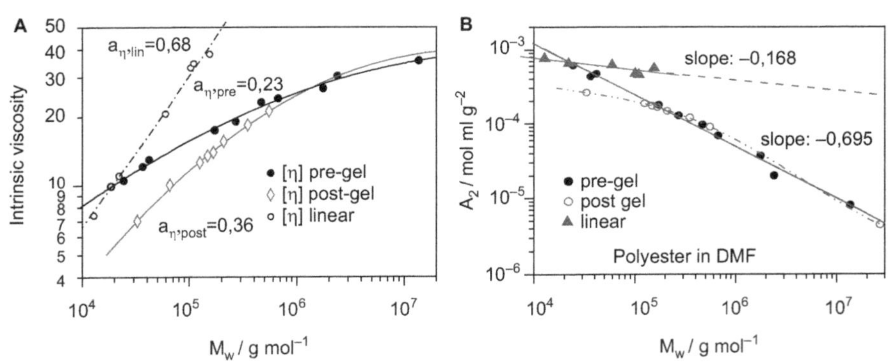

Figure 8.19 Molar mass dependence of the intrinsic viscosity from a linear polyester (phenyl-glycidyl ether/phathalic anhydride)[39] chains and from the corresponding (anhydride cured epoxy) resins from the pre-gel and post-gel domains.[76–78]

Table 8.4 Observed exponents re-calculated ν_{Rg} applying the scaling relationships for the hyperbranched starch fragments and the linear amylose.[24,25]

Starch fragments				Linear amylose			
	ν_x	ν'	d_f		ν_x	ν'	d_f
ν_{R_g}	0.40	0.40	2.50	ν_{R_g}	0.61	0.61	1.64
ν_{R_h}	0.47	0.47	2.17	ν_{R_h}	0.61	0.61	1.64
a_η	0.34	0.45	2.22	a_η	0.84	0.61	1.64
a_{A_2}	−0.66	0.38	2.27	a_A	−0.16	0.61	1.64

Table 8.5 Experimentally observed exponents and re-calculated ν' values and fractal dimension d_f, of an epoxy cured (cross-linked) polyester from the pre-gel and post-gel domain, compared with the corresponding data for the linear chain.

Cross-linked polyester pre-gel				Cross-linked polyester post-gel				Linear polyester			
	ν_x	ν'	d_f		ν_x	ν'	d_f		ν_x	ν'	d_f
ν_{R_g}	0.59	0.59	1.69	ν_{R_g}	0.52	0.52	1.92	ν_{R_g}	—	—	—
ν_{R_h}	0.58	0.58	1.78	ν_{R_h}	0.55	0.55	1.81	ν_{R_h}	0.49	0.49	2.04
a_η	0.23	0.41	2.44	a_η	0.36	0.45	2.22	a_η	0.65	0.55	1.82
a_{A_2}	−0.65	0.45	2.22	a_{A_2}	−0.48	0.51	1.97	a_{A_2}	−0.17	0.61	1.54

An excellent validation of the scaling conception is obtained with the *linear* amylose chain and a fairly good agreement also with the *hyper-branched* starch fragments, with average fractal dimensions of $d_f = 1.64$ for the linear chain and $d_f = 2.29$ for the branched fragments. Similar clear conclusions cannot be drawn from the data of the cross-liked polyester. There is a fairly good agreement between ν from the direct measurements of the radii R_g and R_h with an average of the fractal dimension $d_f = 1.73 \pm 0.06$ in the pre-gel state and $d_f = 1.86 \pm 0.06$ for the post-gel state, which within experimental error may be considered as a reasonable agreement.

A very different impression is obtained with the cross-linked polyesters and the end-linked three-arm star-molecules (Table 8.5). The ν' exponents derived from the data based on *volumes* of the particle, gave similar data with $d_f = 2.33 \pm 0.10$ for the pre-gel sate and $d_f = 2.10 \pm 0.12$ for the post-gel state, not really distinguishable within experimental error. Such agreement is expected for the *randomly branched* samples. However, a significant and contradicting *difference* was obtained with the fractal dimensions obtained from measurements of the *molecular radii*. The data from static and dynamic LS measurements for R_g and R_h are characteristic for *linear* structures in a good solvent, but the other data obtained from the *volume* of the particles correspond to expected *fractal behaviour* of *branched* samples in a *marginally* good solvent.

This discrepancy was noticed also by other colleagues who looked for an explanation. Here the percolation study by Stauffer *et al.*[74,75] gave a decisive hint for the reason of this apparent disagreement which shortly will be commented in the next section.

8.1.8 Percolation on a Lattice

The phenomena of gelation can also be mimicked by simulations on lattices. The concept is simple: one adds units onto empty places of lattice sizes step by step. After a certain 'conversion' the occupied lattice sites start to touch each other; this fact is taken as a bond formation. In this manner growing clusters are formed if the occupation of free lattice sites is increased. Finally, a situation is obtained where one cluster has reached a size which touches the walls of the vessel. (The word 'percolation' reminds us of filtering coffee in which, if too large amount of milled coffee is chosen, it ceases to be *percolated* by the water being used. This is called the percolation threshold, and it is equivalent to the phenomenon of gelation or network formation).

Stauffer was mainly interested in the cluster formation near this threshold, because in the critical regime a special universal critical behaviour is predicted, which he confirmed. Stauffer expressed most of his observations in terms of a critical regime denoted by $|p–p_c|$ where p_c indicates the critical state site occupation of gelation. This critical conversion is difficult to determine accurately, but it has been done in several cases in our laboratories.[39,43,59,76–78]

Near the critical state Stauffer determined the number of different cluster sizes and obtained in this way a number distribution which could be described by a power law curve:

$$w(M) \propto M^{1-\tau} \tag{8.25}$$

with $\tau = 2.2$ similar to the Stockmayer number-distribution eqn (8.24) near the gel point but with an exponent of $\tau_{\text{Stockmayer}} = 2.5$. Weissmüller and Burchard[43,59] checked the Stauffer prediction and determined the weight average molar mass distribution at different extents of cross-linking three-arm stars by diisocyanate. The result is shown in Figure 8.20. The findings can be summarized as follows:

1. At low extent of cross-linking the Stockmayer distribution is well obeyed in which the exponential decay (cut-off function) dominates.
2. With increasing extent of cross-linking the asymptote of the Stauffer relationship is slowly approached, but at high molar mass the exponential decay still remains effective.
3. Only very close to the gel-point is power law behaviour obtained with an exponent of $1 - \tau = -1.36$ or $\tau = 2.36$, slightly higher than predicted by Stauffer. Also in this last curve the exponential decay is no longer noticeable.

The final exponential decay as found by Weissmüller is not in disagreement with Stauffer's prediction. Stauffer emphasized that in the end a cut-off function should be present, because the clusters in a real system can never become infinitely large. He simply set a sharp cut when the clusters in his simulations reached just the size of the chosen lattice.

The next question was, Has this τ-value an influence on the exponent in the molar mass dependence of the radius of gyration, and if so, can this then be corrected? If in experiments the slope in the power law behaviour is determined one always plots the mean square radius of gyration as a function of the molar mass. For a non-fractionated system this is a plot of the *z-average* mean square radius of gyration $\langle R_g^2 \rangle_z$ against the *weight average* molar mass M_w, but to find the true fractal dimension of the particle one has to compare

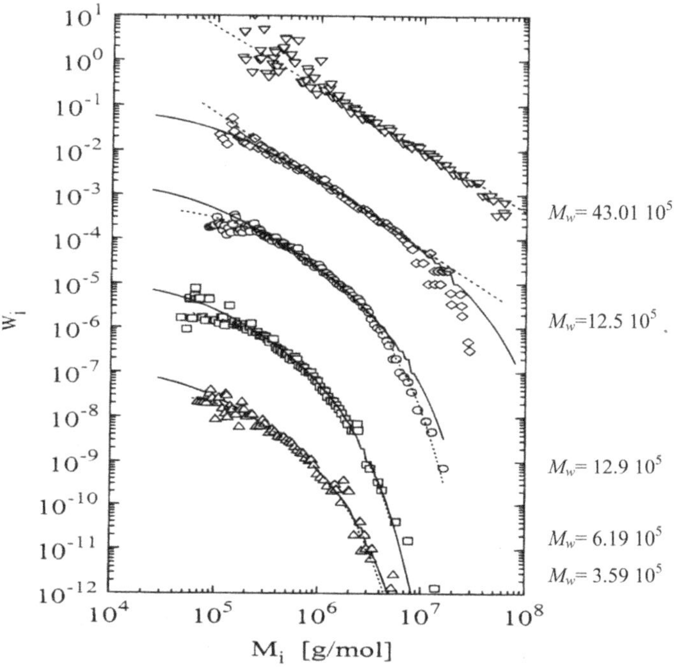

Figure 8.20 Molar distribution function obtained by SEC from OH-terminated A_3 stars end-linked by diisocyanate. The full lines correspond to the Stockmayer distribution, the dashed lines to the Stauffer prediction complemented by a cut-off function.[37,43]
Reprinted with permission from ref. 37. Copyright © 1997 SCI.

the weight average $\langle R_g^2 \rangle_w$ with that of the molar mass M_w. This requires the determination of the ratio M_z/M_w which is the next issue. The required conversion of M_w into M_z could be performed from the distribution data in term of the exponent τ with the result

$$M_w \propto M_z^{3-\tau} \tag{8.26}$$

8.1.9 Characteristics of Size Exclusion Chromatography Diagrams

To solve this problem with experimental data one at first has to make clear that actually the molar mass M_i is recorded as a function of the elution volume ν_e, which commonly is $\nu_e = K_1 - K_2 \log(M_i)$; this means that the equal spacing of the elution volume is not linearly related to the molar mass, which leads to a distorted molar weight distribution. To correct this distortion Weissmüller and Burchard[37,43] converted at first the measured diagram into a cumulant distribution and determined the required weight distribution by numerical differentiation. The result of this manipulation is shown in Figure 8.21.

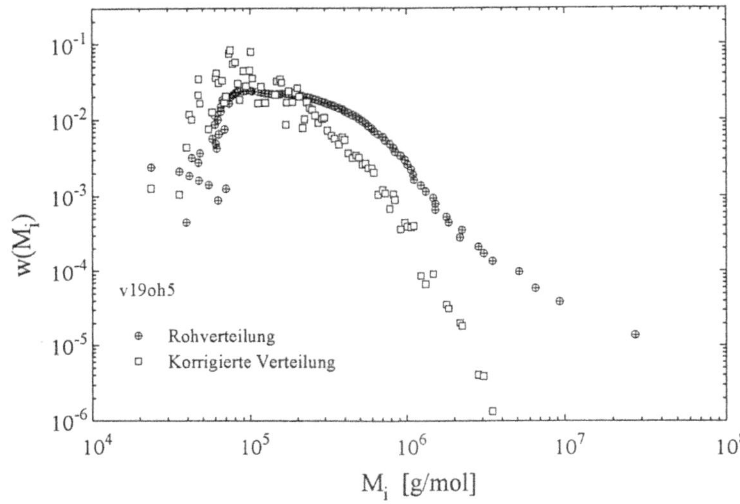

Figure 8.21 Molar mass distribution (raw diagram, filled symbols) as derived from SEC without correction for linearity and the corrected curve obtained after differentiation of the cumulant distribution (open symbols).[37] Adapted with permission from ref. 37. Copyright © 1997 SCI.

The maximum in the corrected graph corresponds to the point where the cut-off function starts to act. The maximum of such curves is considered to represent the mean position of the *z*-average molar mass. Weissmüller and Burchard[37] analysed all SEC curves and plotted the weight-average molar mass M_w against M_{max} and from the slope the correction factor M_w/M_z. The result is shown in Figure 8.22. The slope is 0.69 from which with eqn (8.26) a value of $\tau = 2.31$ is found. A value that lies between the Stockmayer distribution with $\tau = 2.5$ and the percolation value of $\tau = 2.18$. With the data from the end-linked three-arm stars the polydispersity parameter $\tau = 2.31 \pm 0.05$ was obtained. For the polyester Trappe[39] applied a somewhat different route and found a value $\tau = 2.18 \pm 0.02$ in excellent agreement with the percolation prediction.

We now introduce the notation ν_z and ν_w and the corresponding fractal dimensions $d_{f,z}$ obtained experimentally and $d_{f,w}$, in order to distinguish the exponents obtained from a *z*-average mean square radius of gyration and the exponents obtained after correction for the weight average. These two parameters are correlated as

$$\nu_w = \nu_z(3-\tau) \quad \text{or} \quad d_{f,w} = d_{fz}/(3-\tau) \qquad (8.27a,b)$$

The results for the end-linked thee-arm stars and the polyester are given in Figure 8.22.

The corrected $d_{f,w}$ values now correlate much better with *branched* structures. The data are closer to those from the intrinsic viscosity in particular for the samples in the pre-gel domain, but for the post-gel domain a better agreement is obtained with the *z-average* radius of gyration. The latter is of

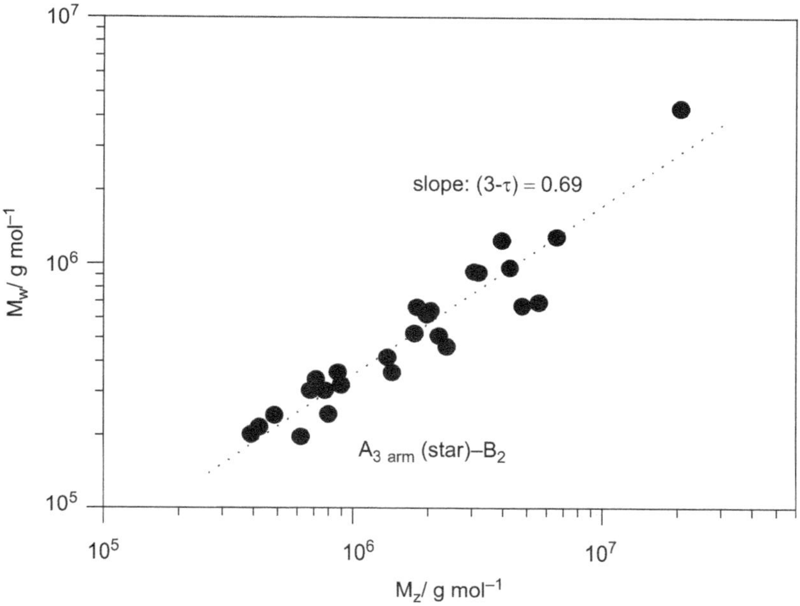

Figure 8.22 Correlation between weight average and *z*-average molecular weight derived from SEC fractionating curves from end-three-arm star molecules.

Table 8.6 The ν-exponents for the molar mass dependence of the radius of gyration from scattering data ν_z and ν_w and from the intrinsic viscosity ν_η, from randomly cross-linked polyesters in the pre- and post-gel regimes, compared with the data from the hyperbranched starch fragments and the linear amylose and the corresponding fractal dimensions. The indexes *z* and *w* refer to the *z*-average and corrected weight average of the-mean square radius of gyration the index η the corresponding value derived from the intrinsic viscosity.

Exponent	Polyester pre-gel	Polyester post-gel	Starch fragments	Amylose linear
ν_z	0.575	0.538	0.48	0.61
ν_w	0.471	0.414	—	0.61
ν_η	0.430	0.510	0.45	0.61
$d_{f,z}$	1.74	1.86	2.34	1.64
$d_{f,w}$	2.13	2.67	—	1.64
$d_{f,\eta}$	2.33	1.97	2.22	1.64

interest because the same is obtained from the hyperbranched samples. The results derived from the intrinsic viscosity have to be taken with some scepticism because of the draining effect which is not yet understood for branched structures. The somewhat lower fractal dimension found for amylose is probably caused by a noticeable chain stiffness (Table 8.6).

One topic remains unclear: this is the physical relation between the τ-parameter and the ν-exponent which arises from excluded volume interaction. According to the derivation from the percolation theory the τ-exponent is nothing else than a correction parameter for the molar mass distributions from the *z*- to a *weight* molar mass distribution. This cannot be the complete answer, because it does not give an explanation of why deviations from the Stockmayer distribution occur. Stockmayer's derivation is based on kinetic reactions, *i.e.* on constant kinetic reaction constants. If these kinetic parameters are not constant but enclose the conversion dependence this would have an additional effect on the probability of reaction in the course of conversion. Such behaviour cannot be treated in the scheme of a mean field approximation. A supposition is that an overcrowding effect causes a slowing-down of branching, and thus a weaker decrease of the amount of detectable cluster should be obtained. The question remains, what is the difference between segmental overcrowding and excluded volume interaction? In principle such overcrowding can be mimicked by a so-called substitution effect which could partially be solved by the kinetic theory of branching in a treatment by Galina *et al.*,[79,80] Hendriks *et al.*,[81] and Ziff and Stell.[82] The aspects of the kinetic theory have not been considered in this book, because no specific results can be derived on conformational properties. We are in the process of testing how such a substitution effect can be treated with the statistical branching theory.

As pointed out by Daoud and Cotton[83] with star-branched macromolecules overcrowding occurs predominantly in the local domain, *i.e.* when more than two branches come into contact, and the probability of such events is clearly enlarged in the close neighbourhood of a branching point. Such local events become noticeable only at very large *q*-values ($q > 0.2$ nm^{-1}) and requires careful SANS measurements in that regime. By 'local' we mean an effect that the reaction of a monomer unit with three or more than two functional groups can be impeded if two of the units have already reacted. We also may recall that most of the derivations were tacitly made with the Debye approximation, *i.e.* the grainy structure of polymers is fully neglected when the structures are represented by integrals. If the value of q comes in a range where $1/q \approx b$ which is the bond length between two neighboured repeat units an additional interference has to be taken into account. Finally, the size of the monomer repeat units no longer can be neglected. Mostly this effect is still considered as small such that it contributes only as a point-like scatterer, but this has already a significant contribution. In the following section two examples from static light scattering (SLS) and one SANS resulting from an AB_2 hyperbranched polyester are presented.

8.1.10 Angular Dependence of Scattering Curves from Measurements and Theoretical Prediction

We have emphasized several times that the mean square radius of gyration is the key parameter in all conformational properties. This is true not only because $\langle R_g^2 \rangle$ is a parameter that can be directly measured by scattering techniques without knowing anything of the origin or synthesis of the sample, but equally important is that the radius of gyration is a unique scaling parameter for the angular dependence such that the scattering curve can in a generalized manner be described by the use of the dimensionless parameter qR_g.

Well-defined macromolecules of really large size could be prepared only in a few cases. Mostly the hyperbranched samples are smaller than $R_g = 12$–15 nm such that only in fortunate cases can the radius of gyration be measured by visible light scattering. Randomly branched samples are exceptions, because it is easy to prepare samples up to conversions close to the gel point where the molar mass increases beyond all limits.

Two examples were studied in the Freiburg laboratory, which are the already discussed cross-linked polyesters and the three-arm star-macromolecules end-linked by diisocyanate over OH end groups on the chain ends from the polystyrene stars. Figure 8.23 shows the scattering curves from five cross-linked polyesters of molar masses between 177 000 g mol^{-1} and 27×10^6 g mol^{-1}. The normalized Kratky plots were chosen, *i.e.* $(qR_g)^2 P(q)$ as a function of qR_g. This special type of representation was used for two reasons.

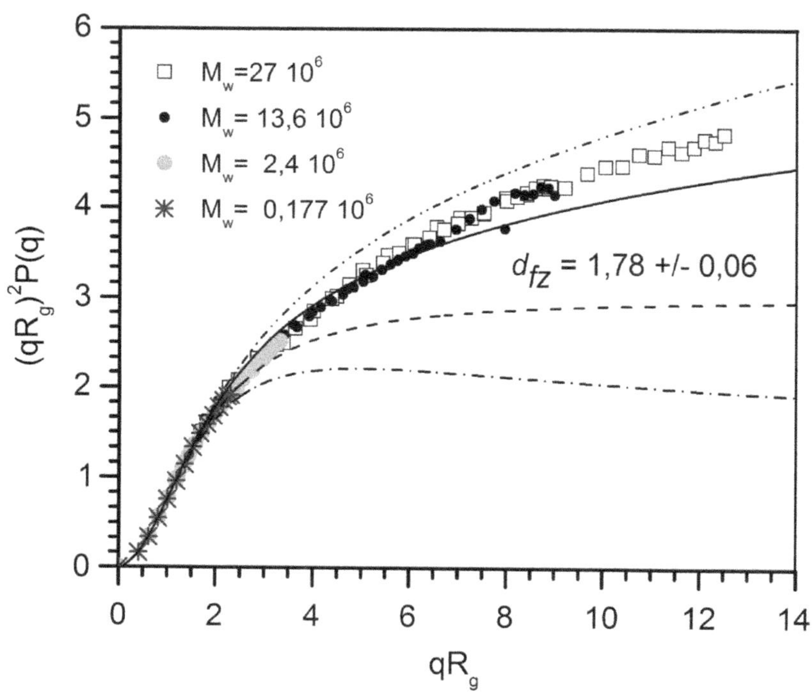

Figure 8.23 Normalized Kratky plot from four diglicidyl phthalic anhydride cross-linked polyester in DMF at 20 °C.[39] The dashed lines correspond to the Sinha curves for $d_{fz} = 1.7$; 1.8, 2.0 and 2.2, respectively (from the top of the figure).

The first is that $P(q)=i(q)/i(q=0)$ and qR_g are dimensionless, and only in this normalized form can results from different laboratories be compared. The other reason is that in a linear plot of $P(q)$ as a function of $(qR)^2$ the scattering function has already decayed to values of about 4%, *i.e.* into a region in which the behaviour from different structures are difficult to discern. By multiplication with $(qR_g)^2$ the asymptotic regime is highly amplified, and even small but characteristic differences in the scattering curves can be made visible.

The different samples in Figure 8.23 form one common curve and display self-similar behaviour. For instance, for unperturbed uniform linear chains the Kratky plot leads to a constant plateau of 2 in height and 3 for polydisperse linear chains. Despite the wide spread in the molar mass all five samples form one common curve and could be fitted with a Sinha curve $d_{fz}=1.78$ in good agreement with the molar mass dependence of $\langle R_g^2\rangle_z^{1/2}$.

A very different behaviour was found with the starch fragments, which are natural hyperbranched samples (Figure 8.24). Obviously the samples are not self-similar to each other, despite the same chemical structure with the anhydro-glucose as the repeat unit. This fact is characteristic for hyperbranched samples which consist of a linear chain contribution and another of convoluted two linear chains, which causes the branching. Without going into details the scattering curve can be calculated on the basis of the mean field approximation with the result

$$P(q)=m_{\text{lin}}P(qR_{\text{g,lin}})_{\text{lin}}+m_{\text{br}}P(qR_{\text{g,br}}) \tag{8.28}$$

$$m_{\text{lin}}=2\frac{\alpha}{1-\alpha}\frac{1}{m}; \qquad m_{\text{br}}=\text{DB}\left(\frac{\alpha}{1-\alpha}\right)^2\frac{1}{m}; \tag{8.28a}$$

with $m=m_{\text{lin}}+m_{\text{br}}$ and DB = degree of branching.

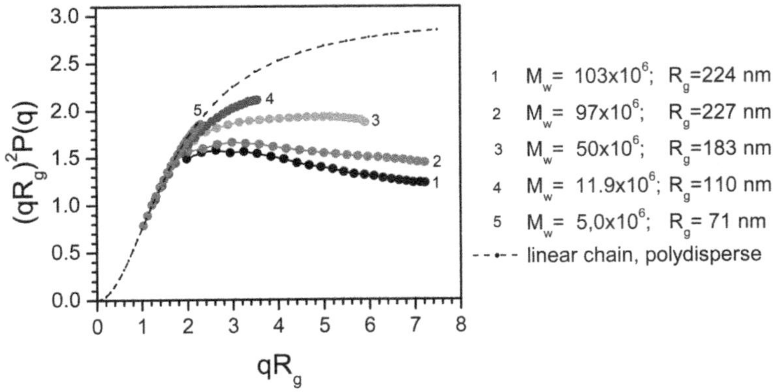

Figure 8.24 Kratky plots from five starch fragments of molar masses as indicated in the graph.[24,25] Evidently the curves cannot be described by *one* common Sinha curve, but a fit is still possible with two different types of Sinha curve.[84]

Eqn 8.28 shows that the mass fraction m_{br} increases quadratically with the conversion α while the linear mass fraction m_{lin} grows only linearly. Thus a change from the scattering behaviour of a linear chain to that of the branched chains should become apparent. For both parts different fractal behaviour has to be assumed. The fit by Sinha's equation is not uniquely possible if the degree of branching is not known. This value has to be determined separately. The common NMR technique could not be applied to these high molar masses. On the other hand the degree of branching can be obtained from the average length of the linear chains between two branching points, which was found 37 units in length (see Figure 8.13 and the corresponding comments). With this value a DB $\approx$ 0.054 is obtained, and the scattering curve from the unperturbed chain conformations can be estimated. The corresponding fit with the Sinha function has not yet been performed.

The wavelength of the used light in water is with $\lambda = 632/1.331$ nm, more than two times larger than the radius of gyration of the largest sample used in this example such that the initial and intermediate regime of the angular dependence could be covered.

A very different situation occurred with a synthesized hyperbranched polyester prepared by Lederer *et al.*[10] with a radius of gyration $R_g = 8.4$ nm. Only the molar mass and the hydrodynamic radius could be measured by static light scattering. SANS measurements had to be performed, and this was done in cooperation with Lindner, Schweins and co-workers at the ILL in Grenoble.[84] Figure 8.25 shows the scattering curve in the log($P(q)$) against

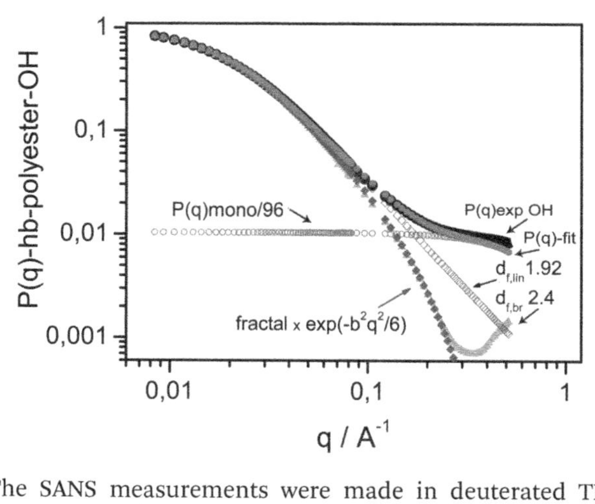

Figure 8.25 The SANS measurements were made in deuterated THF. The filled black circles show the actual measured curve. The open circles refer to measurements of the monomer. After subtraction, the light grey curve is obtained. The open squares represent the fit with the indicated fractal dimensions using the Sinha expression for the linear and branched contributions. The data points represented by the filled squares were obtained after adding the interference from single bonds. When adding the contribution of the monomer the curve with the filled dark grey circles was obtained.[84]

log(q) obtained after extrapolation of the measurements at 5 concentrations towards $c=0$. The measurements were extended to large q-values corresponding to distances of 2 Å, which means that the local structure was probed in this domain. The scattering behaviour of the monomer was measured separately and the contribution of the monomer units where subtracted. The resulting curve then was fitted with the two fractals of the linear chain and that for the convoluted chains which represent the branched contribution. After the successful fit the contribution of the monomer was added again, which gave the curve labelled '$P(q)$ fit' in Figure 8.25. An almost perfect description has been possible. The obtained fractal dimension $d_f=1.94$ would indicate almost unperturbed linear chains whereas for the convoluted chains, the fractal dimension of $d_f=2.4$ is decisive for branched chains again near unperturbed conformations. These data could be interpreted as following marginally good solvent behaviour but probably is the effect that such small molar mass samples are not yet in the asymptotic region of fractal behaviour.

8.2 Addendum

Before summarizing the content of this chapter we wish to add a few comments on the observation with charged dendrimers. The dendrimers were not discussed in detail in Chapters 6–8, and the same holds for linear chains, for the following reasons. The behaviour of linear chains has been taken as being well known and did not require a separate discussion. Certainly, this cannot be said for dendrimers, which have been studied by groups of experts. For a polymer scientist, dendrimers are very small particles with sizes around 1 nm, which, for the application of dynamic light scattering, cause serious problems. Careful SANS measurements gave reliable results and a deeper insight into these particles. In two detailed articles[85,86] the progress obtained by this technique was comprehensively published in combinations with simulations, and these review articles give sufficient information, with one exception: this is the behaviour of charged dendrimers at different degrees of ionic dissociation.

The SANS measurements from charged dendrimers at different pH values disclosed a unique and unexpected behaviour; the radius of gyration remained constant and not affected by the pH.[87,88] In the neutral state more or less the common behaviour of colloids of spherical shape is obtained with a compact core of not ordered segments but complemented by a fairly diffuse segmental arrangement in the outer shell which could be described by the following equation:[88]

$$P(q)=\left\{\frac{3}{(qR)^3}[\sin(qR)-qR\cos(qR)]\exp\left(-\frac{q^2\sigma^2}{4}\right)\right\}^2+a_bP_{\text{fluc.}}(qR) \qquad (8.29)$$

The term in the squared brackets[89] represents the common scattering amplitude for a hard sphere multiplied by an exponential function which corresponds to the Debye–Bueche approach to take into account an inhomogeneous (fuzzy) surface. The second element in the sum of eqn (8.29) is more involved and arises from internal interferences. For further information the paper by Chen *et al.*[88] should be consulted. With eqn (8.29) the radius of gyration is obtained:

$$R_g^2(R,\sigma) = \frac{3}{5}R^2 + \frac{3}{2}\sigma^2 \tag{8.30}$$

in which R is the radius of a hard sphere and σ represents the thickness of a fuzzy shell.[89] The authors were able to determine the changes in the sphere radius and the thickness of the fuzzy shell as a function of the pH with the result that the radius of gyration remained not affected, but the radius of the particle increased weakly. The second term in eqn (8.29) has a dramatic effect on the shape of the scattering curve and causes a maximum at a certain q-value which slowly is building up with the ionic dissociation to a pronounced and fairly narrow maximum. Such a maximum refers to interferences caused by the highly ordered segments in the various generations of the dendrimers. The final conclusion of the scattering phenomena is that on ionic dissociation of the tertiary amino end-groups a well organized outer shell is formed and the repulsive ionic interaction force causes well-ordered arrangements of the units in the various generations.

These remarkable properties are emphasized because a similar but much less pronounced effect may be obtained also with *charged hyperbranched* samples, in particular if a charged shell is attached to a common non-polar hyperbranched core.

8.3 Summary

In this chapter the view of an experimentalist is combined with theoretical predictions. In the first part the global structure properties are considered once again:

- The scaling rules, which were developed for linear chains, turned out not to be applicable to branched polymers. Similarly, the generalized ratio of two volumes $A_2M/[\eta]$ failed to develop universal behaviour.
- The special structure dependence is concluded to result from the hydrodynamic draining function Φ which for branched structures is not yet sufficiently well explored, and the same holds for the segment clouds interpenetrating function Ψ when two particles come into contact.
- In contrast, the ratio $R_g/R_h \equiv \rho$ proved to be a useful parameter to predict the degree of branching (DB). This ratio was derived in Chapter 7 by application of the general branching theory, modified to polymers for unperturbed conformation as well as for particles with excluded volume interaction.

- The angular dependence of scattering curves shows fractal behaviour which is well described by Sinha's equation. However, this fractal exponent does not represent the fractal dimension of the particles. The difference arises, at least partly, from the fact that a *z-average* mean square radius of gyration $\langle R_g^2 \rangle_z$ is compared with a *weight average* molar mass M_w. After correction of the radius of gyration to the weight average a much better agreement with the intrinsic viscosity behaviour is obtained.

References

1. P. G. DeGennes, *Scaling Concepts in Polymer Physics*, Cornell University Press, Ithaka, London, 1979.
2. P. L. Auer and C. S. Gardner, *J. Chem. Phys.*, 1955, **23**, 1545.
3. J. G. Kirkwood, R. W. Zwanzig and R. J. Plock, *J. Chem. Phys.*, 1955, **23**, 213.
4. B. H. Zimm, W. H. Stockmayer and M. Fixman, *J. Chem. Phys.*, 1953, **21**, 1716.
5. M. Fixman, *J. Chem. Phys.*, 1955, **23**, 1656.
6. M. Fixman, *J. Chem. Phys.*, 1961, **35**, 889.
7. E. F. Casassa, *J. Polym. Sci., Part A: Gen. Pap.*, 1965, **3**, 605.
8. K. F. Freed, *Renormalization group theory of macromolecules*, Wiley, New York, 1987.
9. R. K. Kainthan, E. B. Muliawan, S. G. Hatzikiriakos and D. E. Brooks, *Macromolecules*, 2006, **39**, 7708.
10. A. Lederer, W. Burchard, A. Khalyavina, P. Lindner and R. Schweins, *Angew. Chem., Int. Ed.*, 2013, **52**, 4659.
11. S. Bywater, *Physical Chemistry*, Springer, Berlin Heidelberg, 1979, ch. 2, vol. 30, p. 89.
12. M. Kurata, H. Yamakawa and E. Teramoto, *J. Chem. Phys.*, 1958, **28**, 785.
13. K. Akasaka, Y. Nakamura, T. Norisuye and A. Teramoto, *Polym. J.*, 1994, **26**, 1387.
14. J. F. Douglas, J. Roovers and K. F. Freed, *Macromolecules*, 1990, **23**, 4168.
15. J. Roovers, L. L. Zhou, P. M. Toporowski, M. van der Zwan, H. Iatrou and N. Hadjichristidis, *Macromolecules*, 1993, **26**, 4324.
16. J. E. L. Roovers and S. Bywater, *Macormolecules*, 1972, **5**, 384.
17. J. E. L. Roovers and S. Bywater, *Macormolecules*, 1974, **7**, 443.
18. J. Roovers, N. Hadjichristidis and L. J. Fetters, *Macromolecules*, 1983, **16**, 214.
19. K. Huber, W. Burchard and L. J. Fetters, *Macromolecules*, 1984, **17**, 541.
20. W. Burchard, in *Branched Polymers II*, ed. J. Roovers, 1999, vol. 143, p. 113.
21. A. Lederer, W. Burchard, T. Hartmann, J. S. Haataja, N. Houbenov, A. Janke, P. Friedel, R. Schweins and P. Lindner, *Angew. Chem. Int. Ed.*, 2015, DOI: 10.1002/anie.201504059.
22. I. B. Rietveld and J. A. M. Smit, *Macromolecules*, 1999, **32**, 4608.
23. P. Lang, W. Burchard, M. S. Wolfe, H. J. Spinelli and L. Page, *Macromolecules*, 1991, **24**, 1306.

24. G. Galinsky and W. Burchard, *Macromolecules*, 1995, **28**, 2363.
25. G. Galinsky, PhD, University of Freiburg, 1996.
26. A. Lederer, T. Hartmann and H. Komber, *Macromol. Rapid Commun.*, 2012, **33**, 1440.
27. A. Einstein, *Ann. Phys.*, 1911, **34**, 591.
28. A. Einstein, *Ann. Phys.*, 1906, **19**, 289.
29. J. G. Kirkwood and J. Riseman, *J. Chem. Phys.*, 1948, **16**(6), 565.
30. P. J. Flory and T. G. J. Fox, *J. Am. Chem. Soc.*, 1951, 73.
31. W. Burchard, in *Physical Techniques for the Study of Food Biopolymers*, ed. Ross-Murphy, Chapman and Hall, London, 1994.
32. Y. Oono and M. Kohmoto, *J. Chem. Phys.*, 1983, **78**, 520.
33. Y. Oono, P. R. Baldwin and T. Ohta, *Phys. Rev. Lett.*, 1984, **53**, 2149.
34. Y. Oono, *J. Chem. Phys.*, 1983, **79**, 4629.
35. P. J. Flory, *Principles in Polymer Chemistry*, Cornell University Press, Ithaca, New York, 1953.
36. S. Erlander and D. French, *J. Polym. Sci.*, 1956, **20**, 7.
37. M. Weissmüller and W. Burchard, *Polym. Int.*, 1997, **44**, 380.
38. M. Weissmuller, V. Trappe and W. Burchard, *Abstracts of Papers of the American Chemical Society*, 1996, **211**, 107-PMSE.
39. V. Trappe, PhD, University of Freiburg, 1996.
40. J. A. Semlyen and P. J. Flory, *Trans. Faraday Soc.*, 1966, **62**, 2622.
41. J. A. Semlyen, *Cyclic Polymers*, Elsevier Applied Science Publishers, London and New York, 1986.
42. S. B. Ross-Murphy and F. T. Stepto, in *Large Ring Molecules*, ed. J. A. Semlyen, Wiley, Chichester, New York, 1996.
43. M. Weissmüller, PhD, University of Freiburg, 1996.
44. C. Paulo and J. E. Puskas, *Macromolecules*, 2001, **34**, 734.
45. A. J. Heidenreich and J. E. Puskas, *J. Polym. Sci., Part A: Polym. Chem.*, 2008, **46**, 7621.
46. J. E. Puskas, W. Burchard, A. J. Heidenreich and L. D. Santos, *J. Polym. Sci., Part A: Polym. Chem.*, 2012, **50**, 70.
47. A. H. E. Müller, D. Yan and M. Wulkow, *Macromolecules*, 1997, **30**, 7015.
48. W. Radke, G. Litvinenko and A. H. E. Müller, *Macromolecules*, 1998, 31.
49. D. Yan, Z. Zhou and A. H. E. Müller, *Macromolecules*, 1999, **32**, 245.
50. P. F. W. Simon and A. H. E. Müller, *Macromolecules*, 2001, **34**, 6206.
51. M. Gauthier and M. Möller, *Macromolecules*, 1991, **24**, 4548.
52. M. Gauthier, M. Moller and W. Burchard, *Abstracts of Papers of the American Chemical Society*, 1993, **205**, 6-POLY.
53. M. Gauthier, M. Moller and W. Burchard, *Macromol. Symp.*, 1994, **77**, 43.
54. W. Burchard, in *Light Scattering from Polymers*, Springer, Berlin, Heidelberg, 1983, ch. 1, vol. 48, p. 1.
55. A. Z. Akcasu, M. Benmouna and C. C. Han, *Polymer*, 1980, **21**, 866.
56. B. H. Zimm and W. H. Stockmayer, *J. Chem. Phys.*, 1949, **17**, 1301.
57. J. D. Fox and J. F. Robyt, *Carbohydr. Res.*, 1992, **227**, 163.
58. M. Kurata, M. Abe, M. Iwama and M. Matsushima, *Polym. J.*, 1972, **3**, 729.
59. M. Weissmüller and W. Burchard, *Acta Polym.*, 1997, **48**, 571.

60. W. H. Stockmayer and M. Fixman, *Ann. N. Y. Acad. Sci.*, 1953, **57**, 334.
61. W. Burchard and R. Schweins, manuscript in preparation.
62. A. Blumen, A. Jurjiu, T. Koslowski and C. von Ferber, *Phys. Rev. E: Stat., Nonlinear, Soft Matter Phys.*, 2003, 67.
63. A. Blumen, C. von Ferber, A. Jurjiu and T. Koslowski, *Macromolecules*, 2004, **37**, 638.
64. M. Dolgushev and A. Blumen, *J. Chem. Phys.*, 2009, **131**, 044905.
65. M. Dolgushev and A. Blumen, *Macromolecules*, 2009, **42**, 5378.
66. M. Dolgushev and A. Blumen, *J. Chem. Phys.*, 2010, **132**, 124905.
67. M. Dolgushev, G. Berezovska and A. Blumen, *Macromol. Theory Simul.*, 2011, **20**, 621.
68. A. Johner, F. Thalmann, J. Baschnagel, H. Meyer, S. Obukhov and J. P. Wittmer, *J. Stat. Mech.: Theory Exp.*, 2014, **2014**, P04024.
69. J. P. Wittmer, *personal communication*, 2014.
70. B. B. Mandelbrot, *The Fractal Geometry of Nature*, Freeman & Co., New York, 1983.
71. P. Debye and A. M. Bueche, *J. Appl. Phys.*, 1949, **20**, 518.
72. G. Ross, *Opt. Acta*, 1968, **15**, 451.
73. G. Porod, in *Small Angle X-Ray Scattering*, ed. O. Glatter and O. Kratky, Academic Press London, New York, 1982, p. 17.
74. D. Stauffer, *Introduction to Percolation Theory*, Taylor and Francis, London, 1985.
75. D. Stauffer, A. Coniglio and M. Adam, *Adv. Polym. Sci.*, 1982, **44**, 103.
76. V. Trappe, W. Burchard and B. Steinmann, *Macromolecules*, 1991, **24**, 4738.
77. V. Trappe, W. Richtering and W. Burchard, *J. Phys.*, 1992, **2**, 1453.
78. V. Trappe, W. Burchard and B. Steinmann, *Makromol. Chem., Macromol. Symp.*, 1991, **45**, 63.
79. H. Galina and J. Lechowicz, in *Grafting/Characterization Techniques/Kinetic Modeling*, ed. H. Galina, Y. Ikada, K. Kato, R. Kitamaru, J. Lechowicz, Y. Uyama and C. Wu, Springer, Berlin, Heidelberg, 1998, ch. 4, vol. 137, p. 135.
80. H. Galina, J. B. Lechowicz and M. Walczak, *J. Macromol. Sci., Phys.*, 2005, **B44**, 925.
81. E. M. Hendriks, M. H. Ernst and R. M. Ziff, *J. Stat. Phys.*, 1983, **31**, 519.
82. R. M. Ziff and G. Stell, *J. Chem. Phys.*, 1980, **73**, 3492.
83. M. Daoud and J. P. Cotton, *J. Phys.*, 1982, **43**, 531.
84. W. Burchard, A. Khalyavina, P. Lindner, R. Schweins, P. Friedel, M. Wiemann and A. Lederer, *Macromolecules*, 2012, **45**, 3177.
85. M. Ballauff and C. N. Likos, *Angew. Chem., Int. Ed.*, 2004, **43**, 2998.
86. S. Rosenfeldt, N. Dingenouts, M. Ballauff, P. Lindner, C. N. Likos, N. Werner and F. Vogtle, *Macromol. Chem. Phys.*, 2002, **203**, 1995.
87. G. Nisato, R. Ivkov and E. J. Amis, *Macromolecules*, 2000, **33**, 4172.
88. W. R. Chen, L. Porcar, Y. Liu, P. D. Butler and L. J. Magid, *Macromolecules*, 2007, **40**, 5887.
89. J. S. Pedersen, *J. Chem. Phys.*, 2001, **114**, 2839.

CHAPTER 9

Correlation of Molecular and Bulk Properties

9.1 Bulk Properties in Flow and in Thin Films

Most of the applications of hyperbranched polymers are in bulk materials, thus proper interpretation of their bulk properties in correlation to molecular parameters is essential for deliberate materials design. Therefore, knowledge of the properties on a molecular level as is the case for a single molecule in dilute solution has to be precisely used for understanding the behaviour in bulk, where a single molecule is entirely exposed to interactions with other macromolecules.[1]

The particular globular shape of dendritic polymers provides them with a special rheological profile in melt which strongly differs from that of linear polymers. Thus, hyperbranched polymers are frequently used to modify the processing properties of linear polymer systems.[2] Detailed knowledge about the effect of the molecular architecture on the rheology and processing enables one to take advantage of hyperbranched polymers for such applications. Still, theoretical explanation of the dendritic polymer rheology is a challenge. A number of investigations on hyperbranched polymers have been carried out showing strong deviations from the meanwhile well-understood behaviour for linear and star or H-shaped polymers, which have been the subject of theoretical and experimental investigations for several decades.[3–7] These deviations are the result of their significantly denser branching topology which prevents formation of entanglements.[8,9] As a result, hyperbranched polymers exhibit the typical properties of unentangled objects with a Newtonian behaviour. Parameters such as degree of branching, topology and end groups have a strong influence on their rheology. The influence of different degrees of branching was repeatedly reported.[10–13]

RSC Polymer Chemistry Series No. 16
Hyperbranched Polymers: Macromolecules in between Deterministic Linear Chains and Dendrimer Structures
By Albena Lederer and Walther Burchard

Published by the Royal Society of Chemistry, www.rsc.org

Polyethyenes with different branching topologies,[14] from dendritic to nearly linear, show that dendritic structures possess Newtonian flow behaviour in contrast to the typical shear thinning in linear polymers. Examples such as linear and branched poly(methylmethacrylate), highly branched polystyrenes and hyperbranched polyesters[12,15,16] lead to the conclusion that high branching density leads to a behaviour between that of polymers and that of soft colloids, and inter- and intramolecular entanglement is strongly limited.

Parallel with the branching density the high number of end functionalities in dendritic systems significantly influences chain mobility.[17] Due to their high number in dendritic systems, H-bonding formation could be amplified, as shown in the work of McKee *et al.*[13] on linear and randomly branched poly(alkyl methacrylates). Storage modulus and dynamic viscosity measurements show that the H-bonding between end groups dominates the melt rheology response.

As a limiting case of bulk behaviour, thin films of dendritic systems have to be considered. The influence of the molecular properties under the confined conditions in thin films could be easily correlated to the local molecular characteristics, while interaction with solvent molecules, *e.g.* in swelling processes are of significant interest.

Multifunctional dendritic architectures are fascinating candidates for various thin film applications such as humidity detectors or chemical sensors,[18,19] as well for diagnostics and in multifunctional coatings.[18,20–22] In the confinement of thin film geometry the large number of functional groups dominates surface properties and enables specific interactions with different molecules. The architectural parameters in this confined geometry with a thickness of only a few molecular layers play a special role. Ellipsometry and AFM measurements show that the type of *backbone* on the density of the films is of immense importance.[23,24] Direct dependence of the density of the films on the monomer flexibility was found (Figure 9.1) which is completely consistent with the dilute solution behaviour of these polymers shown in Chapter 5. Depending on the monomer geometry different swelling kinetics has been observed. Higher swelling was found for flexible backbone geometries, in which more free volume is available. In contrast, stiff aromatic structures correspond to lower swelling. In this context the more flexible structures possess higher surface hydrophobicity, which is amplified after annealing above the glass transition enabling the macromolecules to undergo reorientation and reorganization of the polymer segments.[25,26]

End-group type can strongly influence the swelling properties and the sensitivity of thin films to analyte molecules too.[26,27] As an example, hyperbranched aromatic polyesters with hydroxy (–OH), carboxylic (–COOH), and acetoxy (–OAc) end groups provide different response to air humidity with variation of the film thickness.[27] Similar sensitivity of the thin films toward alcohols or freons was indicated by reflectometric interference spectroscopy.[29] Indeed, a comparison of the branched and linear polymers with similar type of end groups with non-polar, modified hyperbranched

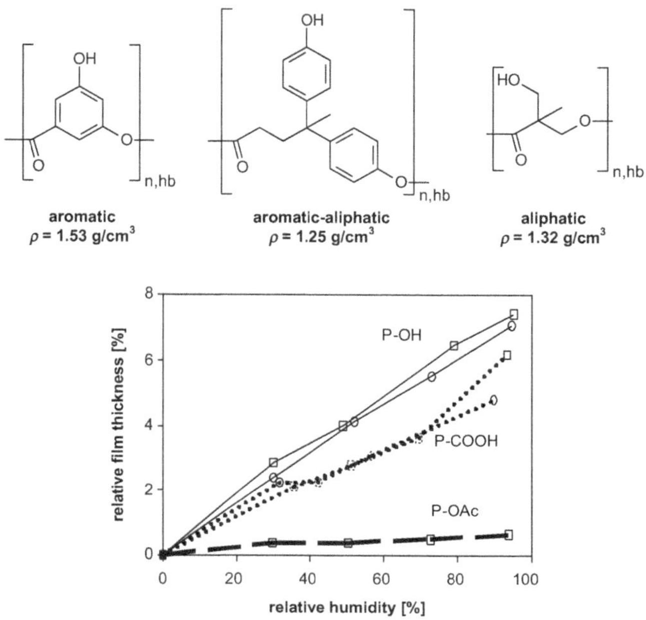

Figure 9.1 Hyperbranched polyesters with different backbones and thin-film density at 50 nm film thickness[24] (compare Figure 5.15). The graph shows the film thickness dependence on the humidity for aromatic hyperbranched polyesters with hydroxy, carboxylic (measurements on two different samples show high reproducibility), and acetoxy end groups.[27] Reprinted with permission from ref. 28. Copyright © 2009 American Chemical Society.

polymers and linear analogous shows clearly that the dominating feature in adsorption processes on thin films is not the branching topology but the large number of functional groups.[24]

Thus, adjusting the thin film properties can be mainly controlled by tuning the type and number of end groups.

In this chapter the mechanical properties and melt rheology will not be discussed in detail. Here we will focus on the correlation between molecular structure and molecular dynamics depending on the temperature as this property is directly coupled to the molecular structure itself, but within a bulk environment.

9.2 Thermal Properties

Thermal stabilities of hyperbranched polymer materials are similar to those of linear polymers and depend largely on the chemical structure; aromatic materials are more stable than aliphatic. Branching strongly influences the thermal characteristics of polymers, like melting point and glass transition temperature. Branching reduces the crystallization ability, and branching topology and length entirely determine the extent of this effect. In the case of hyperbranched polymers crystallization is completely prevented due to the short distances between the branching points. There are only a few examples where crystallization has been observed either due to modification of the end groups with long chains[17] or due to branching dilution leading to linear segments with sufficient length for formation of crystalline domains,[30] and just rarely liquid crystal formation was reported.[31] Therefore the T_g is the main thermal parameter, which defines bulk properties of hyperbranched materials and is strongly influenced by the molecular parameters as degree of branching, molar mass and especially end groups.

9.2.1 Influence of the Degree of Branching

The glass transition temperature (T_g) depends on the mobility of the polymer segments and on the end groups and their interactions, which in turn reflect in the free volume of the molecules. Increasing degree of branching and branching density leads to decreased intramolecular mobility, which should result in higher T_g. However, comparison between polymers with different degrees of branching but similar chemical character show controversial tendencies.[32–34] Glass transition slightly depending on branching was observed for linear hyperbranched aromatic polyesters and their dendrimers with identical repeating unit, as well as number and nature of functional groups.[35] Similar observations were made on linear, branched and 'kinked' PET[36] and linear and branched poly(butylenes isophthalates).[37] This behaviour could be a result of end group effects, explained in the next section or of partial crystallinity, which increases the T_g due to restriction in the segmental motion of the macromolecular segments.

Other studies show that branching causes a decrease of the glass transition temperature as in the case of polyetherimides[38] or polyimides.[39] Zhu *et al.*[33] reported on amorphous poly[3-methyl-3-(hydroxymethyl)oxetane] which shows a dynamic behaviour of T_g, passing through a maximum at degree of branching of 27% and then decreasing. They explain this observation with restriction of the segmental mobility with the degree of branching and simultaneous increase of end-group number which becomes the dominating effect after a certain degree of branching.

Investigations on aromatic aliphatic polyesters with a systematic variation in the degree of branching described in Chapter 3 show indeed a minor, but linear decrease with increasing branching (Figure 9.2). This constant tendency, however, is strongly influenced by the type of end groups. The linear behaviour mentioned was observed for non-polar end groups, while OH end groups lead to increased T_g which does not show constant dependency on the degree of branching.

9.2.2 Influence of the Molar Mass

The glass transition is a property which is strongly influenced by molar mass and this dependency is well understood for linear polymers. The glass

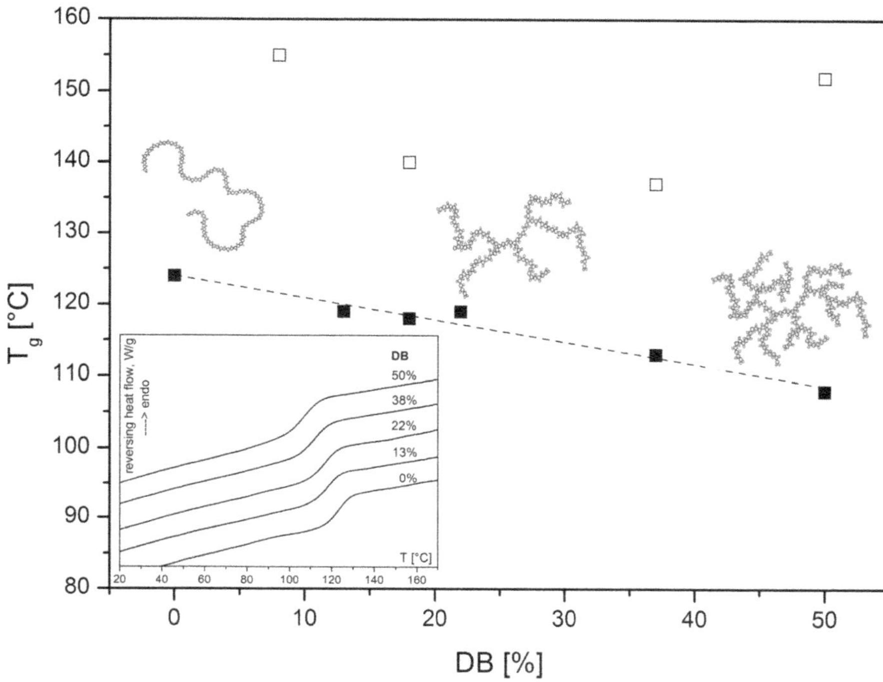

Figure 9.2 Glass transition temperature dependence on the degree of branching for OH-terminated (open symbols) and end-capped, SY-terminated (full symbols) aliphatic–aromatic polyesters. The inset represents the DSC traces of the SY-terminated samples.
Reprinted with permission from ref. 40. Copyright © 2012 Elsevier.

transition as a result of segmental motion of a linear chain depends on its length and the distribution of chain lengths, *e.g.* the molar mass distribution. T_g increases rapidly with molar mass and reaches a plateau at higher values. The relation between glass transition temperature and the molar mass was determined by Fox and Flory[41] as follows:

$$T_{g,\text{linear}} = T_{g,\infty} - \frac{K}{M} \tag{9.1}$$

The $T_{g,\infty}$ corresponds to the plateau value at very high molar masses and is individual for each polymer, so is the constant, K.

The determination of T_g for densely branched statistical polymers is not as easy as for linear polymers due to a multitude of inhomogeneities, *e.g.* molar mass, branching and end-group distributions. Wooley *et al.*[35] succeeded in establishing a relationship for dendrimers, which are essentially monodispers, with well-defined branching and chain-end number allowing for precise T_g investigations. These investigations led to the following dependence of the T_g on the molar mass, taking into account the number of end groups n_e:

$$T_{g,\text{dendrimer}} = T_{g,\infty} - \frac{n_e K}{M} \tag{9.2}$$

However, a zero value of the term $\frac{n_e K}{M}$ was never reached experimentally.

The question arises, Which model of the glass transition temperature depending on the molar mass would better describe the behaviour of hyperbranched polymers? Therefore polymers with low molar mass distributions and variation in the molar masses are needed. For this purpose, fractionation of samples could lead such portfolio of narrowly distributed samples as it was performed by Khalyavina *et al.*[40] (see also Chapter 4). In general, it was found that the development of the function $M(T_g)$ resembles the behaviour of the linear polymers, reaching a plateau at high molar mass values. This observation was done on the basis of polymers with different statistical degree of branching and was compared to a linear polymer with identical chemical structure (see Chapter 3.2.2). This dependency is shown in Figure 9.3.

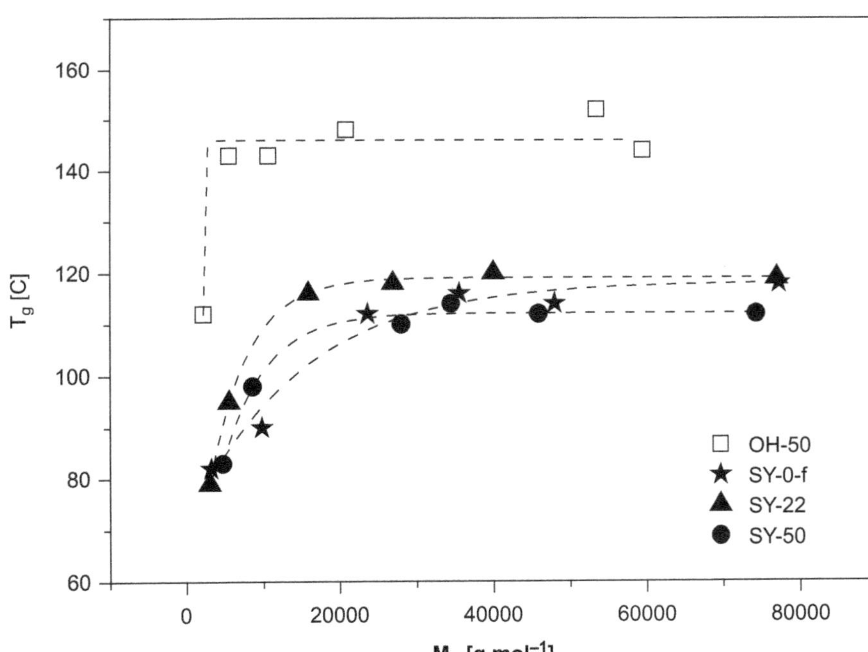

Figure 9.3 Development of the glass transition temperature with molecular weight for fractions of OH-terminated hyperbranched polyesters ($DB = 50$, open symbols) and end-capped, SY-terminated polyesters with $DB = 0$, 22 and 50% (full symbols).
Adapted with permission from ref. 40. Copyright © 2012 Elsevier.

Another parameter, the type of the functional groups, very strongly influences the values of T_g. In case where non-interacting end groups are available, the branching does not significantly influence the T_g/M relationship. It was found that T_g increases slowly and reaches a plateau at a number average molecular weight of approx. 25 000 g mol^{-1}. Yet, by exchanging the end groups with OH-functions capable of H-bonding, the T_g increases sharply at low molar masses (5500 g mol^{-1}) and remains constant with the molar mass. Obviously, the OH-group effect has a very strong influence on the T_g, which is stronger than the impact of both degree of branching and molar mass.

As mentioned above, the plot of the T_g dependence towards molecular weight resembles those of linear systems. The dendrimer behaviour found by Wooley *et al.*[35] is less represented, since there is a difference in the development of functional groups during synthesis in dendrimers and hyperbranched polymers. While for dendrimers this number increases with the generation [eqn (1.3)], the number of functional groups in hyperbranched structures in statistical branching is equal to the degree of polymerization plus 1. Hence, the term n_e/M remains constant and the dependence can be expressed similarly as for linear polymers [eqn (9.1)].

9.2.3 Influence of the End Groups

The end groups play a significant role in manipulating thermal properties. Longer linear parts in hyperbranched polymers are supposed to play the role of plasticizers[17] while polar end groups could significantly increase the T_g. In a similar manner, the decrease of T_g with the introduction of a protective group was observed in the work of Mikhaylova,[42] where the T_g values (227 °C) of OH-terminated hyperbranched aromatic polyesters drop to 158 °C when protecting end groups are introduced. This observation leads to the conclusion that hydrogen bonds have a strong influence on glass transition temperature. For OH-terminated polyesters H-bonding interactions have been confirmed by FTIR measurements.[25,43] This is supported by the results of Wooley *et al.*[34] where similar T_g values were obtained for fully aromatic OH-terminated linear, hyperbranched and dendritic polyesters with comparable amounts of phenolic groups.

Several other examples indicate that the end group effect is obviously very strong in bulk systems predominantly influencing the T_g and only slightly depending on the degree of branching. The results shown in Figures 9.2 and 9.3 were collected from polymers with different degree of branching in which the effect of end groups is excluded by two ways: (1) the number of end groups is linearly dependent on the degree of polymerization and independent on the degree of brancing; and (2) the end groups are end-capped and non-interactive (Chapter 3.2.2). In this way the effect of the degree of branching can be clearly extracted, but the effect of the type and number of end groups cannot be systematically followed, though showing deviations if OH groups come into play. Systematic investigation of this parameter in systems with variations in the degree of branching and number of end groups sheds light on this question using gradually branched polyesters by co-polymerization of AB_2 and AB monomers (Section 3.2.1).[44]

In contrast to Figure 9.2, T_g for these systems increases with increasing degree of branching very sharply (Figures 9.4 and 9.5). This is explained by the complex overlapping of different parameters such as hydrogen bonding, branching and different chemistry of the bifunctional AB monomer, which is added in different ratios to the AB_2 comonomer for branching variation.

The pronounced tendency has to be ascribed to the influence of the end groups on the thermal behaviour. With decreasing degree of branching the number of the OH-functional groups decreases from one OH group at 50% degree of branching to 0.3 OH groups per monomer unit at 30% degree of branching (Figure 9.5). By this decrease, the formation of H-bonds between the macromolecules in bulk is strongly reduced, the flexibility of the molecules becomes higher and T_g decreases.

At 29% degree of branching even a melting region was observed, indicating partial crystallinity although the T_g of this polymer is the lowest in this series (Figure 9.4). This fact shows that the H-bonding obviously not only overrides the branching but also prevents the crystallinity.

The effect of the dendritic branching on the glass transition cannot simply be explained by the particular branching topology. It is defined by the combination of the monomer chemical character and the number and type

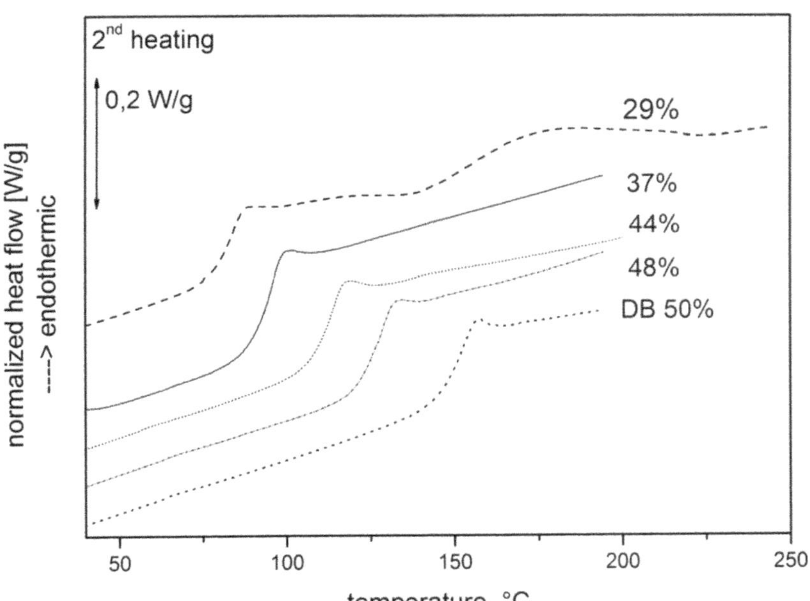

Figure 9.4 T_g dependency on the degree of branching for $AB_2 + AB$ co-polyesters. Reprinted with permission from ref. 44 Copyright © 2008 Wiley-VCH Verlag GmbH & Co. KGaA, Weinheim.

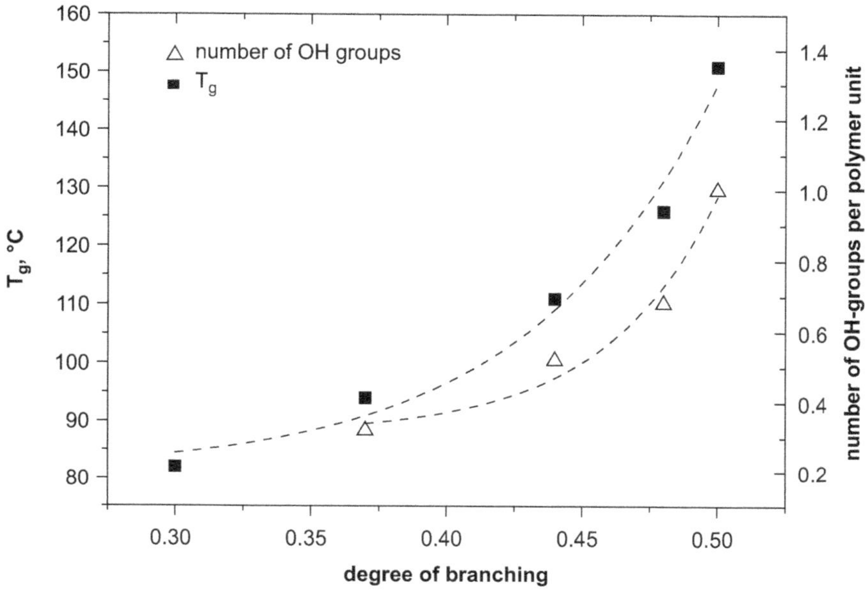

Figure 9.5 Dependence of T_g and number of OH groups on the degree of branching for an $AB_2 + AB$ polyester with varying content of AB co-monomer.[44]

of functional groups in the polymer. This also holds when the solution properties of these polymers are considered. However, the influence of H-bonding in dilute solution seems not to be as pronounced as in bulk, where the branching effect could be completely overruled by the H-bonding effect.

References

1. W. W. Graessley and J. Roovers, *Macromolecules*, 1979, **12**, 5.
2. B. Voit, *J. Polym. Sci., Part A: Polym. Chem.*, 2005, **43**, 267.
3. T. C. B. McLeish, in *Macromolecular Engineering: Precise Synthesis, Materials, Properties, Applications*, ed. K. Matyjaszewski, Y. Gnanou and L. Leibler, Wiley-VCH, Weinheim, 2007, vol. 2, p. 1605ff.
4. T. C. B. McLeish and S. T. Milner, *Adv. Polym. Sci.*, 1999, **143**, 195.
5. S. B. Kharchenko, R. M. Kannan, J. Cernohous and S. Venkataramani, *Macromolecules*, 2003, **36**, 399.
6. P. G. DeGennes, *J. Phys., Lett.*, 1975, **36**.
7. J. Roovers, *Macromol. Symp.*, 1997, **121**, 89.
8. P. J. Farrington, C. J. Hawker, J. M. J. Fréchet and M. E. Mackay, *Macromolecules*, 1998, **31**, 5043.
9. I. Sendijarevic and A. J. McHugh, *Macromolecules*, 2000, **33**, 590.
10. J. P. Sheth, S. Unal, E. Yilgor, I. Yilgor, F. L. Beyer, T. E. Long and G. L. Wilkes, *Polymer*, 2005, **46**, 10180.

11. R. Patil, R. H. Colby, D. J. Read, G. Chen and Z. Guan, *Macromolecules*, 2005, **38**, 10571.
12. P. F. W. Simon, A. H. E. Müller and T. Pakula, *Macromolecules*, 2001, **34**, 1677.
13. M. G. McKee, C. L. Elkins, T. Park and T. E. Long, *Macromolecules*, 2005, **38**, 6015.
14. Z. Ye and S. Zhu, *Macromolecules*, 2003, **36**, 2194.
15. S. B. Kharchenko and R. M. Kannan, *Macromolecules*, 2002, **36**, 407.
16. Suneel, D. M. A. Buzza, D. J. Groves, T. C. B. McLeish, D. Parker, A. J. Keeney and W. J. Feast, *Macromolecules*, 2002, **35**, 9605.
17. D. Schmaljohann, L. Häußler, P. Pötschke, B. I. Voit and T. J. A. Loontjens, *Macromol. Chem. Phys.*, 2000, **201**, 49.
18. B. Voit, D. Beyerlein, K.-J. Eichhorn, K. Grundke, D. Schmaljohann and T. Loontjens, *Chem. Eng. Technol.*, 2002, **25**, 7.
19. M. Vollprecht, F. Dieterle, S. Busche, G. Gauglitz, K.-J. Eichhorn and B. Voit, *Anal. Chem.*, 2005, **77**, 5542.
20. M. Johansson, E. Malmström and A. Hult, *J. Polym. Sci., Part A: Polym. Chem.*, 1993, **31**, 619.
21. A. Sidorenko, X. W. Zhai, A. Greco and V. V. Tsukruk, *Langmuir*, 2002, **18**, 3408.
22. A. Sidorenko, X. W. Zhai, S. Peleshanko, A. Greco, V. V. Shevchenko and V. V. Tsukruk, *Langmuir*, 2001, **17**, 5924.
23. M. A. Elrehim, B. Voit, B. Bruchmann, K.-J. Eichhorn, K. Grundke and C. Bellmann, *J. Polym. Sci. Part A: Polym. Chem.*, 2005, **43**, 3376.
24. S. Reichelt, K.-J. Eichhorn, D. Aulich, K. Hinrichs, N. Jain, D. Appelhans and B. Voit, *Colloids Surf., B*, 2009, **69**, 169.
25. Y. Mikhaylova, G. Adam, L. Häussler, K.-J. Eichhorn and B. Voit, *J. Mol. Struct.*, 2006, **788**(1–3), 80.
26. Y. Mikhaylova, E. Pigorsch, K. Grundke, K. J. Eichhorn and B. Voit, *Macromol. Symp.*, 2004, **210**, 271.
27. D. Beyerlein, G. Belge, K.-J. Eichhorn, G. Gauglitz, K. Grundke and B. Voit, *Macromol. Symp.*, 2001, **164**, 117.
28. B. I. Voit and A. Lederer, *Chem. Rev.*, 2009, **109**, 5924.
29. G. Belge, D. Beyerlein, C. Betsch, K.-J. Eichhorn, G. Gauglitz, K. Grundke and B. Voit, *Anal. Bioanal. Chem.*, 2002, **374**, 403.
30. A. Burgath, A. Sunder and H. Frey, *Macromol. Chem. Phys.*, 2000, **201**, 782.
31. M. Marcos, R. Martin-Rapun, A. Omenat and J. L. Serrano, *Chem. Soc. Rev.*, 2007, **36**, 1889.
32. M. G. McKee, S. Unal, G. L. Wilkes and T. E. Long, *Prog. Polym. Sci.*, 2005, **30**, 507.
33. Q. Zhu, J. Wu, C. Tu, Y. Shi, L. He, R. Wang, X. Zhu and D. Yan, *J. Phys. Chem. B*, 2009, **113**, 5777.
34. K. L. Wooley, J. M. J. Fréchet and C. J. Hawker, *Polymer*, 1994, **35**, 4489.
35. K. L. Wooley, C. J. Hawker, J. M. Pochan and J. M. J. Frechet, *Macromolecules*, 1993, **26**, 1514.

36. M. Jayakannan and S. Ramakrishnan, *J. Polym. Sci., Part A: Polym. Chem.*, 1998, **36**, 309.
37. L. Finelli, N. Lotti and A. Munari, *J. Appl. Polym. Sci.*, 2002, **84**, 2001.
38. L. J. Markoski, J. S. Moore, I. Sendijarevic and A. J. McHugh, *Macromolecules*, 2001, **34**, 2695.
39. J. Peter, A. Khalyavina, J. Kříž and M. Bleha, *Eur. Polym. J.*, 2009, **45**, 1716.
40. A. Khalyavina, L. Häußler and A. Lederer, *Polymer*, 2012, **53**, 1049.
41. T. G. J. Fox and P. J. Flory, *J. Polym. Sci.*, 1954, **14**, 315.
42. Y. Mikhaylova, PhD Thesis, TU Dresden, 2006.
43. E. Žagar and J. Grdadolnik, *J. Mol. Struct.*, 2003, **658**(3), 143.
44. F. Schallausky, M. Erber, H. Komber and A. Lederer, *Macromol. Chem. Phys.*, 2008, **209**, 2331.

Subject Index

$A_2M_w/[\eta]$ ratio, of molecular parameters, 235–239
$\rho = R_g/R_h$ ratio, of molecular parameters, 239–244
Ψ-parameter, 230

$AB_2 + AB$ approach, and degree of branching tuning, 42–47
$AB_2 + ABB^*$ approach, and degree of branching tuning, 47–50
AB_2 hyperbranched samples, and branched polymers, 207–208
absolute molar mass, determination of, 69–76
addendum, 264–265
angular dependence of scattered light, and branched polymers, 204–210
  AB_2 hyperbranched samples, 207–208
  degree of polymerization, 208–209
  and global structure parameters, 208–209
  hydrodynamic radius, 206–207
  overview, 204–206
  radius of gyration, 206
  unperturbed conformations, 209–210
angular dependence of scattering molecular properties, 260–264

Baker–Williams fractionation, 58
branched co-polymers, 202–204
branched homopolymers, 199–201
branched macromolecules
  and dynamic light scattering, 176–177
  and radius of gyration, 220–224
branched polymers, 3–13
  dendrimers, 6–8
  dendritic hybrids, 11–13
  hyperbranched polymers, 8–11
  overview, 3–6
branching, 20–28
  contraction factors, 22–24
  degree of (*See* degree of branching (DB))
  overview, 20–22
Brownian motion, and dynamic light scattering, 167–168
bulk properties, of hyperbranched polymers, 269–271

capillary viscometry, 89, 94, 139–140
chemical structure influence, and solution viscosity, 107–115
  and backbone character, 107–110
  of functionality, 110–115
cluster formation, and dynamic light scattering, 179–180
compound distribution, and branched polymers, 199
concentration dependence, 159–161, 183–185

conformational properties, of branched polymers
angular dependence of scattered light, 204–210
AB_2 hyperbranched samples, 207–208
degree of polymerization, 208–209
and global structure parameters, 208–209
hydrodynamic radius, 206–207
overview, 204–206
radius of gyration, 206
unperturbed conformations, 209–210
and general branching theory, 194–204
basic properties of, 196
probable distributions, 196–199
randomly branched polymers, 199–204
remarks on, 194–196
overview, 193–194
and volume exclusion, 210–228
description, 210–211
Flory expansion factor, 211–212
particle scattering factor *P(q)*, 213–220
perturbed conformations, 223–224
size distribution of real chains, 212–213
unperturbed conformations, 222–223
conformation of dendritic polymers
control over molar mass, 56–61
degree of branching tuning, 42–56
$AB_2 + AB$ approach, 42–47
$AB_2 + ABB^*$ approach, 47–50
pseudo-dendrimers by AB_2 modification, 50–56
overview, 39–42
variation of end-group functionality, 61–66
contraction parameters, and molecular properties, 244–249
convolution, and branched polymers, 198–199

DB. *See* degree of branching (DB)
Debye equation, 216
degree of branching (DB)
and contraction factors, 22–23, 27–28
description, 24–27
determination of, 28–35
side reactions and, 33–35
spectroscopic approaches, 28–30
synthetic approach, 30–33
and hyperbranched polymers, 81–83
thermal properties, 271–272
tuning of, 42–56
$AB_2 + AB$ approach, 42–47
$AB_2 + ABB^*$ approach, 47–50
pseudo-dendrimers by AB_2 modification, 50–56
degree of polymerization, and branched polymers, 208–209
dendrimers, 6–8
solution viscosity of, 97–99
dendritic hybrids, 11–13
dendritic polymers
conformation of
control over molar mass, 56–61
degree of branching tuning, 42–56
overview, 39–42
variation of end-group functionality, 61–66

dendritic polymers (*continued*)
separation methods for, 78–86
according to degree of branching, 81–83
and elution properties, 83–86
solution viscosity of, 97–107
dendrimers, 97–99
dendronized polymers, 105–107
hyperbranched polymers, 99–102
pseudo-dendrimers, 102–105
dendronized polymers, solution viscosity of, 105–107
determination of degree of branching, 28–35
side reactions and, 33–35
spectroscopic approaches, 28–30
synthetic approach, 30–33
dimension $\lambda/20$ particles, and static light scattering, 145–151
dynamic light scattering, 164–185
Brownian motion and stochastic processes, 167–168
and cluster formation, 179–180
concentration dependence, 183–185
development of optics, 165–167
effect of external forces, 169–171
electrophoretic, 180–183
linear and branched macromolecules, 176–177
overview, 164–165, 185
and polydispersity, 179–180
and proteins, 177–179
time correlation functions, 168–169, 171–175

electrophoretic dynamic light scattering, 180–183
elution properties, and hyperbranched polymers, 83–86
end-group functionality, and dendritic polymers, 61–66
end groups, and hyperbranched polymers, 275–277
external forces effect, and dynamic light scattering, 169–171

Flory expansion factor, 211–212
Flory–Fox equation, 24, 72, 73, 90, 91, 92, 104, 120, 128, 251
fractal approach, and molecular properties, 251–255
fractal behaviour, 158–159
functionality influence, on solution viscosity, 110–115

Gaussian distribution, 170, 189
gel point, 8
general branching theory, and branched polymers, 194–204
basic properties of, 196
probable distributions, 196–199
compound distribution, 199
convolution, 198–199
definition, 196–197
moments of distribution, 197–198
randomly branched polymers, 199–204
branched co-polymers, 202–204
branched homopolymers, 199–201
remarks on, 194–196
generalized ratios, of molecular parameters, 234–244
ratio $A_2M_w/[\eta]$, 235–239
ratio $\rho = R_g/R_h$, 239–244
star molecules, 243–244
global parameters, characterization, 154–163
concentration dependence, 159–161
fractal behaviour, 158–159

mean square radius of gyration, 154–157
second virial coefficient, 161–162
self-similarity, 162–163
global structure parameters, and branched polymers, 208–209

heterodyne light scattering, 185–186
Huggins constant, 89
Huygens' principle, 141
hydrodynamic radius
and branched macromolecules, 220–224
and branched polymers, 206–207
hyperbranched polymers, 8–11
absolute molar mass, determination of, 69–76
bulk properties in flow and in thin films, 269–271
separation methods for dendritic polymers, 78–86
according to degree of branching, 81–83
and elution properties, 83–86
solution viscosity of, 99–102
suitability of separation technique, 76–78
thermal properties of, 271–277
degree of branching influence, 271–272
end groups influence, 275–277
molar mass influence, 272–275

Kuhn-Mark-Houwink-Sakurada (KMHS) equation, 89–90
and solution viscosity, 93–97

LAC. *See* liquid adsorption chromatography (LAC)
LCCC. *See* liquid chromatography at the critical conditions (LCCC)
light application, and structure characterization, 140–141
linear macromolecules, and dynamic light scattering, 176–177
liquid adsorption chromatography (LAC), 77
liquid chromatography at the critical conditions (LCCC), 77

MALS. *See* multi-angle light scattering detector (MALS)
mean square radius of gyration, 154–157
microscopy, and structure characterization, 138–139
modern theory of segmental dynamics, 186–189
molar mass
control over, dendritic polymers, 56–61
and static light scattering, 151–153
and thermal properties, of hyperbranched polymers, 272–275
molar mass distribution, and molecular properties, 249–250
molecular parameters, 228–234
generalized ratios of, 234–244
ratio $A_2M_w/[\eta]$, 235–239
ratio $\rho = R_g/R_h$, 239–244
star molecules, 243–244
monodisperse colloidal particles, 229
Ψ-parameter, 230
polymers and soft spherical particles, 230
viscosity draining parameter, 230–234
molecular properties
angular dependence of scattering, 260–264
contraction parameters, 244–249

molecular properties (*continued*)
fractal approach, 251–255
molar mass distribution, 249–250
overview, 227–228
percolation on lattice, 255–257
problems to be solved, 251
size exclusion chromatography, 257–260
moments of distribution, and branched polymers, 197–198
monodisperse colloidal particles, 229
multi-angle light scattering detector (MALS), 71

non-invasive structure characterization, 137–141
application of light, 140–141
microscopy, 138–139
rheology, 139–140
sedimentation by ultracentrifugation, 140

particle scattering factor *P(q)*, and volume exclusion, 213–215
fractal approach to measure, 215–220
percolation on lattice, and molecular properties, 255–257
percolation threshold, 256
perturbed conformations, and volume exclusion, 223–224
phantom chains, 215
point-like scattering units, and static light scattering, 142–144
polarization effect, and static light scattering, 144–145
polydispersity, and dynamic light scattering, 179–180
power spectrum, 165
probable distributions, and branched polymers, 196–199
compound distribution, 199
convolution, 198–199
definition, 196–197
moments of distribution, 197–198
proteins, and dynamic light scattering, 177–179
pseudo-dendrimers
by AB_2 modification, 50–56
solution viscosity of, 102–105, 124–132

radius of gyration
and branched macromolecules, 220–224
and branched polymers, 206
mean square, 154–157
and static light scattering, 151–153
randomly branched polymers, 199–204
branched co-polymers, 202–204
branched homopolymers, 199–201
Renormalization Group Theory, 247
rheology, and structure characterization, 139–140

Schulz–Blaschke constant, 89
second virial coefficient, 161–162
sedimentation, and structure characterization, 140
segmental density influence, on solution viscosity, 115–132
and pseudo-dendrimers, 124–132
and statistical branching, 115–124
self-avoiding walk (SAW) simulations, 213
self-similarity, and structure characterization, 162–163
separation methods, for dendritic polymers, 78–86
according to degree of branching, 81–83
and elution properties, 83–86

shielding ratio, 101
side reactions, and degree of branching, 33–35
size distribution of real chains, and volume exclusion, 212–213
size exclusion chromatography, 257–260
soft spherical particles, 230
solution viscosity
  of dendritic polymers, 97–107
    dendrimers, 97–99
    dendronized polymers, 105–107
    hyperbranched polymers, 99–102
    pseudo-dendrimers, 102–105
  influence of chemical structure, 107–115
    and backbone character, 107–110
    of functionality, 110–115
  influence of segmental density, 115–132
    and pseudo-dendrimers, 124–132
    and statistical branching, 115–124
  and Kuhn–Mark–Houwink–Sakurada dependency, 93–97
  overview, 88–93
spectroscopic approaches, and degree of branching, 28–30
star molecules, 243–244
static light scattering, 141–154
  and molar mass, 151–153
  overview, 154
  from particles of dimensions λ/20, 145–151
  from point-like scattering units, 142–144
  and polarization effect, 144–145
  and radius of gyration, 151–153
statistical branching, and solution viscosity, 115–124
stochastic processes, and dynamic light scattering, 167–168
structure characterization
  dynamic light scattering, 164–185
    Brownian motion and stochastic processes, 167–168
    and cluster formation, 179–180
    concentration dependence, 183–185
    development of optics, 165–167
    effect of external forces, 169–171
    electrophoretic, 180–183
    linear and branched macromolecules, 176–177
    overview, 164–165, 185
    and polydispersity, 179–180
    and proteins, 177–179
    time correlation functions, 168–169, 171–175
  global parameters, 154–163
    concentration dependence, 159–161
    fractal behaviour, 158–159
    mean square radius of gyration, 154–157
    second virial coefficient, 161–162
    self-similarity, 162–163
  heterodyne light scattering, 185–186
  modern theory of segmental dynamics, 186–189
  non-invasive, 137–141
    application of light, 140–141
    microscopy, 138–139

structure characterization (*continued*)
 rheology, 139–140
 sedimentation by ultracentrifugation, 140
 static light scattering, 141–154
  and molar mass, 151–153
  overview, 154
  from particles of dimensions λ/20, 145–151
  from point-like scattering units, 142–144
  and polarization effect, 144–145
  and radius of gyration, 151–153
substitution effect, 260
synthetic approach, and degree of branching, 30–33

thermal properties, of hyperbranched polymers, 271–277
 degree of branching influence, 271–272
 end groups influence, 275–277
 molar mass influence, 272–275
time correlation functions, 168–169, 171–175
tuning of degree of branching, 42–56
 $AB_2 + AB$ approach, 42–47
 $AB_2 + ABB^*$ approach, 47–50
 pseudo-dendrimers by AB_2 modification, 50–56

ultracentrifugation, and structure characterization, 140
universal calibration, 91
universal constant, 90–91
unperturbed conformations, and branched polymers, 209–210, 222–223

viscosity draining parameter, 230–234
volume exclusion, and branched polymers, 210–228
 description, 210–211
 Flory expansion factor, 211–212
 particle scattering factor *P(q)*, 213–215
  fractal approach to measure, 215–220
 perturbed conformations, 223–224
 size distribution of real chains, 212–213
 unperturbed conformations, 222–223